Current Topics in Behavioral Neurosciences

Volume 38

Current Topics in Behavioral Neurosciences provides critical and comprehensive discussions of the most significant areas of behavioral neuroscience research, written by leading international authorities. Each volume offers an informative and contemporary account of its subject, making it an unrivalled reference source. Titles in this series are available in both print and electronic formats.

With the development of new methodologies for brain imaging, genetic and genomic analyses, molecular engineering of mutant animals, novel routes for drug delivery, and sophisticated cross-species behavioral assessments, it is now possible to study behavior relevant to psychiatric and neurological diseases and disorders on the physiological level. The *Behavioral Neurosciences* series focuses on "translational medicine" and cutting-edge technologies. Preclinical and clinical trials for the development of new diagnostics and therapeutics as well as prevention efforts are covered whenever possible.

More information about this series at http://www.springer.com/series/7854

Eric Vermetten • Dewleen G. Baker •
Victoria B. Risbrough
Editors

Behavioral Neurobiology of PTSD

Springer

Editors
Eric Vermetten
Department Psychiatry
Leiden University Medical Center
Leiden, The Netherlands

Dewleen G. Baker
Department of Psychiatry
UC San Diego School of Medicine
San Diego, California
USA

Victoria B. Risbrough
Department of Psychiatry
University of California San Diego
La Jolla, California
USA

ISSN 1866-3370 ISSN 1866-3389 (electronic)
Current Topics in Behavioral Neurosciences
ISBN 978-3-030-06920-9 ISBN 978-3-319-94824-9 (eBook)
https://doi.org/10.1007/978-3-319-94824-9

This Springer imprint is published by the registered company Springer Nature Switzerland AG.
The registered company address is: Gewerbestrasse 11, 6330 Cham, Switzerland

Preface

This volume is a collection of papers in the series *Current Topics in Behavioral Neuroscience* dedicated to post-traumatic stress disorder (PTSD). The volume reflects the state of the art in the behavioral neurosciences of PTSD. It is not intended as a comprehensive survey of behavioral neuroscience of PTSD, but is aimed at bringing together preclinical and clinical scientific studies that address current topics of relevance to trauma and PTSD. With this series of review articles, we provide an issue that summarizes the latest studies in the long-term consequences of trauma. Also, we provide an issue that maximizes the reader's insight into the mechanisms of risk and resilience to stress and that addresses both theoretical and applied aspects of developing intervention tools for trauma-related disorders.

The planning of the volume started early in 2015 and has 3 years later resulted in a comprehensive selection of topics that reflect research on novel approaches in the domain of behavioral neuroscience of PTSD. We have selected a compendium of papers to reflect excellence in both research and recognition of novel domains in PTSD research. The papers range across a wide array of topics that include development, epigenetics, animal models and neurocognition, a Research Domain Criteria (RDoC) approach to PTSD, emotion regulation, anhedonia, PTSD comorbidity with TBI, as well as novel approaches to therapy such as exposure-based treatment with ketamine.

The first chapter by Stevens et al. on developmental contributors to trauma response provides an in-depth overview of early factors that interact with development to contribute to later trauma responses. The authors describe development of neural substrates that have been associated with PTSD and identify developmental shifts around critical ages that may contribute to vulnerability. They underscore the need in this field for longitudinal research from middle childhood through early adulthood. They also emphasize that we need to move away from lumping early trauma as one factor altogether.

The second chapter comprehensively reviews the first wave of epigenetic studies that has been performed in an attempt to identify the molecular underpinnings of the chronic effects of trauma exposure. The attention is on the role of microRNAs in moderating or mediating the impact of severe stress and trauma. Adding to the epigenetics of methylation, the mRNA domain is a novel field of study in PTSD research. The authors stress the need for new research comprising translational and cross-species approaches that use longitudinal designs for studying trajectories of change contrasting susceptible and resilient subjects.

Because PTSD is precipitated by a definite traumatic experience, animal models can simulate the induction of PTSD and test causal factors with longitudinal designs. The third paper reviews the widely used animal models of PTSD in rodents and gives a comprehensive overview of their strengths and weaknesses in terms of face, construct, and predictive validity. This is important since many symptoms can be modeled using behavioral tests as long as strengths and weaknesses of the model are identified.

In the chapter by Schmidt and Vermetten, the core findings in neurobiological PTSD research are matched to the RDoC research domains and units of analysis. Several of the core findings in PTSD such as amygdalar overactivity have been linked to all RDoC domains without further specification of their distinct role in the pathophysiological pathways associated with these domains. Still incomplete are the cellular and molecular processes that are decisive for regulation of psychic processes and for the expression of psychopathological symptoms. While RDoC can be valuable, there are critical domains for PTSD missing, and the authors have suggested stress regulation and maintenance of consciousness to be added.

Jak et al. highlight the key role that cognitive functioning plays in both the development and maintenance (or exacerbation) of PTSD symptoms, and their paper focuses on neurocognitive changes following treatment for PTSD as well. They show that neurocognition in PTSD is predictive of school achievement, obtaining and maintaining employment, job advancement, and maintaining relationships and that improved cognition predicted better health and quality of life.

The sixth chapter discusses the role of emotion regulation circuitry implicated in stress-related psychopathology from a developmental and transdiagnostic perspective. VanTieghem and Tottenham highlight converging evidence suggesting that multiple forms of early adverse experiences impact the functional development of amygdala-prefrontal circuitry and make important suggestions for future longitudinal and translational research to better elucidate the mechanisms and facilitate development of interventions that can attenuate risk for psychopathology in youth exposed to early life stress.

The chapter by Veen et al. explores the use of ketamine as an interesting candidate for targeting emotional memories. In a novel approach to PTSD, it is hypothesized that a single intravenous infusion of a subanesthetic dose of ketamine can be considered as a viable augmentation strategy for trauma-focused psychotherapy in patients with PTSD. The chapter also uses a "question-based drug development plan" and reviews topics that need to be addressed concerning the psychotherapeutic approach and phase orientation of this novel ketamine/ pharmacological assisted psychotherapy.

The chapter by Spadoni et al. provides a summary of some of the most common and the most innovative neuroimaging approaches used to characterize the neural circuits associated with PTSD, as well as TBI, and their comorbidity. They summarize the state of the science for each disorder and describe the few studies that have explicitly attempted to characterize the neural substrates of their shared and dissociable influence. They argue that future studies should exploit innovative neuroimaging approaches and longitudinal designs.

It is well known that fear learning is critical in the development and maintenance of PTSD and that safety signal learning and extinction are necessary for recovery. Strauss et al. review studies that suggest that sleep disruption plays a role in safety signal learning and extinction, and thus the development and maintenance of PTSD symptoms, thereby presenting an important modifiable target in PTSD treatment.

Glenn et al. discuss the methodology associated with broad categories of how contextual fear learning is manipulated in imaging studies and highlight findings for the primary neural circuitry involved in each paradigm. They also offer methodological recommendations for human studies of contextual fear acquisition, including using stimuli that distinguish configural learning from discrete cue associations and clarifying how context is experimentally operationalized.

In a comprehensive research update, van Huijstee and Vermetten list and discuss the promises and pitfalls in current research studies on the dissociative subtype of PTSD. The authors argue that inclusion of the novel dissociative subtype of PTSD in the *DSM-5* stimulates research on the prevalence, symptomatology, and neurobiology of the dissociative subtype of PTSD and poses a challenge to improve treatment outcome in PTSD patients with dissociative symptoms.

The last chapter by Risbrough et al. asks the question whether anhedonia predicts PTSD. Whereas childhood trauma has long been associated with increased anhedonia and increased subsequent risk for trauma-related disorders in adulthood, in this chapter the focus is on a novel, emerging, direct contributor to anhedonia in both rodents and humans: fragmented, chaotic environmental signals during critical periods of development. Preliminary longitudinal data are reviewed that aims to show that fragmented chaotic environmental signals are associated with increased anhedonia in adolescence, with further evidence that anhedonia in adolescence/young adulthood may be a risk factor for PTSD and depression.

In this volume, all 12 chapters illustrate that there is a need for neurodevelopmental and cohort approaches to PTSD. And, all papers illustrate that there has been an enormous effort to capture risk and resilience factors, as well as to identify neuroscientific biomarkers of expressed illness. The field is advancing and maturing rapidly, yet research efforts will need to be sustained over the next decades to consolidate knowledge in the field.

We are grateful that these authors have provided us with these comprehensive reviews of current topics. They have a value to the understanding of the complex interface of trauma exposure and chronic psychopathology. Lastly, a special thanks to all the reviewers for this special issue.

Leiden, The Netherlands
San Diego, CA, USA
La Jolla, CA, USA
May 2018

Eric Vermetten
Dewleen G. Baker
Victoria B. Risbrough

Contents

Developmental Contributors to Trauma Response: The Importance of Sensitive Periods, Early Environment, and Sex Differences

Jennifer S. Stevens, Sanne J.H. van Rooij, and Tanja Jovanovic

Abstract This review considers early factors that interact with development to contribute to later trauma responses, including developmental sensitive periods, the effects of early environment, and the emergence of sex differences. We also describe development of neural substrates that have been associated with posttraumatic stress disorder and specifically focus on fear behavior and circuitry. Emerging evidence suggests that there may be developmental shifts around age 10 in these underlying circuits that may contribute to vulnerability. We also discuss age-related changes in the importance of caregiver availability as positive buffering factors. Hormonal changes later in development with onset during puberty appear to further shape development trajectories toward risk or resilience. We highlight these recent findings as well as the great need for further longitudinal research from middle childhood through early adulthood.

Keywords Amygdala • Child development • Early environment • Fear conditioning • Sensitive periods • Sex differences • Social buffering • Trauma

Contents

J.S. Stevens, S.J.H. van Rooij, and T. Jovanovic (✉)
Department of Psychiatry and Behavioral Sciences, Emory University School of Medicine, 49 Jesse Hill Jr Dr, Suite 331, Atlanta, GA 30303, USA
e-mail: tjovano@emory.edu

Curr Topics Behav Neurosci (2018) 38: 1–22
DOI 10.1007/7854_2016_38
Published Online: 22 March 2018

Many of the most potent risk factors for posttraumatic stress symptoms and other negative responses to trauma begin early in life, as the early environment interacts with the developing brain. Childhood adversity is one of the most potent and well-known risk factors for the development of posttraumatic stress disorder (PTSD). It is well established that early deprivation, for example, institutional rearing, but also childhood abuse, or exposure to other traumatic events, such as witnessing violence, can have a long-lasting effect on the individual. However, other factors such as risk genotype, brain development, and emergence of pubertal sex differences also influence the later risk for developing PTSD. A great deal of research has established retrospective links between these early factors and psychiatric and medical outcomes in adulthood. Further research is now needed to address why early life is such an important period and to outline the specific windows of development that interact with environmental factors that increase risk for negative outcomes.

Here, we highlight the importance of developmental timing as a key factor influencing the impacts of early experiences on risk for trauma-related pathology. We focus on the development of fear learning systems, as deficits in fear inhibition and extinction have been demonstrated to be critical mediators of the trauma response in adults (Jovanovic et al. 2010; Jovanovic and Ressler 2010). We begin by outlining sensitive periods in development for trauma exposure and emergence of psychopathology, and early development in the neurobiology of fear, as well as windows of divergence in brain development: toward risk-related or healthy outcomes. We then highlight protective aspects of the early rearing environment, which may buffer against the negative effects of early-life stress. Finally, we end with the developmental timing of sex differences in risk for trauma-related pathology. While there is an emerging literature on genes that may contribute to vulnerability for trauma-related mental illness, here we focus primarily on environmental factors during development.

1 Sensitive Periods

1.1 Sensitive Periods for Trauma Exposure and Related Pathology

The prevalence of anxiety disorders has been shown to increase during late childhood and early adolescence, suggesting that this period may be developmentally critical in identifying individuals at risk for adult psychopathology (Beesdo et al. 2009; Cohen et al. 1993). In a replication of the National Comorbidity Survey, anxiety diagnoses were found to be highly prevalent at 28.8% with the earliest disorder emerging at a median age of 11 (Kessler et al. 2005). Separation anxiety and specific phobias emerge at the earliest age, followed by social phobia in adolescence and generalized anxiety, which can emerge in adulthood (Beesdo et al. 2009; Cohen et al. 1993). Several longitudinal studies of children and adolescents found no sex differences in childhood (prior to age 12), but a highly significant increase in anxiety disorders in girls relative to boys in adolescence (Velez et al. 1989). PTSD can be difficult to diagnose in children using adult criteria; therefore several modifications have been made to account for developmental stage, such as using trauma reenactment during play and frightening dreams as symptom presentations (Makley and Falcone 2010). The most common PTSD-eliciting event in children is injury or motor vehicle accident, with prevalence rates of about 25%. Physical or sexual abuse can result in rates as high as 58% (Makley and Falcone 2010). Because onset of PTSD is dependent on the timing of the traumatic event, it is difficult to determine developmentally sensitive periods for the emergence of symptoms in childhood; however some studies have diagnosed children as early as preschool age (Cohen and Scheeringa 2009). Longitudinal studies of PTSD in children have found it to be a stable diagnosis over 2 years, indicating the need for treatment intervention with children (Cohen and Scheeringa 2009).

Exposure to trauma during childhood has long been recognized as a significant predictor of PTSD in adulthood in addition to other mental disorders, such as depression (Nemeroff et al. 2006). Adversity includes several different negative experiences, such as child maltreatment and poverty. Child maltreatment is a pervasive public health problem as more than three million children received intervention from Child Protective Services in 2012 (Administration on Children, Youth and Families and US Department of Health and Human Services 2012). Poverty is even more prevalent; currently over half of all students in US public schools come from low-income families according to the National Center for Education Statistics (Suitts 2015). The long-lasting impact of adversity during early life on the brain has been well established over the last several decades in animal research (Sanchez et al. 2001) and in human imaging studies (see (Bick and Nelson 2016) for recent review).

Although there is clear evidence of the impact of childhood trauma on brain development, specific age-related sensitive periods have not been well defined. Yet, the developmental timing of trauma exposure can have a significant impact on risk for altered brain development and psychopathology (Tottenham and Sheridan 2009). While large cohort studies of child abuse using the Adverse Childhood Experiences (ACEs) in over 80,000 individuals underscore the tremendous negative impact of childhood trauma on adult health (Schüssler-Fiorenza Rose et al. 2014), most do not examine timing of trauma exposure very specifically, but rather report ACEs prior to age 18. While there are a limited number of prospective studies that have examined the effects of trauma exposure in childhood, a recent study using retrospective recall identified age 10 as the period when trauma severity recall was greatest (Pechtel et al. 2014). In addition, a survey of US families recently reported that cumulative stressors prior to age 13 significantly increase the odds of psychological distress in adulthood (Björkenstam et al. 2015). Early adolescence is marked with increased independence resulting in higher likelihood of exposure to trauma. It has been recognized for decades that middle school children from low-income urban neighborhoods experience high levels of community violence (Hill et al. 1996). Further, PTSD symptoms in children aged 8–13 are highly correlated with trauma exposure within the last year (Bevans et al. 2009). However, longitudinal developmental studies defining age-related sensitive windows of trauma exposure and PTSD symptoms are still lacking in the literature.

1.2 Sensitive Periods for Development of Fear Neurobiology

Adult neurobiological models of risk for negative outcomes following trauma focus primarily on hyper-reactivity of the amygdala (Shin and Liberzon 2010), a part of the limbic system located in the medial temporal lobe of the brain and an integral component of mammalian fear circuitry (Davis et al. 1993; Fanselow 1994; LeDoux 1992). Studies of humans with brain damage have also found that the amygdala modulates the fear response: temporal lobectomy in patients results in loss of fear-conditioned startle (Funayama et al. 2001). Amygdala hyper-reactivity is observed in patients with chronic PTSD (Fonzo et al. 2010; Rauch et al. 2000; Shin et al. 2005; Stevens et al. 2013), and, more importantly, amygdala reactivity prior to trauma exposure may predict later PTSD severity (Admon et al. 2009, 2013). Amygdala reactivity is moderated by inhibitory connections from the ventromedial prefrontal cortex (vmPFC), which appear to be abnormal in PTSD (Shin et al. 2005; Stevens et al. 2013). Pediatric PTSD is marked by the same underlying neurobiology, i.e., children and youth with PTSD show heightened amygdala reactivity and impaired amygdala-prefrontal connectivity (Keding and Herringa 2015; Wolf and Herringa 2016). The development of fear circuitry may thus contribute to risk for later PTSD and other trauma-related pathology.

Animal models using fear conditioning have suggested very robust age effects on fear-related learning. For example, early postnatal development is associated with different neural circuitry underlying fear regulation compared to the juvenile period of development, such that the prelimbic cortex is involved in expression of learned fear in juvenile but not infant rats (Li et al. 2012). In humans, early studies using structural magnetic resonance imaging (MRI) outline a pattern of similarly prolonged development in the brain regions supporting fear learning, from early childhood through early adulthood. For example, amygdala volume increased in males from ages 4 to 18, while hippocampal volume increased in females in the same age range (Giedd et al. 1996). On the other hand, cerebral gray matter development follows an inverted U-shape, showing early increases in volume and thickness that peak in late childhood, followed by decreased volume and density after adolescence (Giedd et al. 1999, 2015; Gogtay et al. 2004; Shaw et al. 2006) with the medial prefrontal cortex (mPFC) showing the longest developmental trajectories (Kolb et al. 2012). White matter integrity in the uncinate and cingulum, the primary tracts connecting the amygdala and hippocampus with the PFC, show steep linear increases from childhood through early adulthood, not peaking until after age 35 (Lebel et al. 2012). Such prolonged patterns of development for the amygdala and prefrontal cortex, and connections between these regions, indicate that there is a large window spanning childhood and adolescence during which experience, particularly traumatic experience, can shape trajectories of brain development.

In addition to the aforementioned studies suggesting that gray matter volumes for the prefrontal and temporal lobes peak from ages 12 to 16 (Giedd et al. 1999, 2015); a recent study found a developmental shift in functional connectivity between the amygdala and the mPFC during the viewing of fearful faces (Gee et al. 2013a). The cross-sectional study included children from 4 years of age to adults and found that these areas were positively connected prior to age 10 years and negatively connected after age 10 years (Gee et al. 2013a). The observed negative functional connectivity continued to increase from adolescence to adulthood. Earlier studies using similar methods found that adolescents showed greater amygdala reactivity to fearful faces than adults (Guyer et al. 2008). Fear conditioning studies comparing adolescents to adults have found that adolescents show greater fear-conditioned responses compared to adults (Shechner et al. 2015), suggesting blunting of fear responses with adulthood. Together, these structural and functional data point to developmental decreases in activation in limbic subcortical structures in response to fear-related cues from childhood to adulthood.

Similar developmental effects on inhibition have been observed using acoustic startle responses in children and adolescents. One study examined habituation of startle responses in 7–9-year-olds compared to 10–12-year-olds in children with anxiety disorders and controls (Waters et al. 2008). This study found that startle increased with age only in the at-risk group and was already higher in the younger children with anxiety disorders. While this study suggested that risk phenotypes emerge between 9 and 10 years of age, it was based on a relatively small sample size. A more recent study of 40 healthy children between 8 and 13 years old used fear conditioning methods and found that adult-like patterns in fear-potentiated startle emerged around 10 years of age (Glenn et al. 2011). One of the key findings of this study was the inability of younger children to inhibit fear to safety signals, suggesting that this ability develops in healthy children around 10 years of age. Data from our studies at the Grady Trauma Project (GTP) support a developmental shift in safety signal processing around age 10, in that participants who were older than 10 years of age showed higher fear-potentiated startle and significant discrimination between the CS+ (danger signal) and the CS- (safety signal) (Jovanovic et al. 2014). In contrast, children up to 9 years of age showed deficits in fear inhibition – this trait has been associated with PTSD in adults (Jovanovic and Norrholm 2011). Of note, developmental studies that have used skin conductance response (SCR) as a physiological measure have had mixed results. A recent study that compared age groups between 5- and 10-year-old healthy children did not find age-related differences in fear conditioning or extinction (Michalska et al. 2016). In our studies in children from the Grady Trauma Project, we have found that SCR to the CS+ was greatest in children under 10 with high anxiety (Jovanovic et al. 2014). We also found that SCR to the CS+ was associated with fear-related symptoms of PTSD in children (Gamwell et al. 2015).

The abovementioned sensitive periods in brain development and physiology also point to the importance of middle childhood as a sensitive period in which environmental insults such as early adverse experiences can have long-lasting neurobiological impacts. Studies by Tottenham and colleagues investigated fear circuitry in children adopted from institutional care with high rates of neglect. These studies found larger amygdala volumes in children with prolonged maternal deprivation early in life (Tottenham et al. 2010). In an MRI study of orphaned children, those that were adopted prior to 15 months of age had similar amygdala volumes to controls, whereas children adopted after 15 months of age showed increased amygdala volumes later in childhood (tested around 10 years of age). Although this early trauma may increase risk for anxiety disorders in children, the MRI results in the study were not directly related to anxiety, since the relationship remained significant even after exclusion of children with anxiety (Tottenham et al. 2010). However, the study did find that amygdala volume was positively correlated with internalizing and anxiety symptoms in the children. In addition to increased amygdala volume, amygdala reactivity to fearful faces and functional connectivity between the prefrontal cortex and the amygdala are altered in children and adolescents with early-life stress. Moreover, normal developmental changes appear to be disrupted (Tottenham and Sheridan 2009; Gee et al. 2013a). A recent study using retrospective data from adults recalling childhood trauma exposure found a dose response between trauma severity and amygdala volume, with a significant increase in volume with trauma recalled between ages 10 and 11 (Pechtel et al. 2014). Finally, studies of pediatric PTSD found that PTSD was associated with a smaller PFC and hippocampus and that symptom severity correlated with the decrease in these areas (Keding and Herringa 2015; De Bellis et al. 2002). On the other hand, pediatric PTSD was also associated with increased dorsal cingulate reactivity to threatening cues in children ages 8–18, with age 10 and 11 reflecting highest activity (Wolf and Herringa 2016). Taken together, these studies indicate that timing of trauma exposure during development has significant consequences on PTSD symptoms as well as its neural underpinnings and that middle childhood around age 10 may be a sensitive period for these effects.

2 Early Environmental Factors Influencing Risk for PTSD

2.1 Rearing Environment

Growing up in an adverse environment is a known risk factor for the emergence of mental disorders; however, a positive rearing environment may exert protective or beneficial effects on the development of psychiatric disorders. The effect of rearing condition is often investigated in nonhuman primates by comparing monkeys reared by their mother with monkeys reared by age-matched peers or nursery-reared monkeys. Peer- or nursery-reared monkeys show more behavioral fear (Clarke and Snipes 1998), lower baseline cortisol levels (Shannon et al. 1998), and larger stress-sensitive brain regions (Spinelli et al. 2009) compared to mother-reared monkeys, indicating that rearing condition may have long-term effects on stress responses and related neurobiology. The availability of the mother may have some specific effects by means of maternal buffering, which are discussed in a separate section below.

Environmental enrichment (EE) methods have been used in order to investigate the effect of a positive rearing environment. EE is thought to improve cognitive, motor, and sensory functions compared to standard housing (Nithianantharajah and Hannan 2006). In a typical EE study, laboratory rodents are placed in single-sex housing in groups with unlimited access to food and water. EE cages, however, also consist of toys, nesting materials, tunnels, running wheels, and ladders, which are regularly changed to bring novelty and complexity to the rodents' environment (Takuma et al. 2011). In the first study, the protective effects of EE against stress were investigated. Rats were tested in an EE 2 weeks prior to an inescapable foot shock procedure, which was used to induce PTSD-like anxious behavior. However, the study found that EE did not protect against the effect of this foot shock procedure (Hendriksen et al. 2010). Second, in a follow-up study, the rats that underwent the foot shock procedure were placed in an EE after the shock, which reversed the PTSD-like anxiety behavior and increased cell proliferation in the hippocampus (Hendriksen et al. 2010). Third, the effects of EE as treatment for prenatal chronic stress and early-life stress were tested. Prenatal chronic stress was induced by means of unpredictable foot shocks (Nowakowska et al. 2014; Yang et al. 2006, 2007), and early-life stress was induced by housing newborn rats (postnatal day (PND) 2–21) in cages with limited nesting/bedding materials (Cui et al. 2006). Both resulted in depressive-like behavior, impaired spatial learning and memory, and impaired hippocampal long-term potentiation in young adulthood (PND 53–57). EE treatment during childhood and adolescence (PND 22–52) nullified the negative effects of prenatal and early-life stress (Nowakowska et al. 2014; Yang et al. 2006, 2007; Cui et al. 2006). Furthermore, EE during the peripubertal period reversed the negative effects of postnatal maternal separation on endocrine and behavioral stress responses (Francis et al. 2002). It can be concluded that EE does not protect against the immediate effects of stress, but EE treatment can reverse the negative effects of prenatal and early-life stress.

Investigating the effects of rearing environment in a human sample is far more challenging than in a laboratory setting. However, studies investigating children who were raised in institutional care and were adopted early in life have provided some insights in effects of rearing conditions. In a groundbreaking randomized trial of foster care, the Bucharest Early Intervention Project (BEIP) assigned children who were raised in Romanian orphanages to either a care-as-usual or foster care intervention (Humphreys et al. 2015). A recent follow-up of the study found that foster care attenuated many of the symptoms of mental illness that developed in the institutionalized children. Several other studies from the same project showed improved cognitive development (Nelson et al. 2007), emotional responses, attachment, a better psychiatric outcome (Bos et al. 2011), and improved neural activity (McLaughlin et al. 2011; Moulson et al. 2009) in children who were adopted compared to children who remained in orphanages.

Another line of research on internationally adopted children from orphanages outside the USA has also yielded data on the effects of early environment. In most cases, institutionalized children were adopted by high-functioning and high-socioeconomic-status (SES) parents, who could provide an enriched environment for their children. Although the effects of parental care cannot be separated from the effect of EE, this situation is the best opportunity to study EE as an intervention in humans. When children were compared to never-institutionalized children, early deprivation was found to have long-lasting effects on brain and behavior. Previously institutionalized children showed difficulties in emotion regulation, increased amygdala volumes (Tottenham et al. 2010), and reactivity (Tottenham et al. 2011), as well as early maturation of amygdala-prefrontal connectivity (Gee et al. 2013b). In sum, even though adoption improves many aspects of cognitive and emotional development, EE does not completely counteract the negative effects of the early environment and therefore does not seem to have the same effects as observed in rats.

2.2 *Maternal Buffering*

Parental availability has a particularly large stress-reducing effect on offspring (Callaghan and Tottenham 2016). In species where the primary caregiver is the mother, this phenomenon is called maternal buffering (Kikusui et al. 2006). Nonhuman primate studies have shown that maternal separation induces significant stress in the infants, indicated by increased behavioral stress and cortisol responses. Reunion with the mother, or having access to visual or auditory stimuli associated with the mother, significantly decreases cortisol levels of the offspring (Bayart et al. 1990; Coe et al. 1978; Levine et al. 1985). In rat pups, pairing a shock with a neutral odor typically results in avoidance of the odor in the future; however, fear learning does not occur when the mother is present during odor conditioning (Barr et al. 2009; Moriceau and Sullivan 2006). It is thought that this mechanism prevents the pups from learning to fear their mother, even when she (accidently) induces pain by, for example, stepping on the pups. This promotes attachment to the mother and thus survival in these pups that are dependent on her for care (Sullivan et al. 2000). With increasing age, the effects of maternal buffering on cortical activity in the rat decrease (Sarro et al. 2014).

A dampened fear response in the presence of the mother has also been observed in human studies. The cortisol response to a social stressor in children, but not in adolescents, is eliminated by maternal support (Hostinar et al. 2015). In an fMRI study, children showed dampened amygdala activation to pictures of their mother compared to pictures of a stranger, whereas there was no difference in adolescents (Gee et al. 2014). In another study that used fear conditioning of startle responses, children showed an attenuated fear response to a safety signal when the mother was available during conditioning (van Rooij et al. 2016). However, when the mother was not in the same room, children were not able to discriminate danger and safety signals. Importantly, the effect of maternal buffering was only observed in children and not in young adolescents (van Rooij et al. 2016). This is most likely explained by the functional emergence of the prefrontal cortex, which is important for learning to discriminate fear and danger and inhibiting the fear response when appropriate (Lissek et al. 2014). This postulation is supported by an fMRI study in which more adult-like connectivity between the prefrontal cortex and the amygdala was observed when viewing maternal versus stranger pictures (Gee et al. 2014). Again, the effects of maternal buffering were only observed in children and not in adolescents (Gee et al. 2014).

In contrast to childhood where the individual is dependent on the parent, adolescence is a time of increasing independence. The reduced effects of maternal buffering in this phase might reflect this transition. Although adolescents or adults do not directly benefit from maternal presence, a study with mother- versus nursery-reared rhesus macaques showed that maternal presence during childhood may have long-term effects on the ability of adult monkeys to benefit from social support later in life (Winslow et al. 2003).

2.3 Social Support

Social support is thought to influence responses to stress and trauma during development or later in life. The importance of social support as a protective factor against psychiatric disorders has been demonstrated in several preclinical and human studies. In one rodent study, mice that were isolated from conspecifics showed higher anxiety-like responses and lower basal plasma corticosterone, but a larger corticosterone increase to novel or stressful situations compared to mice exposed to social or enriched conditions (Ros-Simó and Valverde 2012). Similarly, socially isolated rats show increased freezing behavior after fear conditioning, whereas socially partnered rats enhance their grooming behavior (DaSilva et al. 2011). These findings underscore the importance of social interaction for a healthy behavioral and neuronal development in response to stress and trauma.

In a human study, social support was found to protect against depression in a genetically at-risk group of maltreated children (Kaufman et al. 2006). In an adult sample, baseline social support was found to protect patients with chronic illness from the development of PTSD (Dinenberg et al. 2014). Likewise, low levels of post-deployment social support were associated with increased PTSD and depression symptoms in a military cohort (Pietrzak et al. 2009, 2010). The positive effects of social support on trauma survivors may explain the therapeutic effects of some innovative drugs that promote social affiliation, such as oxytocin (Koch et al. 2016) and methylenedioxymethamphetamine (Mithoefer et al. 2016).

The studies reviewed here suggest that the environmental factors of rearing environment, maternal availability, and social support influence the response to stress and trauma and may protect against the development of mental disorders. Being able to successfully control mild-to-moderate stressors can result in the development of an adaptive stress response, which may protect the individual from negative effects of future uncontrollable stressors. Enriched environments and maternal or other social support during development promote adaptive stress responses. A recent finding that maternally reported warmth did not impact the effects of maternal availability on fear conditioning in children with trauma exposure (van Rooij et al. 2016) suggests that the quality of the mother-child relationship may not be as important as maternal availability. It is clear that childhood is a particularly sensitive period for environmental influences. As was observed in most of the maternal buffering studies, maternal presence did not impact adolescents older than 10 years of age, only children. Considering the steep increase in psychiatric disorders during adolescence, it is important to advance our understanding of environmental influences on the more plastic period before age 10.

3 Developmental Emergence of Sex Differences in Risk for PTSD

From a public health perspective, sex differences in the response to trauma are a critical factor for further investigation, given that women are at greater risk than men for developing trauma-related mental health disorders Women are nearly twice as likely as men to be diagnosed with PTSD and mood disorders (Dell'Osso et al. 2013; Kessler et al. 1995, 2005; Kline et al. 2013; Steven et al. 2013). Interestingly, women do not experience a greater number of traumas than men; in fact, it has been estimated that a woman is only 77% as likely as a man to experience trauma (Tolin and Foa 2006). However, women and men show differences in specific types of traumas they are most likely to experience. For example, women are more likely than men to have experienced forms of trauma that are particularly severe, such as sexual assault or abuse, starting in childhood (Tolin and Foa 2006). More importantly for informing prevention and early intervention strategies, there may also be sex-dependent vulnerabilities that predate trauma. Studies of child development have the potential to reveal mechanisms that produce the striking sex differences in trauma-related psychopathology in adults.

3.1 Sex-Dependent Risk Factors for Trauma-Related Pathology

Although it is clear that women are at greater risk for PTSD than men, research on preexisting vulnerability in women is only in very early phases, with promising recent findings pointing to risk factors in brain circuits, peripheral autonomic physiology, and genetics. Exaggerated reactivity of the amygdala (Fonzo et al. 2010; Admon et al. 2009) and an impairment in its connections with the rostral and subgenual anterior cingulate cortex (ACC) (Stevens et al. 2013; Sripada et al. 2012) appear to be key risk factors for trauma-related psychopathology, particularly hyperarousal and other anxiety symptoms (Stevens et al. 2013). Notably, both the amygdala and prefrontal cortex are sensitive to the effects of gonadal steroid hormones (Clark et al. 1988; MacLusky et al. 1986; Roselli et al. 2001). Meta-analysis of the neuroimaging literature indicates that women show greater amygdala reactivity to negative emotional stimuli than men (Stevens and Hamann 2012), which may increase their risk for trauma-related pathology.

Evidence suggests that estrogen levels in women may influence physiological and neural reactivity to threat, possibly providing one mechanism for women's increased PTSD risk. In an fMRI study of fear conditioning and extinction, naturally cycling women during a high-estrogen phase showed greater activation of fear-processing regions than men or women with low estrogen levels (Hwang et al. 2015). Similarly, pregnancy may be a particularly vulnerable time for women, as circulating estrogens and the stress hormone cortisol increase over the course of pregnancy. A recent study showed that pregnant women reported greater levels of hyperarousal symptoms than nonpregnant women and showed greater fear-potentiated startle to a safety signal, reflecting impaired fear inhibition (Michopoulos et al. 2014). Interestingly, low estrogen levels may also increase risk: trauma-exposed women with low estrogen levels (both naturally cycling and postmenopausal) showed impaired fear inhibition (Glover et al. 2013) and impaired fear extinction (Glover et al. 2012) in a fear-potentiated startle paradigm. Women with low (versus high) estrogen also showed greater connectivity between the amygdala and dorsal ACC (Engman et al. 2016), a region associated with increased fear expression and arousal. Further studies are needed in order to investigate the possibility of an inverted U-shaped dose-response relationship between estrogen and risk for PTSD and to outline whether there are specific circumstances (e.g., pregnancy, menopause) in which high versus low levels of estrogens may promote risk.

Rodent work has demonstrated a possible mechanism by which estrogen levels may interact with stressors to increase fear reactivity in females. Ovariectomized female rats who received exogenous estrogen replacement during stress exposure showed an increase in dendrite length and density in an inhibitory pathway between the prefrontal cortex and amygdala, whereas those without estrogen replacement showed no such increase (Shansky et al. 2010). In contrast, male rats exposed to stress showed no dendritic remodeling in the same pathway (Shansky et al. 2009). A human analog of this pathway has been shown to be impaired in women with PTSD; relative to trauma-exposed women without PTSD, traumatized women with PTSD showed less connectivity between the amygdala and a prefrontal region that regulates amygdala activity, the subgenual anterior cingulate cortex (Stevens et al. 2013).

Genetic variation may also contribute to PTSD risk in women. For example, a data-driven analysis of genetics associated with fear in both humans and mice identified a sex-specific risk factor for PTSD (Ressler et al. 2011). Genetic polymorphisms associated with PTSD in humans were examined for overlap with genes whose expression in the mouse amygdala showed large changes before and after a fear-conditioning experiment. In this study, 17 overlapping genes were identified, with the strongest effects for genes coding for pituitary adenylate cyclase-activating polypeptide (PACAP) and its receptor PAC1R. High peripheral levels of PACAP and a genetic polymorphism in *PAC1R* gene were associated with PTSD in women but not men (Ressler et al. 2011). In women, the *PAC1R* polymorphism was also associated with greater dark-enhanced startle, a psychophysiological correlate of anxiety, and less discrimination between danger and safety signals with fear-potentiated startle (Ressler et al. 2011). Further, the same risk allele was associated with greater amygdala reactivity to threat and less functional coupling of the amygdala with regulatory regions including the hippocampus and rostral ACC (Pohlack et al. 2015; Stevens et al. 2014).

3.2 Emergence of Sex-Dependent Risk Factors over Development

In order to understand how sex differences influence fear neurocircuitry and interact with trauma response, it is critical to study the emergence of sex differences over development. Although most sexually dimorphic brain changes occur with the increase in gonadal hormone activity during puberty, it is possible that even before puberty environmental influences and maturational factors differentially impact girls and boys. An emerging literature in humans points to sex differences in emotional neurophysiology before puberty. Psychophysiological measures of emotional arousal such as heart rate and skin conductance response indicate that reactivity to negative stimuli is greater in prepubertal girls than boys (McManis et al. 2001). In a recent study of prepubertal sex differences in fear conditioning, girls showed less discrimination between danger and safety signals compared to boys, a phenotype that has been associated with PTSD (Gamwell et al. 2015). Further, the skin conductance response to the conditioned danger signal was correlated with different PTSD symptoms in girls and boys, suggesting sex-specific patterns in physiological underpinnings of trauma-related pathology (Gamwell et al. 2015). Finally, in girls, but not boys, early-life stress was associated with increased basal cortisol, and this predicted later impaired resting state connectivity between the amygdala and mPFC (Burghy et al. 2012).

In order to examine whether the *PAC1* genotype has sex-specific effects on anxiety prior to puberty, we examined dark-enhanced startle in 28 boys and 22 girls aged 10 ± 0.2. Dark-enhanced startle was greater in children with the risk (CC) genotype when girls and boys were considered together, with no effect of sex (Jovanovic et al. 2013). This indicated that the sex-dependent effects of this genotype on anxiety are not observed prior to the activational effects of estrogens postpuberty. However, it is notable that PAC1 genotype was associated with heightened startle in boys (Jovanovic et al. 2013), but not in adult men (Ressler et al. 2011), pointing to an intriguing possibility that postpubertal hormonal changes may confer a protective effect upon men with the risk genotype. This idea is supported by preclinical research showing protective effects of testosterone in males. For example, in a study of fear conditioning in which female mice showed greater freezing behavior than males after conditioning to a tone, estrogen levels and even ovariectomy in females had no effect on their freezing behaviors. Instead, males showed an *increase* in freezing after orchidectomy, which returned again to a lower level after testosterone administration (Chen et al. 2014). Additional experiments that manipulate hormone levels in males are needed to replicate these findings and to examine links with *PAC1* genotype.

Important sex differences appear to occur with the increase in gonadal hormones during puberty. For example, it appears that sex differences in amygdala reactivity may not emerge until after puberty. A small study comparing adolescents aged 9–17 and adults aged 25–36 showed greater amygdala reactivity in female versus male adults, but not adolescents (McClure et al. 2004). These findings were replicated in larger studies showing no sex differences in the amygdala response to emotional stimuli in children aged 7–13 (Guyer et al. 2008; Pagliaccio et al. 2013). In addition, a study of regional volumes in the amygdala and hippocampal complex in a large sample of children aged 4–18 showed no sex differences before puberty, but with increasing pubertal maturity, the volume of hippocampal complex structures decreased in boys and increased in girls (Hu et al. 2013). Girls and boys did not differ in patterns of amygdala volume development in this study. Further research specifically targeting late adolescence and early adulthood is needed to determine the specific developmental period during which sex differences in amygdala function emerge.

By adolescence, children exposed to stress already show similar patterns of brain activation as trauma-exposed women and men: in adolescent girls, but not boys, childhood maltreatment was associated with reduced functional connectivity between the amygdala and subgenual ACC (Herringa et al. 2013). Recently, a large study was conducted to examine how early stress exposure might interact with genetic polymorphisms in genes regulating HPA axis function, in girls and boys aged 7–12 (Pagliaccio et al. 2015). Pagliaccio and colleagues found that the experience of stressful life events was associated with greater amygdala reactivity to negative emotional stimuli, consistent with previous studies. However, genetic factors showed an interaction with sex and pubertal status: polymorphisms in HPA axis-related genes predicted greater amygdala reactivity to fearful stimuli in pubertal girls and greater amygdala reactivity to *neutral* stimuli in pubertal boys. Interestingly, amygdala reactivity may be a more stable trait across development in boys than in girls: a study following 4-month-old infants into adulthood showed that infant boys with a "high-reactive" pattern of behavior (vigorous motor activity, crying to unfamiliar stimuli) also showed greater amygdala reactivity to novel face stimuli as adults, relative to men who were "low-reactive" as infants (Schwartz et al. 2012), and this was not the case in women. It is possible that amygdala function in women is more responsive to environmental and hormonal effects than in men.

4 Summary and Conclusions

Several important developmental factors contribute to risk for PTSD, including childhood adversity and hormonal activation. The neuroplasticity of the developing neural circuitry of fear responses leads to putative sensitive periods when trauma exposure may be particularly detrimental. It is important to fully understand the timing of such sensitive periods in order to apply optimal intervention and prevention strategies for PTSD. Figure 1 shows the putative timing for the primary factors that increase risk for PTSD during development such as early deprivation, child abuse and trauma exposure, and activation of gonadal hormones. In contrast, factors

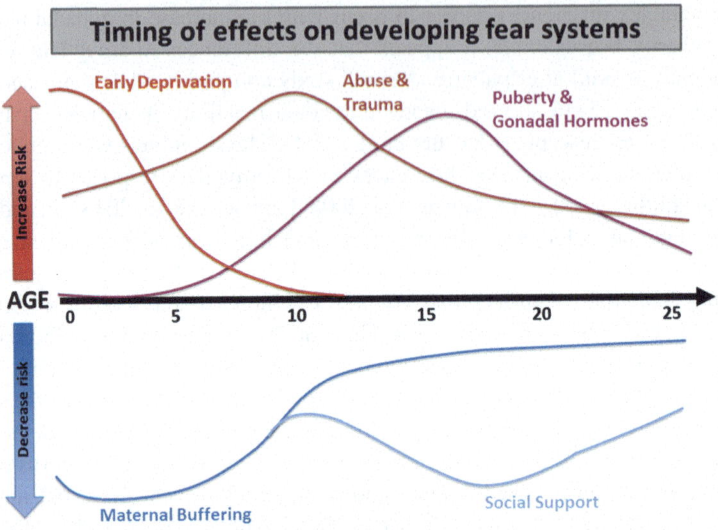

Fig. 1 The figure depicts the factors that increase and decrease risk for PTSD and the putative timing during development when these factors exert their greatest influence on underlying brain circuitry

that decrease risk for PTSD include maternal buffering and social support. In the aftermath of early deprivation, enriched environments have been shown to potentially attenuate negative risk.

Fear learning behavior continues to change and develop over a long window of development, with the ability to differentiate threatening and safe cues strengthening in middle childhood around age 10 (e.g., (Jovanovic et al. 2014)). Similarly, the amygdala and its connections with the vmPFC continue to develop into young adulthood. During this long window of developmental change, trauma can shift the trajectory of development toward outcomes associated with risk, such as the greater amygdala reactivity and volume observed in previously institutionalized children (Tottenham et al. 2010, 2011). Interestingly, retrospective studies of adults reporting their childhood experiences suggest that the impacts of trauma on amygdala structure are greatest in middle to late childhood (Pechtel et al. 2014). However, the specific boundaries of sensitive periods during which trauma has its greatest impacts have not yet been defined. Longitudinal studies following participants through childhood and adolescence are needed to address this question.

During this same window of early childhood through middle childhood, the brain also appears to be sensitive to positive effects of the early environment. Aspects of the rearing environment can buffer against the negative effects of early-life stress, including trauma (Bos et al. 2011). In addition, social buffering provides a powerful protective influence against stress and changes with development such that maternal presence provides greatest buffering influence through middle childhood and less so during adolescence (van Rooij et al. 2016). Importantly, the potential to benefit from social support in adulthood is thought to depend on maternal availability during childhood (Winslow et al. 2003).

Finally, the increase in gonadal steroid hormones during the pubertal period in adolescence appears to be a later window that shapes the brain toward risk or resilience. Developmental studies of amygdala reactivity suggest that sex differences in amygdala reactivity emerge only after puberty, with the increase in levels of steroid hormones (Chen et al. 2014; McClure et al. 2004), and additional studies targeting late adolescence are needed in order to define a specific developmental window. However, it is notable that there are also intriguing findings pointing to prepubertal sex differences in physiological measures of arousal and HPA axis responses to stressors (Gamwell et al. 2015; McManis et al. 2001; Burghy et al. 2012).

In summary, development is a highly plastic period that is influenced by environmental factors. Neural development points to several sensitive periods during development, including an early childhood period prior to age 3, a late childhood period around age 10, and an adolescent period associated with puberty. Figure 1 depicts the putative periods when environmental factors can exert their influences. Future studies should carefully define these and other sensitive periods in order to provide critical windows of opportunity for intervention and even prevention of trauma-related pathology such as PTSD.

References

Administration on Children, Youth and Families, US Department of Health and Human Services (2012) Child maltreatment

Admon R, Lubin G, Stern O, Rosenberg K, Sela L, Ben-Ami H, Hendler T (2009) Human vulnerability to stress depends on amygdala's predisposition and hippocampal plasticity. Proc Natl Acad Sci 106(33):14120–14125. doi:10.1073/pnas.0903183106

Admon R, Lubin G, Rosenblatt JD, Stern O, Kahn I, Assaf M, Hendler T (2013) Imbalanced neural responsivity to risk and reward indicates stress vulnerability in humans. Cereb Cortex 23 (1):28–35. doi:10.1093/cercor/bhr369

Barr GA, Moriceau S, Shionoya K, Muzny K, Gao P, Wang S, Sullivan RM (2009) Transitions in infant learning are modulated by dopamine in the amygdala. Nat Neurosci 12(11):1367–1369 http://www.nature.com/neuro/journal/v12/n11/suppinfo/nn.2403_S1.html

Bayart F, Hayashi KT, Faull KF, Barchas JD, Levine S (1990) Influence of maternal proximity on behavioral and physiological responses to separation in infant rhesus monkeys (Macaca mulatta). Behav Neurosci 104(1):98–107

Beesdo K, Knappe S, Pine DS (2009) Anxiety and anxiety disorders in children and adolescents: developmental issues and implications for DSM-V. Psychiatr Clin North Am 32(3):483–524 http://dx.doi.org/10.1016/j.psc.2009.06.002

Bevans K, Cerbone AB, Overstreet S (2009) The interactive effects of elevated mid-afternoon cortisol and trauma history on PTSD symptoms in children: a preliminary study. Psychoneuroendocrinology 34(10):1582–1585. doi:10.1016/j.psyneuen.2009.04.010

Bick J, Nelson CA (2016) Early adverse experiences and the developing brain. Neuropsychopharmacology 41(1):177–196

Björkenstam E, Burström B, Brännström L, Vinnerljung B, Björkenstam C, Pebley AR (2015) Cumulative exposure to childhood stressors and subsequent psychological distress. An analysis of US panel data. Soc Sci Med 142:109–117 http://dx.doi.org/10.1016/j.socscimed.2015.08.006

Bos K, Zeanah CH, Fox NA, Drury SS, McLaughlin KA, Nelson CA (2011) Psychiatric outcomes in young children with a history of institutionalization. Harv Rev Psychiatry 19(1):15–24. doi:10.3109/10673229.2011.549773

Burghy CA, Stodola DE, Ruttle PL, Molloy EK, Armstrong JM, Oler JA, Fox ME, Hayes AS, Kalin NH, Essex MJ (2012) Developmental pathways to amygdala-prefrontal function and internalizing symptoms in adolescence. Nat Neurosci 15(12):1736–1741

Callaghan BL, Tottenham N (2016) The neuro-environmental loop of plasticity: a cross-species analysis of parental effects on emotion circuitry development following typical and adverse caregiving. Neuropsychopharmacology 41(1):163–176

Chen LS, Tzeng WY, Chuang JY, Cherng CG, Gean PW, Yu L (2014) Roles of testosterone and amygdaloid LTP induction in determining sex differences in fear memory magnitude. Horm Behav 66(3):498–508. doi:10.1016/j.yhbeh.2014.07.008

Clark AS, Maclusky NJ, Goldman-Rakic PS (1988) Androgen binding and metabolism in the cerebral cortex of the developing rhesus monkey. Endocrinology 123(2):932–940. doi:10.1210/endo-123-2-932

Clarke AS, Snipes M (1998) Early behavioral development and temperamental traits in mother- vs peer-reared rhesus monkeys. Primates 39(4):433–448

Coe CL, Mendoza SP, Smotherman WP, Levine S (1978) Mother-infant attachment in the squirrel monkey: adrenal response to separation. Behav Biol 22(2):256–263. doi:10.1016/S0091-6773(78)92305-2

Cohen JA, Scheeringa MS (2009) Post-traumatic stress disorder diagnosis in children: challenges and promises. Dialogues Clin Neurosci 11(1):91–99

Cohen P, Cohen J, Kasen S, Velez CN, Hartmark C, Johnson J, Rojas M, Brook J, Streuning EL (1993) An epidemiological study of disorders in late childhood and adolescence--I. Age- and gender-specific prevalence. J Child Psychol Psychiatry 34(6):851–867

Cui M, Yang Y, Yang J, Zhang J, Han H, Ma W, Li H, Mao R, Xu L, Hao W, Cao J (2006) Enriched environment experience overcomes the memory deficits and depressive-like behavior induced by early life stress. Neurosci Lett 404(1–2):208–212. doi:10.1016/j.neulet.2006.05.048

DaSilva JK, Husain E, Lei Y, Mann GL, Tejani-Butt S, Morrison AR (2011) Social partnering significantly reduced rapid eye movement sleep fragmentation in fear-conditioned, stress-sensitive Wistar-Kyoto rats. Neuroscience 199:193–204

Davis M, Falls WA, Campeau S, Kim M (1993) Fear-potentiated startle: a neural and pharmacological analysis. Behav Brain Res 58:175–198

De Bellis MD, Keshavan M, Shifflett H, Iyengar S, Beers SR, Hall J, et al (2002) Brain structures in pediatric maltreatment-related PTSD: a sociodemographically matched study. Biol Psychiatry 52:1066–1078

Dell'Osso L, Carmassi C, Massimetti G, Stratta P, Riccardi I, Capanna C, Akiskal KK, Akiskal HS, Rossi A (2013) Age, gender and epicenter proximity effects on post-traumatic stress symptoms in L'Aquila 2009 earthquake survivors. J Affect Disord 146(2):174–180

Dinenberg RE, McCaslin SE, Bates MN, Cohen BE (2014) Social support may protect against development of posttraumatic stress disorder: findings from the heart and soul study. Am J Health Promot 28(5):294–297. doi:10.4278/ajhp.121023-QUAN-511

Engman J, Linnman C, Van Dijk KR, Milad MR (2016) Amygdala subnuclei resting-state functional connectivity sex and estrogen differences. Psychoneuroendocrinology 63:34–42. doi:10.1016/j.psyneuen.2015.09.012

Fanselow MS (1994) Neural organization of the defensive behavior system responsible for fear. Psychon Bull Rev 1:429–438

Fonzo GA, Simmons AN, Thorp SR, Norman SB, Paulus MP, Stein MB (2010) Exaggerated and disconnected insular-amygdalar blood oxygenation level-dependent response to threat-related emotional faces in women with intimate-partner violence posttraumatic stress disorder. Biol Psychiatry 68(5):433–441. doi:10.1016/J.Biopsych.2010.04.028

Francis DD, Diorio J, Plotsky PM, Meaney MJ (2002) Environmental enrichment reverses the effects of maternal separation on stress reactivity. J Neurosci 22(18):7840–7843

Funayama ES, Grillon C, Davis M, Phelps EA (2001) A double dissociation in the affective modulation of startle in humans: effects of unilateral temporal lobectomy. J Cogn Neurosci 13(6):721–729

Gamwell K, Nylocks M, Cross D, Bradley B, Norrholm SD, Jovanovic T (2015) Fear conditioned responses and PTSD symptoms in children: sex differences in fear-related symptoms. Dev Psychobiol 57(7):799–808. doi:10.1002/dev.21313

Gee DG, Humphreys KL, Flannery J, Goff B, Telzer EH, Shapiro M, Hare TA, Bookheimer SY, Tottenham N (2013a) A developmental shift from positive to negative connectivity in human amygdala–Prefrontal circuitry. J Neurosci 33(10):4584–4593. doi:10.1523/jneurosci.3446-12.2013

Gee DG, Gabard-Durnam LJ, Flannery J, Goff B, Humphreys KL, Telzer EH, Hare TA, Bookheimer SY, Tottenham N (2013b) Early developmental emergence of human amygdala-prefrontal connectivity after maternal deprivation. Proc Natl Acad Sci U S A 110 (39):15638–15643. doi:10.1073/pnas.1307893110

Gee DG, Gabard-Durnam L, Telzer EH, Humphreys KL, Goff B, Shapiro M, Flannery J, Lumian DS, Fareri DS, Caldera C, Tottenham N (2014) Maternal buffering of human amygdala-prefrontal circuitry during childhood but not during adolescence. Psychol Sci 25 (11):2067–2078. doi:10.1177/0956797614550878

Giedd JN, Vaituzis AC, Hamburger SD, Lange N, Rajapakse JC, Kaysen D, Vauss YC, Rapoport JL (1996) Quantitative MRI of the temporal lobe, amygdala, and hippocampus in normal human development: ages 4–18 years. J Comp Neurol 366(2):223–230

Giedd JN, Blumenthal J, Jeffries NO, Castellanos FX, Liu H, Zijdenbos A, Paus T, Evans AC, Rapoport JL (1999) Brain development during childhood and adolescence: a longitudinal MRI study. Nat Neurosci 2(10):861–863

Giedd JN, Raznahan A, Alexander-Bloch A, Schmitt E, Gogtay N, Rapoport JL (2015) Child psychiatry branch of the National Institute of Mental Health longitudinal structural magnetic resonance imaging study of human brain development. Neuropsychopharmacology 40 (1):43–49. doi:10.1038/npp.2014.236

Glenn CR, Klein DN, Lissek S, Britton JC, Pine DS, Hajcak G (2011) The development of fear learning and generalization in 8–13 year-olds. Dev Psychobiol 54(7):675–684. doi:10.1002/dev.20616

Glover EM, Jovanovic T, Mercer KB, Kerley K, Bradley B, Ressler KJ, Norrholm SD (2012) Estrogen levels are associated with extinction deficits in women with posttraumatic stress disorder. Biol Psychiatry 72(1):19–24. doi:10.1016/j.biopsych.2012.02.031

Glover EM, Mercer KB, Norrholm SD, Davis M, Duncan E, Bradley B, Ressler KJ, Jovanovic T (2013) Inhibition of fear is differentially associated with cycling estrogen levels in women. J Psychiatry Neurosci 38(3):120129

Gogtay N, Giedd JN, Lusk L, Hayashi KM, Greenstein D, Vaituzis AC, Nugent TF, Herman DH, Clasen LS, Toga AW, Rapoport JL, Thompson PM (2004) Dynamic mapping of human cortical development during childhood through early adulthood. Proc Natl Acad Sci U S A 101(21):8174–8179. doi:10.1073/pnas.0402680101

Guyer AE, Monk CS, McClure-Tone EB, Nelson EE, Roberson-Nay R, Adler AD, Fromm SJ, Leibenluft E, Pine DS, Ernst M (2008) A developmental examination of amygdala response to facial expressions. J Cogn Neurosci 20(9):1565–1582. doi:10.1162/jocn.2008.20114

Hendriksen H, Prins J, Olivier B, Oosting RS (2010) Environmental enrichment induces behavioral recovery and enhanced hippocampal cell proliferation in an antidepressant-resistant animal model for PTSD. PLoS One 5(8):e11943. doi:10.1371/journal.pone.0011943

Herringa RJ, Birn RM, Ruttle PL, Burghy CA, Stodola DE, Davidson RJ, Essex MJ (2013) Childhood maltreatment is associated with altered fear circuitry and increased internalizing symptoms by late adolescence. Proc Natl Acad Sci U S A 110(47):19119–19124. doi:10.1073/pnas.1310766110

Hill HM, Levermore M, Twaite J, Jones LP (1996) Exposure to community violence and social support as predictors of anxiety and social and emotional behavior among African American children. J Child Fam Stud 5(4):399–399

Hostinar CE, Johnson AE, Gunnar MR (2015) Parent support is less effective in buffering cortisol stress reactivity for adolescents compared to children. Dev Sci 18(2):281–297. doi:10.1111/desc.12195

Hu S, Pruessner JC, Coupe P, Collins DL (2013) Volumetric analysis of medial temporal lobe structures in brain development from childhood to adolescence. Neuroimage 74:276–287. doi:10.1016/j.neuroimage.2013.02.032

Humphreys KL, Gleason MM, Drury SS, Miron D, Nelson 3rd CA, Fox NA, Zeanah CH (2015) Effects of institutional rearing and foster care on psychopathology at age 12 years in Romania: follow-up of an open, randomised controlled trial. Lancet Psychiatry 2(7):625–634. doi:10.1016/S2215-0366(15)00095-4

Hwang MJ, Zsido RG, Song H, Pace-Schott EF, Miller KK, Lebron-Milad K, Marin M-F, Milad MR (2015) Contribution of estradiol levels and hormonal contraceptives to sex differences within the fear network during fear conditioning and extinction. BMC Psychiatry 15(1):295

Jovanovic T, Norrholm SD (2011) Neural mechanisms of impaired fear inhibition in posttraumatic stress disorder. Front Behav Neurosci:5. doi:10.3389/fnbeh.2011.00044

Jovanovic T, Ressler KJ (2010) How the neurocircuitry and genetics of fear inhibition may inform our understanding of PTSD. Am J Psychiatry 167(6):648–662. doi:10.1176/appi.ajp.2009.09071074

Jovanovic T, Norrholm SD, Blanding NQ, Davis M, Duncan E, Bradley B, Ressler KJ (2010) Impaired fear inhibition is a biomarker of PTSD but not depression. Depress Anxiety 27 (3):244–251. doi:10.1002/da.20663

Jovanovic T, Norrholm SD, Davis J, Mercer KB, Almli L, Nelson A, Cross D, Smith A, Ressler KJ, Bradley B (2013) PAC1 receptor (ADCYAP1R1) genotype is associated with dark-enhanced startle in children. Mol Psychiatry 18(7):742–743. doi:10.1038/mp.2012.98

Jovanovic T, Nylocks KM, Gamwell KL, Smith A, Davis TA, Norrholm SD, Bradley B (2014) Development of fear acquisition and extinction in children: effects of age and anxiety. Neurobiol Learn Mem 113:135–142. doi:10.1016/j.nlm.2013.10.016

Kaufman J, Yang B-Z, Douglas-Palumberi H, Grasso D, Lipschitz D, Houshyar S, Krystal JH, Gelernter J (2006) Brain-derived neurotrophic factor–5-HTTLPR Gene interactions and environmental modifiers of depression in children. Biol Psychiatry 59(8):673–680 http://dx.doi.org/10.1016/j.biopsych.2005.10.026

Keding TJ, Herringa RJ (2015) Abnormal structure of fear circuitry in pediatric post-traumatic stress disorder. Neuropsychopharmacology 40(3):537–545. doi:10.1038/npp.2014.239

Kessler RC, Sonnega A, Bromet E, Hughes M, Nelson CB (1995) Posttraumatic stress disorder in the National Comorbidity Survey. Arch Gen Psychiatry 52(12):1048–1060. doi:10.1001/archpsyc.1995.03950240066012

Kessler RC, Berglund P, Demler O, Jin R, Merikangas KR, Walters EE (2005) Lifetime prevalence and age-of-onset distributions of DSM-IV disorders in the National Comorbidity Survey Replication. Arch Gen Psychiatry 62(6):593–602

Kikusui T, Winslow JT, Mori Y (2006) Social buffering: relief from stress and anxiety. Philos Trans R Soc Lond B Biol Sci 361(1476):2215–2228. doi:10.1098/rstb.2006.1941

Kline A, Ciccone DS, Weiner M, Interian A, St. Hill L, Falca-Dodson M, Black CM, Losonczy M (2013) Gender differences in the risk and protective factors associated with PTSD: a prospective study of National Guard troops deployed to Iraq. Psychiatry 76(3):256–272

Koch SB, van Zuiden M, Nawijn L, Frijling JL, Veltman DJ, Olff M (2016) Intranasal oxytocin normalizes amygdala functional connectivity in post-traumatic stress disorder. Neuropsychopharmacology. doi:10.1038/npp.2016.1

Kolb B, Mychasiuk R, Muhammad A, Li Y, Frost DO, Gibb R (2012) Experience and the developing prefrontal cortex. Proc Natl Acad Sci 109(Supplement 2):17186–17193. doi:10.1073/pnas.1121251109

Lebel C, Gee M, Camicioli R, Wieler M, Martin W, Beaulieu C (2012) Diffusion tensor imaging of white matter tract evolution over the lifespan. Neuroimage 60(1):340–352

LeDoux J (1992) Brain mechanisms of emotion and emotional learning. Curr Opin Neurobiol 2:191–197

Levine S, Johnson DF, Gonzalez CA (1985) Behavioral and hormonal responses to separation in infant rhesus monkeys and mothers. Behav Neurosci 99(3):399–410

Li S, Kim JH, Richardson R (2012) Differential involvement of the medial prefrontal cortex in the expression of learned fear across development. Behav Neurosci 126:217–225

Lissek S, Bradford DE, Alvarez RP, Burton P, Espensen-Sturges T, Reynolds RC, Grillon C (2014) Neural substrates of classically conditioned fear-generalization in humans: a parametric fMRI study. Soc Cogn Affect Neurosci 9(8):1134–1142. doi:10.1093/scan/nst096

MacLusky NJ, Naftolin F, Goldman-Rakic PS (1986) Estrogen formation and binding in the cerebral cortex of the developing rhesus monkey. Proc Natl Acad Sci U S A 83(2):513–516

Makley AT, Falcone Jr RA (2010) Posttraumatic stress disorder in the pediatric trauma patient. Semin Pediatr Surg 19(4):292–299. doi:10.1053/j.sempedsurg.2010.06.006

McClure EB, Monk CS, Nelson EE, Zarahn E, Leibenluft E, Bilder RM, Charney DS, Ernst M, Pine DS (2004) A developmental examination of gender differences in brain engagement during evaluation of threat. Biol Psychiatry 55(11):1047–1055. doi:10.1016/j.biopsych.2004.02.013

McLaughlin KA, Fox NA, Zeanah CH, Nelson CA (2011) Adverse rearing environments and neural development in children: the development of frontal electroencephalogram asymmetry. Biol Psychiatry 70(11):1008–1015. doi:10.1016/j.biopsych.2011.08.006

McManis MH, Bradley MM, Berg WK, Cuthbert BN, Lang PJ (2001) Emotional reactions in children: verbal, physiological, and behavioral responses to affective pictures. Psychophysiology 38(2):222–231

Michalska KJ, Shechner T, Hong M, Britton JC, Leibenluft E, Pine DS, Fox NA (2016) A developmental analysis of threat/safety learning and extinction recall during middle childhood. J Exp Child Psychol 146:95–105. doi:10.1016/j.jecp.2016.01.008

Michopoulos V, Rothbaum AO, Corwin E, Bradley B, Ressler KJ, Jovanovic T (2014) Psychophysiology and posttraumatic stress disorder symptom profile in pregnant African-American women with trauma exposure. Arch Womens Ment Health 18(4):639–648

Mithoefer MC, Grob CS, Brewerton TD (2016) Novel psychopharmacological therapies for psychiatric disorders: psilocybin and MDMA. Lancet Psychiatry 3(5):481–488. doi:10.1016/S2215-0366(15)00576-3

Moriceau S, Sullivan RM (2006) Maternal presence serves as a switch between learning fear and attraction in infancy. Nat Neurosci 9(8):1004–1006 http://www.nature.com/neuro/journal/v9/n8/suppinfo/nn1733_S1.html

Moulson MC, Westerlund A, Fox NA, Zeanah CH, Nelson CA (2009) The effects of early experience on face recognition: an event-related potential study of institutionalized children in Romania. Child Dev 80(4):1039–1056. doi:10.1111/j.1467-8624.2009.01315.x

Nelson CA, Zeanah CH, Fox NA, Marshall PJ, Smyke AT, Guthrie D (2007) Cognitive recovery in socially deprived young children: the Bucharest early intervention project. Science 318 (5858):1937–1940. doi:10.1126/science.1143921

Nemeroff CB, Bremner JD, Foa EB, Mayberg HS, North CS, Stein MB (2006) Posttraumatic stress disorder: a state-of-the-science review. J Psychiatr Res 40(1):1–21

Nithianantharajah J, Hannan AJ (2006) Enriched environments, experience-dependent plasticity and disorders of the nervous system. Nat Rev Neurosci 7(9):697–709

Nowakowska E, Kus K, Ratajczak P, Cichocki M, Woźniak A (2014) The influence of aripiprazole, olanzapine and enriched environment on depressant-like behavior, spatial memory dysfunction and hippocampal level of BDNF in prenatally stressed rats. Pharmacol Rep 66 (3):404–411. doi:10.1016/j.pharep.2013.12.008

Pagliaccio D, Luby JL, Gaffrey MS, Belden AC, Botteron KN, Harms MP, Barch DM (2013) Functional brain activation to emotional and nonemotional faces in healthy children: evidence for developmentally undifferentiated amygdala function during the school-age period. Cogn Affect Behav Neurosci 13(4):771–789. doi:10.3758/s13415-013-0167-5

Pagliaccio D, Luby JL, Bogdan R, Agrawal A, Gaffrey MS, Belden AC, Botteron KN, Harms MP, Barch DM (2015) HPA axis genetic variation, pubertal status, and sex interact to predict amygdala and hippocampus responses to negative emotional faces in school-age children. Neuroimage 109:1–11. doi:10.1016/j.neuroimage.2015.01.017

Pechtel P, Lyons-Ruth K, Anderson CM, Teicher MH (2014) Sensitive periods of amygdala development: the role of maltreatment in preadolescence. Neuroimage 97:236–244. doi:10.1016/j.neuroimage.2014.04.025

Pietrzak RH, Johnson DC, Goldstein MB, Malley JC, Southwick SM (2009) Psychological resilience and postdeployment social support protect against traumatic stress and depressive symptoms in soldiers returning from Operations Enduring Freedom and Iraqi Freedom. Depress Anxiety 26(8):745–751. doi:10.1002/da.20558

Pietrzak RH, Johnson DC, Goldstein MB, Malley JC, Rivers AJ, Morgan CA, Southwick SM (2010) Psychosocial buffers of traumatic stress, depressive symptoms, and psychosocial difficulties in veterans of Operations Enduring Freedom and Iraqi Freedom: the role of resilience, unit support, and postdeployment social support. J Affect Disord 120(1–3):188–192

Pohlack ST, Nees F, Ruttorf M, Cacciaglia R, Winkelmann T, Schad LR, Witt SH, Rietschel M, Flor H (2015) Neural mechanism of a sex-specific risk variant for posttraumatic stress disorder in the Type I receptor of the pituitary adenylate cyclase activating polypeptide. Biol Psychiatry 78(12):840–847

Rauch SL, Whalen PJ, Shin LM, McInerney SC, Macklin ML, Lasko NB, Orr SP, Pitman RK (2000) Exaggerated amygdala response to masked facial stimuli in posttraumatic stress disorder: a functional MRI study. Biol Psychiatry 47(9):769–776

Ressler KJ, Mercer KB, Bradley B, Jovanovic T, Mahan A, Kerley K, Norrholm SD, Kilaru V, Smith AK, Myers AJ, Ramirez M, Engel A, Hammack SE, Toufexis D, Braas KM, Binder EB, May V (2011) Post-traumatic stress disorder is associated with PACAP and the PAC1 receptor. Nature 470(7335):492–497. doi:10.1038/nature09856

Roselli CE, Klosterman S, Resko JA (2001) Anatomic relationships between aromatase and androgen receptor mRNA expression in the hypothalamus and amygdala of adult male cynomolgus monkeys. J Comp Neurol 439(2):208–223

Ros-Simó C, Valverde O (2012) Early-life social experiences in mice affect emotional behaviour and hypothalamic-pituitary-adrenal axis function. Pharmacol Biochem Behav 102(3):434–441

Sanchez M, Ladd C, Plotsky P (2001) Early adverse experience as a developmental risk factor for later psychopathology: evidence from rodent and primate models. Dev Psychopathol 13(3):419

Sarro EC, Wilson DA, Sullivan RM (2014) Maternal regulation of infant brain state. Curr Biol 24 (14):1664–1669. doi:10.1016/j.cub.2014.06.017

Schüssler-Fiorenza Rose SM, Xie D, Stineman M (2014) Adverse childhood experiences and disability in U.S. adults. PM R 6(8):670–680. doi:10.1016/j.pmrj.2014.01.013

Schwartz CE, Kunwar PS, Greve DN, Kagan J, Snidman NC, Bloch RB (2012) A phenotype of early infancy predicts reactivity of the amygdala in male adults. Mol Psychiatry 17 (10):1042–1050. doi:10.1038/mp.2011.96

Shannon C, Champoux M, Suomi SJ (1998) Rearing condition and plasma cortisol in rhesus monkey infants. Am J Primatol 46(4):311–321. doi:10.1002/(SICI)1098-2345(1998)46:4<311::AID-AJP3>3.0.CO;2-L

Shansky RM, Hamo C, Hof PR, McEwen BS, Morrison JH (2009) Stress-induced dendritic remodeling in the prefrontal cortex is circuit specific. Cereb Cortex 19(10):2479–2484

Shansky RM, Hamo C, Hof PR, Lou W, McEwen BS, Morrison JH (2010) Estrogen promotes stress sensitivity in a prefrontal cortex-amygdala pathway. Cereb Cortex 20(11):2560–2567. doi:10.1093/cercor/bhq003

Shaw P, Greenstein D, Lerch J, Clasen L, Lenroot R, Gogtay N, Evans A, Rapoport J, Giedd J (2006) Intellectual ability and cortical development in children and adolescents. Nature 440 (7084):676

Shechner T, Britton JC, Ronkin EG, Jarcho JM, Mash JA, Michalska KJ, Leibenluft E, Pine DS (2015) Fear conditioning and extinction in anxious and nonanxious youth and adults: examining a novel developmentally appropriate fear-conditioning task. Depress Anxiety 32 (4):277–288. doi:10.1002/da.22318

Shin LM, Wright CI, Cannistraro PA, Wedig MM, McMullin K, Martis B, Macklin ML, Lasko NB, Cavanagh SR, Krangel TS, Orr SP, Pitman RK, Whalen PJ, Rauch SL (2005) A functional magnetic resonance imaging study of amygdala and medial prefrontal cortex responses to overtly presented fearful faces in posttraumatic stress disorder. Arch Gen Psychiatry 62 (3):273–281. doi:10.1001/archpsyc.62.3.273

Shin LM, Liberzon I (2010) The neurocircuitry of fear, stress, and anxiety disorders. Neuropsychopharmacology 35(1):169–191

Spinelli S, Chefer S, Suomi SJ, Higley JD, Barr CS, Stein E (2009) Early-life stress induces long-term morphologic changes in primate brain. Arch Gen Psychiatry 66(6):658–665. doi:10.1001/archgenpsychiatry.2009.52

Sripada RK, King AP, Garfinkel SN, Wang X, Sripada CS, Welsh RC, Liberzon I (2012) Altered resting-state amygdala functional connectivity in men with posttraumatic stress disorder. J Psychiatry Neurosci 37(4):241–249. doi:10.1503/jpn.110069

Steven BK, Williams G, Najman J, Alati R (2013) Exploring the female specific risk to partial and full PTSD following physical assault. J Trauma Stress 26(1):86

Stevens JS, Hamann S (2012) Sex differences in brain activation to emotional stimuli: a meta-analysis of neuroimaging studies. Neuropsychologia 50(7):1578–1593. doi:10.1016/j.neuropsychologia.2012.03.011

Stevens JS, Jovanovic T, Fani N, Ely TD, Glover EM, Bradley B, Ressler KJ (2013) Disrupted amygdala-prefrontal functional connectivity in civilian women with posttraumatic stress disorder. J Psychiatr Res 47(10):1469–1478. doi:10.1016/j.jpsychires.2013.05.031

Stevens JS, Almli LM, Fani N, Gutman DA, Bradley B, Norrholm SD, Reiser E, Ely TD, Dhanani R, Glover EM, Jovanovic T, Ressler KJ (2014) PACAP receptor gene polymorphism impacts fear responses in the amygdala and hippocampus. Proc Natl Acad Sci U S A 111 (8):3158–3163. doi:10.1073/pnas.1318954111

Suitts S (2015) A new majority: low income students now a majority in the Nation's Public Schools. Southern Education Foundation Report

Sullivan RM, Landers M, Yeaman B, Wilson DA (2000) Neurophysiology: good memories of bad events in infancy. Nature 407(6800):38–39

Takuma K, Ago Y, Matsuda T (2011) Preventive effects of an enriched environment on rodent psychiatric disorder models. J Pharmacol Sci 117(2):71–76. doi:10.1254/jphs.11R07CP

Tolin DF, Foa EB (2006) Sex differences in trauma and posttraumatic stress disorder: a quantitative review of 25 years of research. Psychol Bull 132(6):959

Tottenham N, Sheridan MA (2009) A review of adversity, the amygdala and the hippocampus: a consideration of developmental timing. Front Hum Neurosci 3:68. doi:10.3389/neuro.09.068.2009

Tottenham N, Hare TA, Quinn BT, McCarry TW, Nurse M, Gilhooly T, Millner A, Galvan A, Davidson MC, Eigsti I-M, Thomas KM, Freed PJ, Booma ES, Gunnar MR, Altemus M, Aronson J, Casey BJ (2010) Prolonged institutional rearing is associated with atypically large amygdala volume and difficulties in emotion regulation. Dev Sci 13(1):46–61

Tottenham N, Hare TA, Millner A, Gilhooly T, Zevin JD, Casey BJ (2011) Elevated amygdala response to faces following early deprivation. Dev Sci 14(2):190–204. doi:10.1111/j.1467-7687.2010.00971.x

van Rooij SJH, Cross D, Stevens JS, Vance LA, Kim YJ, Bradley B, Tottenham N, Jovanovic T (2016) Maternal buffering of fear-potentiated startle in children and adolescents with trauma exposure. Soc Neurosci. doi:10.1080/17470919.2016.1164244

Velez CN, Johnson JIM, Cohen P (1989) A longitudinal analysis of selected risk factors for childhood psychopathology. J Am Acad Child Adolesc Psychiatry 28(6):861–864. doi:10.1097/00004583-198911000-00009

Waters AM, Craske MG, Bergman RL, Naliboff BD, Negoro H, Ornitz EM (2008) Developmental changes in startle reactivity in school-age children at risk for and with actual anxiety disorder. Int J Psychophysiol 70(3):158–164

Winslow JT, Noble PL, Lyons CK, Sterk SM, Insel TR (2003) Rearing effects on cerebrospinal fluid oxytocin concentration and social buffering in rhesus monkeys. Neuropsychopharmacology 28(5):910–918

Wolf RC, Herringa RJ (2016) Prefrontal-amygdala dysregulation to threat in pediatric posttraumatic stress disorder. Neuropsychopharmacology 41(3):822–831

Yang J, Li W, Liu X, Li Z, Li H, Yang G, Xu L, Li L (2006) Enriched environment treatment counteracts enhanced addictive and depressive-like behavior induced by prenatal chronic stress. Brain Res 1125(1):132–137. doi:10.1016/j.brainres.2006.10.028

Yang J, Hou C, Ma N, Liu J, Zhang Y, Zhou J, Xu L, Li L (2007) Enriched environment treatment restores impaired hippocampal synaptic plasticity and cognitive deficits induced by prenatal chronic stress. Neurobiol Learn Mem 87(2):257–263. doi:10.1016/j.nlm.2006.09.001

MicroRNAs in Post-traumatic Stress Disorder

Clara Snijders, Laurence de Nijs, Dewleen G. Baker, Richard L. Hauger, Daniel van den Hove, Gunter Kenis, Caroline M. Nievergelt, Marco P. Boks, Eric Vermetten, Fred H. Gage, and Bart P.F. Rutten

Clara Snijders and Laurence de Nijs contributed equally to this work.

C. Snijders, L. de Nijs, G. Kenis, and B.P.F. Rutten (✉)
Department of Psychiatry and Neuropsychology, School for Mental Health and Neuroscience (MHeNS), Faculty of Health, Medicine and Life Sciences, Maastricht University, European Graduate School of Neuroscience, (EURON), Maastricht 6200 MD, The Netherlands
e-mail: b.rutten@maastrichtuniversity.nl

D.G. Baker and R.L. Hauger
Department of Psychiatry, University of California, San Diego, La Jolla, CA 92037, USA

VA Center of Excellence for Stress and Mental Health, San Diego, La Jolla, CA 92037, USA

VA San Diego Healthcare System, San Diego, La Jolla, CA 92037, USA

D. van den Hove
Department of Psychiatry and Neuropsychology, School for Mental Health and Neuroscience (MHeNS), Faculty of Health, Medicine and Life Sciences, Maastricht University, European Graduate School of Neuroscience, (EURON), Maastricht 6200 MD, The Netherlands

Laboratory of Translational Neuroscience, Department of Psychiatry, Psychosomatics and Psychotherapy, University of Würzburg, Würzburg 97080, Germany

C.M. Nievergelt
Department of Psychiatry, University of California, San Diego, La Jolla, CA 92037, USA

VA Center of Excellence for Stress and Mental Health, San Diego, La Jolla, CA 92037, USA

M.P. Boks
Brain Center Rudolf Magnus, University Medical Center Utrecht, Utrecht 3584 CG, The Netherlands

E. Vermetten
Military Mental Health Research Center, Ministry of Defense, P.O. Box 90000, Utrecht 3509 AA, The Netherlands

Department of Psychiatry, Leiden University Medical Center, Leiden 2333 ZA, The Netherlands

Arq Psychotrauma Research Group, Diemen 1112 XE, The Netherlands

F.H. Gage
Laboratory of Genetics, The Salk Institute for Biological Studies, La Jolla, CA 92037, USA

Curr Topics Behav Neurosci (2018) 38: 23–46
DOI 10.1007/7854_2017_32
Published Online: 17 October 2017

Abstract Post-traumatic stress disorder (PTSD) is a psychiatric disorder that can develop following exposure to or witnessing of a (potentially) threatening event. A critical issue is to pinpoint the (neuro)biological mechanisms underlying the susceptibility to stress-related disorder such as PTSD, which develops in the minority of ~15% of individuals exposed to trauma. Over the last few years, a first wave of epigenetic studies has been performed in an attempt to identify the molecular underpinnings of the long-lasting behavioral and mental effects of trauma exposure. The potential roles of non-coding RNAs (ncRNAs) such as microRNAs (miRNAs) in moderating or mediating the impact of severe stress and trauma are increasingly gaining attention. To date, most studies focusing on the roles of miRNAs in PTSD have, however, been completed in animals, using cross-sectional study designs and focusing almost exclusively on subjects with susceptible phenotypes. Therefore, there is a strong need for new research comprising translational and cross-species approaches that use longitudinal designs for studying trajectories of change contrasting susceptible and resilient subjects. The present review offers a comprehensive overview of available studies of miRNAs in PTSD and discusses the current challenges, pitfalls, and future perspectives of this field.

Keywords Brain • Epigenetics • microRNA • Post-traumatic stress disorder • Review

Contents

1 Introduction

Over the last few decades, epigenetic mechanisms have been proposed to be key mediators of the lasting behavioral and molecular effects of traumatic stress exposure (Schmidt et al. 2011). While a first wave of epigenetic studies in this area focused mostly on DNA methylation, epigenetic studies in more recent years have expanded this approach by analyzing the expression of non-coding RNA (ncRNA) species and their impact on gene expression. These RNA molecules of different sizes and forms include non-coding stretches of 20–25 nucleotides named microRNAs (miRNAs). These are increasingly being investigated for their pathophysiological connection to psychiatric disorders including post-traumatic stress disorder (PTSD). More recently, studies have started to focus on the potential use of miRNAs as biomarkers of PTSD.

The present review provides an overview of the current status of the literature on miRNAs in relation to exposure to traumatic stress and its impact on mental health in humans and other mammals. To do so, we briefly describe PTSD-related neurobiological alterations along with the basic concepts of epigenetic mechanisms. Next, an overview of the current scientific evidence on miRNAs in relation to PTSD in humans and PTSD-related symptoms in animals is provided. Finally, current challenges, pitfalls, and future perspectives in studying the potential role of miRNAs in PTSD are discussed.

2 Post-traumatic Stress Disorder

As we know, PTSD is a psychiatric disorder that is triggered by a (potentially) life-threatening traumatic event, i.e., an event capable of producing intense feelings of fear, helplessness, and horror (American Psychiatric Association 2013). Characteristic symptoms include re-experiencing of the traumatic event through intrusive imagery or recurrent nightmares, constant avoidance of reminders of the event, negative mood, and hyperarousal reflected by insomnia and/or hypervigilance. Although these symptoms are often of limited intensity and duration, in a small, susceptible minority of the population they persist longer than 1 month following trauma exposure and create significant distress. Long-term persistence of symptoms is characteristic of PTSD, while the ability to withstand trauma without developing any stress symptoms or rapid recovery from an acute stress reaction without progression to PTSD is referred to as resiliency.

Over the past few decades, PTSD has repeatedly been associated with several neurobiological alterations including decreased hippocampal volume (Smith 2005; Karl et al. 2006; Shin et al. 2006), hyperactivity of the amygdala and hypoactivity of the dorsal and rostral anterior cingulate (AC) cortices and ventromedial prefrontal cortex (vmPFC) (Shin et al. 2006; Etkin and Wager 2007; El Khoury-Malhame et al. 2011). In an attempt to further elucidate the (neuro)biological processes underlying the observed differential susceptibility to traumatic stress, a large number of studies have focused on alterations in the hypothalamus-pituitary-adrenal (HPA) axis. Since the HPA axis is a core component of the mammalian stress response, its (dys) function has been extensively studied in the context of PTSD. In healthy individuals, stressful events trigger neurons of the hypothalamic paraventricular nucleus (PVN) to secrete corticotropin-releasing hormone (CRH) and vasopressin, which causes the release of adrenocorticotropin (ACTH) from the anterior pituitary and finally glucocorticoids from the adrenal cortex (Chrousos and Gold 1992). The activity of the HPA axis is modulated via several brain regions; for example, CRH neurons in the

PVN are inhibited by the hippocampus and PFC and stimulated by areas such as the amygdala (Sherin and Nemeroff 2011). Finally, in order to regulate their own synthesis, glucocorticoids inhibit excessive synthesis and release of CRH and ACTH by controlling hippocampal and PVN neurons, and downregulating CRH_1 receptors and corticotrope function in the anterior pituitary, thereby creating a negative feedback mechanism (Sherin and Charles 2011).

Several studies have found that subjects with PTSD show increased levels of CRH in cerebrospinal fluid (CSF) (Baker et al. 1999), as well as a blunted ACTH response to CRH (Yehuda 2006), a disturbed negative feedback loop (Geracioti et al. 2008), and increased sensitivity of glucocorticoid receptors (GRs) and chronically lowered cortisol levels (Yehuda 2001; Yehuda et al. 2000). Although dysregulation of the HPA axis is well-documented in the context of stress-related disorders and PTSD has repeatedly been associated with reduced cortisol levels, variability in response between individuals remains. The current hypothesis is that cortisol levels depend upon gender and the type of trauma exposure among other factors (Meewisse et al. 2007; Young and Breslau 2004; Lemieux and Coe 1995). To further unravel the molecular regulation of biological mechanisms underlying the onset and course of PTSD, more recent research has also focused on the involvement of epigenetic mechanisms.

3 Epigenetics: The Role of miRNAs

The term epigenetics refers to a variety of heritable but reversible processes involved in the regulation of gene expression under influence of environmental factors without the original genetic code being altered (Peschansky and Wahlestedt 2014). These epigenetic modifications are numerous and include (hydroxy)methylation of DNA cytosine residues, post-translational modifications (PTMs) of histone proteins and ncRNAs (Kouzarides 2007; Venkatesh and Workman 2015). ncRNAs refer to a class of small RNA molecules that are transcribed from genomic DNA without being translated into proteins (Peschansky and Wahlestedt 2014). Instead, these RNAs are directly involved in cellular function and gene regulation. Next to ribosomal and transfer RNAs, ncRNAs include the most commonly studied small interfering RNAs (Zamore 2002), circular RNAs (Memczak et al. 2013), piwi-interacting RNAs (Aravin et al. 2007), and miRNAs.

3.1 Biogenesis and Mode of Action of miRNAs

miRNAs are small (~22 nt in length) ncRNA molecules found in most eukaryotes (Fabian and Sonenberg 2012). Hundreds of different miRNAs are expressed within an organism and are involved in post-transcriptional regulation of gene expression (Pritchard et al. 2012). miRNAs are commonly classified as "intergenic" or "intronic." Intergenic miRNA are transcribed from genomic DNA by RNA polymerase II and/or III (Borchert et al. 2006) and intronic miRNA are processed from intronic regions of heterogeneous nuclear RNA (hnRNA) (Ramalingam et al. 2014). In both cases, a primary miRNA (pri-miRNA) is formed and further cleaved and stabilized by the protein complex microprocessor that includes the ribonuclease III Drosha and its co-factor, DiGeorge syndrome critical region 8 (DGCR8) (Borchert et al. 2006). This process takes place within the nucleus and results in a precursor miRNA (pre-miRNA) of 70–100 nt in length forming a hairpin structure (Issler and Chen 2015; Lee et al. 2003). Following transport to the cytoplasm by the nuclear transport factor Exportin-5, a complex including the RNase III Dicer further processes the pre-miRNA to yield a miRNA duplex containing the final mature miRNA strand and a so-called passenger strand (Fig. 1) (Davis-Dusenbery and Hata 2010).

Binding of the 5′ end of the mature miRNA (i.e., the "seed" sequence) to an almost complementary 6–8 nt seed match sequence in the 3′ UTR of mRNA induces mRNA degradation or translational inhibition (Pritchard et al. 2012;

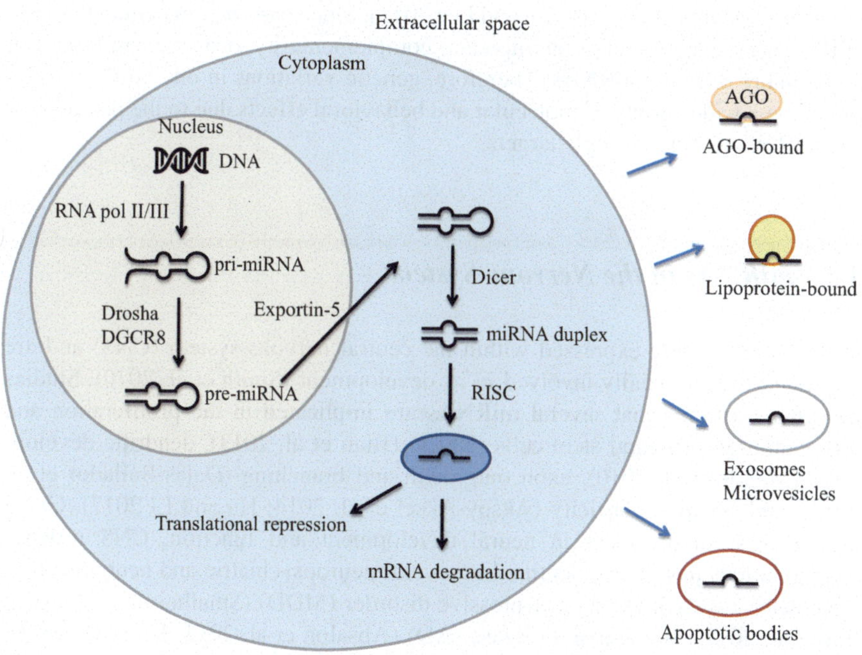

Fig. 1 miRNA biogenesis and cellular locations. miRNAs are transcribed into pri-miRNA by RNA polymerase II and/or III before being further processed by Drosha and DGCR8 to form a cleaved pre-miRNA. After transportation to the cytoplasm by Exportin-5, this pre-miRNA is further digested by a complex including the RNase III Dicer. The mature miRNA is then involved in translational repression and/or mRNA degradation through interaction with the RISC. In the extracellular space, miRNAs are protected from degradation by RNases through binding to RNA-binding proteins (e.g., Ago 1 or 2) or (high-density) lipoproteins, or packaging into exosomes or microvesicles

Davis-Dusenbery and Hata 2010). Thus, miRNAs hold the potential to post-transcriptionally regulate gene expression. Specifically, the mature miRNA triggers the activation of the RNA-induced silencing complex (RISC), a large protein complex containing an Argonaute protein (Ago2) needed for gene silencing, and the mature single-stranded miRNA that leads the complex towards the appropriate mRNA target (Fig. 1) (Fabian and Sonenberg 2012). It was commonly believed that, at this point, only the functional guide strand of the double-stranded miRNA product was incorporated into the RISC and the passenger strand was being degraded (Issler and Chen 2015). However, increasing evidence shows that the passenger strand also has biological functions and target mRNAs (Yang et al. 2013). In either case, depending on the type of Ago protein, the target will be cleaved directly or additional proteins may be needed to achieve silencing. However, exactly how this complex interacts with mRNA strands and which additional proteins are recruited remains unclear.

Currently, it is believed that miRNAs regulate 30–60% of human protein-coding genes (Friedman et al. 2009; Lewis et al. 2005). Several studies have investigated genetic variations such as single nucleotide polymorphisms (SNPs) in the 3′ UTRs of mRNAs (Hanin et al. 2014; Jin and Lee 2013). Since base-pair matching between miRNAs and mRNAs relies on imprecise complementarity, one single miRNA can target many different mRNAs. Therefore, genetic variations in one miRNA target can cause a wide variety of molecular and behavioral effects due to the potential of one miRNA to bind multiple targets.

3.2 miRNAs in the Nervous System

miRNAs are widely expressed within the central nervous system (CNS) and are suggested to be crucially involved in its development (Smith et al. 2010). Studies have demonstrated that several miRNAs are implicated in the proliferation and differentiation of neural stem cells (NSCs) (Bian et al. 2013), dendritic development (Magill et al. 2010), axon outgrowth and branching (Dajas-Bailador et al. 2012), and synaptic plasticity (Aksoy-Aksel et al. 2014; Hu and Li 2017). Given their central involvement in neural development and function, CNS miRNA dysregulations have been identified in several neuropsychiatric and neurodegenerative disorders such as major depressive disorder (MDD) (Smalheiser et al. 2012; Bai et al. 2012), Alzheimer's disease (AD) (Absalon et al. 2013; Hu et al. 2013), and Parkinson's disease (PD) (Wang et al. 2008; Doxakis 2010). Identifying exactly which and how miRNAs within the CNS interact to exert their regulatory effects will be crucial for our understanding of their precise involvement in these and other neurological disorders.

3.3 Circulating miRNAs

While most miRNAs are found inside the cells, a significant number of miRNAs have been observed in extracellular compartments such as biofluids, including blood plasma, serum, saliva, urine, tears, and CSF (Park et al. 2009; Taylor and Gercel-Taylor 2013; Hanke et al. 2010; Weber et al. 2010). These extracellular miRNAs are relatively stable since they are commonly bound to proteins such as Ago1 or 2 and (mostly high density) lipoproteins or packed into vesicles and thus protected from degradation by RNases (Fig. 1) (Taylor and Gercel-Taylor 2013; Turchinovich et al. 2013; Camussi et al. 2011; Valadi et al. 2007; Mitchell et al. 2008; Vickers et al. 2011; Wagner et al. 2013).

Packaging of miRNAs is the most common mechanism used to protect circulating miRNAs. miRNAs can be packaged into apoptotic bodies, shedding vesicles called microvesicles, or exosomes resulting from multivesicular bodies (MVBs) fusing with the plasma membrane (Taylor and Gercel-Taylor 2013; Turchinovich et al. 2013). miRNAs encapsulated within MVBs are believed to arise from the disassembled RISC and are packed along with several RISC-associated components (Gibbings et al. 2009). Once secreted, exosomes translocate easily across cell membranes, thus allowing miRNAs to be taken up by other cells where they hold the potential to actively alter gene expression, among other functions (Wang et al. 2010). Although packaged miRNAs are thought to be specifically involved in RNA-mediated cell-to-cell communication, Ago-bound miRNAs appear to be non-specific residues of cellular activity or cell death (Turchinovich et al. 2013). Indeed, Ago-miRNA complexes have not been found to be actively released or taken up by recipient cells, unlike exosomal miRNAs (Turchinovich et al. 2013). Although several theories have been postulated with regard to extracellular miRNA origin, stability and precise function in recipient cells, many questions remain to be answered. However, circulating miRNAs have several properties that make them interesting relevant candidates to be investigated as biomarkers; they are stable in various biofluids, their sequences are conserved among different species, the expression of some miRNAs is specific to tissues or biological stages, and the level of miRNAs can be easily assessed by various methods, such as small-RNA sequencing, microarrays and quantitative polymerase chain reaction (PCR) (Etheridge et al. 2011). As such, circulating miRNAs in biofluids may reflect miRNA expression and/or dysfunction in the brain.

3.4 Mechanism of miRNA Regulation

In the past few years it has become clear that miRNA expression is regulated by DNA methylation and histone modifications and vice versa (Satrom et al. 2007). Several proteins of the methyl-CpG-binding domain (MBDs) family, i.e. proteins binding to methylated DNA cytosine residues, directly influence miRNA expression (Liu et al.

2010; Chen et al. 2012). Moreover, disturbed methylation patterns arising in promoter regions of miRNA genes have been linked to several human diseases, including neurodegenerative disorders (reviewed in (Van den Hove et al. 2014) for AD). Similarly, histone modifiers have not only been shown to interact with DNA methyltransferases (Dnmts), enzymes involved in maintaining or establishing de novo DNA methylation patterns (Rose and Klose 2014; Raabe and Spengler 2013), but are also suggested to affect miRNA expression levels (Scott et al. 2006). Interestingly, miRNAs themselves have been shown to target histone modifier molecules involved in histone PTM and Dnmt1, 3a, and 3b (Fabbri et al. 2007) through a process termed RNA-directed DNA methylation (Sato et al. 2011). For instance, Dicer-null mouse embryonic stem cells have been shown to express significantly lower levels of Dnmt1, Dnmt3a, and Dnmt3b, further resulting in altered DNA methylation patterns (Sinkkonen et al. 2008). The presence of such epigenetic feedback loops highlights the complex interaction between miRNAs and other epigenetic mechanisms (Schouten et al. 2013).

4 miRNAs, Stress and PTSD

The results of studies examining miRNAs in the context of stress and PTSD in humans or PTSD-related symptoms in animals are described below and summarized in Tables 1 and 2, respectively.

4.1 Evidence from Animal Studies

4.1.1 miRNAs and Fear Conditioning

Patients with PTSD are known to show enhanced fear conditioning and to benefit from exposure-based therapy (Blechert et al. 2007). This therapy is very similar to the fear extinction training used in animals (Norberg et al. 2008). Therefore, the first study to indirectly examine the role of miRNAs in PTSD focused on their involvement in fear extinction (Lin et al. 2011). In this study, the level of miR-128b was increased in the infra-limbic PFC (ILPFC) of mice following fear extinction training, implicating its involvement in fear conditioning (Lin et al. 2011). Previously, proteins involved in miRNA biogenesis had already been shown to play a role in memory formation. Indeed, the deletion of Dicer1 in the forebrain of mice caused a decrease in several miRNAs and enhanced learning and memory strength (Konopka et al. 2010). Several recent animal studies have confirmed that specific miRNAs in several brain regions are involved in fear memory consolidation (Dias et al. 2014; Griggs et al. 2013), contextual fear memory (Vetere et al. 2014), state-dependent fear (Jovasevic et al. 2015), and memory acquisition of trace fear conditioning (Wang et al. 2013).

Table 1 Animal studies examining the role of miRNAs in PTSD

Sex (if mentioned) and species	Model	Tissue	miRNA analyses	Primary finding
Male rats	Surgical traumatic stress	Frontal cortex	TaqMan miRNA assay, qRT-PCR	↑ miR-222 in the frontal cortex 3d following traumatic stress (Zhao et al. 2011)
	Cultured neurons			
Male rats	Auditory FC	Amygdala	qRT-PCR, miRNA microarray, TaqMan miRNA assay, miRNA overexpression	↓ miR-182 1 h following auditory FC. Overexpression in lateral amygdala disrupted long-term memory formation (Griggs et al. 2013)
Male rats	3d of immobilization and tail shock sessions	Serum, amygdala	qRT-PCR, TaqMan miRNA assay	↑ miR-142-5p, miR-19b, miR-1928, miR-223-3p, miR-322*, miR-324, miR-421-3p, miR-463*, miR-674* in serum & amygdala (Balakathiresan et al. 2014)
Male rats	7d CSDS	mPFC, BLA, circulation	miRNA microarray, qPCR	Vulnerability to stress is associated with: Circulation: ↓ miR-24-2-5p, miR-27a-3p, miR-30e-5p, miR-3590-3p, miR-362-3p, miR-532-5p mPFC: ↑ miR-126a-3p, miR-708-5p BLA: 77 dysregulated miRNAs, none associated with vulnerability to stress (Chen et al. 2015)
Male rats	6d of electric FS	Hypothalamus	RT-PCR	Traumatic stress was related to ↑ miR-34c in the hypothalamus (Li et al. 2016)
Mice	Fear extinction training	ILPFC	Lentiviral vector (miR KD/overexpression)	miR-128b is involved in formation of fear extinction memory (Lin et al. 2011)
Male mice	42d of chronic variable stress	Sperm	TaqMan miRNA assay	↑ miR-193*, miR-204, miR-29c, miR-30a, miR-30c, miR-32, miR-375, miR-532-3p, miR-698 in parental sperm (Rodgers et al. 2013)
Male mice	Single electric FS	PFC	Microarray, RT-qPCR	FXT administration in shocked mice causes ↓ mmu-miR-1971 expression (Schmidt et al. 2013)
Male mice	FC	Hippocampus	Lentiviral vector (miR KD), TaqMan miRNA assay	↑ miR-132 30 min after trace FC. Overexpression in hippocampus impairs FC acquisition (Wang et al. 2013)
Male mice	Social defeat stress	Heart	miRNA array	Heart injury following social stress was associated with decreased miR-29b, miR-302a and let-7d levels in one strain (Cho et al. 2014)
Male mice	Auditory FC	BLA	miRNA microarray, luciferase assay	miR-34a is involved in fear memory consolidation (Dias et al. 2014)
Male mice	MSUS	Sperm, serum, brain	Deep sequencing, qRT-PCR	↑ miR-375-3p and -5p, miR-200b-3p, miR-672-5p, miR-466-5p in F1 sperm, serum, hippocampus, and hypothalamus and in F2 serum and hippocampus (Gapp et al. 2014)
Mice	FC	Hippocampus	Lentiviral vector (miR KD), TaqMan miRNA assay	Inhibition of miR-92 in hippocampus impairs contextual fear conditioning (Vetere et al. 2014)
Male mice	10d CSDS	Amygdala	miRNA microarray, qRT-PCR	miR-19b associates with Ago2, regulates Adrb1, and is significantly elevated in amygdala of stressed mice (Volk et al. 2014)
Male mice	Contextual FC	Hippocampus	miRNA microarray	miR-33 regulates GABA-related proteins (Jovasevic et al. 2015)
Male and female mice	Cell cultures	Cortex	qPCR, Luciferase assay, mRNA pulldown assay	miR-511 targets and suppresses FKBP5 mRNA and protein levels (Zheng et al. 2016)

(continued)

Table 1 (continued)

Sex (if mentioned) and species	Model	Tissue	miRNA analyses	Primary finding
Male mice	10d CSDS	Amygdala	miRNA microarray	miR-15a associates with Ago2, increases following chronic stress, and downregulates FKBP51 levels (Volk et al. 2016)

Studies are grouped according to species and listed in chronological order within groups. If the sex of the animals is missing, it was not mentioned in the original study. *FC* fear conditioning, *d* days, *ctrl* non-stressed controls, *qRT-PCR* quantitative reverse transcription polymerase chain reaction, *miR* microRNA, *CSDS* chronic social defeat stress, *mPCF* medial prefrontal cortex, *BLA* basolateral amygdala, *qPCR* quantitative polymerase chain reaction, *ILPFC* infra-limbic prefrontal cortex, *KD* knockdown, *FS* foot shock, *FXT* fluoxetine, *MSUS* unpredictable maternal separation combined with unpredictable maternal stress, *mRNA* messenger RNA

4.1.2 Circulating miRNAs as Biomarkers of PTSD

Over the past few years, fluctuations of miRNA levels in body fluids have been shown to correlate with psychiatric disorders, including MDD (Bocchio-Chiavetto et al. 2013), schizophrenia (Lai et al. 2011), and bipolar disorder (Rong et al., n.d.). These studies suggest potential for the use of circulating miRNAs as diagnostic biomarkers of mental disorders. The first study investigating circulating miRNAs as biomarkers of PTSD-related symptoms found that the expression of nine miRNAs was increased both in the amygdala and serum of rats exposed to 3 days of immobilization and tail shock sessions (Balakathiresan et al. 2014). One of the increased stress-responsive miRNAs, miR-19b, was also found to be involved in the regulation of fear-associated genes. A third lead for miR-19b involvement comes from a study using mice undergoing chronic social defeat stress (CSDS) that reported significant increases in the basolateral amygdala (BLA) following CSDS as compared to non-stressed controls (Volk et al. 2014). Finally, miR-19b was also found associated with Ago2 and to target the amygdalar Adrenergic Receptor Beta 1 (Adrb1).

More recently, the potential of miRNAs to be used as biomarkers of both vulnerability and resilience to stress was examined. In one study, circulating miRNA profiles were examined 3 days before and 24 h following CSDS in rats (Chen et al. 2015). Prior to the stressful event, four miRNAs (miR-4-2-5p, miR-27a-3p, miR-30e-5p, miR-362-3p) were significantly decreased only in those rats that later became vulnerable to stress. Following stress exposure, four different miRNAs (miR-139-5p, miR-28-3p, miR-326-3p, miR-99b-5p) were decreased in resilient animals. These results show that different miRNAs potentially confer vulnerability to future stress or promote sustained resilience. Taken together, these studies show promise for using miRNAs as biomarkers of vulnerability and resiliency to stress.

4.1.3 miRNAs in Transgenerational Inheritance of Early Stress

Several animal studies have shown that ncRNAs are abundantly present in sperm and may be involved in non-Mendelian inheritance of behavioral phenotypes (Rassoulzadegan et al. 2006; Liu et al. 2012). Therefore, to assess the potential role of miRNAs in the transgenerational inheritance of parental stress, Gapp et al. (2014) examined sperm samples of a mouse model of unpredictable maternal separation with unpredictable maternal stress (MSUS). Several miRNAs (among other ncRNAs) were upregulated in F1 MSUS sperm (but not F2 sperm) as compared to the sperm of non-stressed control mice. Several miRNA levels were further altered in serum, hippocampus and hypothalamus of F1 MSUS mice, and in serum and hippocampus of F2 MSUS mice. Interestingly, following injection of RNAs purified from MSUS male sperm into wild-type fertilized mouse oocytes, similar behavioral, metabolic, and molecular effects were obtained as compared to direct exposure to MSUS. Additionally, the offspring of these mice showed depressive-like behaviors. These and other results (Rodgers et al. 2013) provide support for the involvement of RNAs, including miRNAs, in the transgenerational transmission of behavioral phenotypes.

4.1.4 miRNAs Targets the FK506 Binding Protein 5 (FKBP5) Gene

The only stress-related gene that has been suggested to be regulated by miRNAs is *FKBP5*. Genetic variations in *FKBP5* have been extensively studied in the context of gene x environment (GxE) interactions and the influence of early life adversity with regard to PTSD (Binder et al. 2004, 2008; Mehta et al. 2011). The immunophilin FKBP5 is a HSP90 co-chaperone that strongly controls glucocorticoid receptor (GR) sensitivity and signaling by binding to GRs in the cytosol thereby decreasing GR ligand affinity and nuclear translocation (Zannas et al. 2015). Several studies have shown that homozygous genotypes for SNPs in *FKBP5* interact with early life (but not adult) adversity, increasing the risk for later development of PTSD (Binder et al. 2008; Zimmermann et al. 2011). Epigenetic mechanisms have repeatedly been found to contribute to the regulation of *FKBP5* expression (Klengel et al. 2013; Yehuda et al. 2016). Moreover, FKBP51, one of the proteins encoded by *FKBP5*, presents an interesting target for the treatment of stress-related disorders. Increased levels of FKBP51 have been suggested to increase the risk of MDD and PTSD and the deletion of *FKBP5* has been shown to prevent age-related depression-like phenotypes (Sabbagh et al. 2014). However, pharmacologically targeting FKBP51 has proven to be challenging due to the strong sequence similarity between this and other FKBP proteins (Schmidt et al. 2012). Recently, two independent studies have shown that miR-15a and miR-511 affect FKBP51 levels by targeting FKBP5 mRNAs (Zheng et al. 2016; Volk et al. 2016). In the first study, FKBP51 levels were found to be decreased and miR-15a levels significantly increased in the amygdala of mice subjected to CSDS as compared to non-stressed controls (Volk et al. 2016). This same pattern was found in peripheral blood of healthy humans following dexamethasone treatment and in individuals exposed to early life trauma (Volk et al. 2016). In the second study, FKBP5 mRNA and protein levels were found to be decreased by miR-511, which was further shown to be involved in neuronal differentiation (Zheng et al. 2016). These findings indicate that both miRNAs are interesting potential candidates for the treatment of stress-related disorders and set the foundations for further studies to examine the exact roles of both miRNAs in *FKBP5* regulation.

4.2 Evidence from Clinical Studies

Apart from Volk et al. (2016) examining miR-15a profiles in blood samples from patients exposed to childhood trauma and healthy individuals administered with dexamethasone, most human studies researching the link between miRNAs and PTSD so far have focused on miRNAs in relation to immunological dysregulations.

Immune dysfunctions are well documented in PTSD and have been reviewed recently (Neigh and Ali 2016; Michopoulos et al. 2017). PTSD has repeatedly been linked to an excessive inflammatory state, possibly resulting from insufficient counter regulation of PTSD-induced immune activation due to cortisol hyposecretion (Gill et al. 2009; Daskalakis et al. 2016). The first study examining peripheral blood mononuclear cells (PBMCs) of combat veterans diagnosed with PTSD found that alterations in specific miRNAs correlated with immunological changes (Zhou et al. 2014). Specifically, miR-125a and miR-181c were significantly decreased in PTSD patients as compared to healthy controls. Further

Table 2 Human studies examining the role of miRNAs in childhood trauma and PTSD

Population	Sample size (if mentioned)	Tissue	miRNA analyses	Primary finding
Male and female combat veterans with PTSD	52 (30 PTSD, 22 ctrl)	PBMCs	miRNA microarray, RT-PCR	↓ miR-125a, miR-181c (Zhou et al. 2014)
Male and female patients with PTSD and comorbid depression	78 (51 PTSD&dep, 27 ctrl)	Whole blood	qPCR, RNA-seq	↓ miR-3130-5p, ↓ *DICER1* mRNA levels (Wingo et al. 2015)
Male individuals		Whole blood	miRNA microarray	↑ miR-15a following DEX administration or childhood trauma exposure (Volk et al. 2016)
Male combat veterans with PTSD	33 (16 PTSD, 17 ctrl)	PBMCs	qRT-PCR	↓ miR-193a-5p (Bam et al. 2016a)
Male combat veterans with PTSD	48 (24 PTSD, 24 ctrl)	PBMCs	RNA-seq, miRNA microarray, qRT-PCR	190 differentially expressed miRNAs among which 183 downregulated (Bam et al. 2016b)
Male combat veterans with PTSD	24 (15 PTSD, 9 ctrl)	Whole blood	miRNA-seq	8 differentially expressed miRNAs; 4 upregulated, 4 downregulated (Martin et al. 2017)

Studies are listed in chronological order. *DEX* dexamethasone, *PBMCs* peripheral blood mononuclear cells, *PTSD&dep* PTSD with comorbid depression, *(mi)RNA-seq* (mi)RNA-sequencing analyses revealed that miR-125a targeted *IFN-γ* and downregulated the production of the pro-inflammatory cytokine IFN-γ. Therefore, the observed increase in IFN-γ in PBMCs of PTSD patients appears to be, at least in part, epigenetically regulated. Intriguingly, miR-27a-3p, which was downregulated in the circulation of rats vulnerable to future stress (Chen et al. 2015), and miR-19b (Balakathiresan et al. 2014; Volk et al. 2014) and miR-223 (Balakathiresan et al. 2014), which were increased in the serum and the amygdala of stressed rodents, were also dysregulated in the present cohort of combat veterans with PTSD (Zhou et al. 2014). However, it is worth mentioning that, while two independent animal studies found miR-19b levels to be increased in several tissues following stress exposure (Balakathiresan et al. 2014; Volk et al. 2014), one study reported increased levels of miR-223 (Balakathiresan et al. 2014), the same miRNAs were significantly decreased in PBMCs of the human cohort (Zhou et al. 2014).

Following this initial study linking miRNAs and immune dysfunctions in PTSD, two recent studies by the same research group provide further evidence for the epigenetic regulation of inflammation in PTSD (Bam et al. 2016a, b). In addition to IFN-γ, the pro-inflammatory cytokine IL-12 was increased in the same cohort of combat veterans, and miR-193a-5p, which was suggested to target IL-12B, was downregulated (Bam et al. 2016a). These results further suggest that pro-inflammatory gene expression is regulated by miRNAs.

Recently, one study found 8 miRNAs to be differentially expressed (4 upregulated and 4 downregulated) in peripheral blood samples of returning combat veterans as compared to controls (Martin et al. 2017). Pathway analyses revealed that these miRNAs target genes involved in Wnt signaling and axon guidance. However, being limited by a small sample size, this study encourages larger studies to further unravel the involvement of miRNAs in PTSD vulnerability.

5 Current Challenges, Pitfalls, and Future Perspectives

As reflected by the present overview, most studies to date that examine the role of miRNAs in PTSD have used (almost exclusively male) animals. Human studies of this subject are now beginning to emerge and have so far only examined peripheral blood samples. Moreover, most studies have included animals or humans, rarely both, and have focused on susceptible phenotypes only, i.e., those animals and individuals suffering the consequences of trauma exposure. To the best of our knowledge, only one study has examined the potential of miRNAs as biomarkers of both vulnerability and resiliency (Chen et al. 2015). Furthermore, a major limitation of current epigenetic research is the lack of longitudinal studies that would enable identification of dynamic epigenetic changes over time (Fig. 2). For these reasons, future research is critically needed to overcome a few pressing issues.

First, given the tissue specificity of epigenetic alterations and the evident inability to study the brains of living human beings, there is a strong need for researchers to incorporate human postmortem brain analyses in their study design. This approach

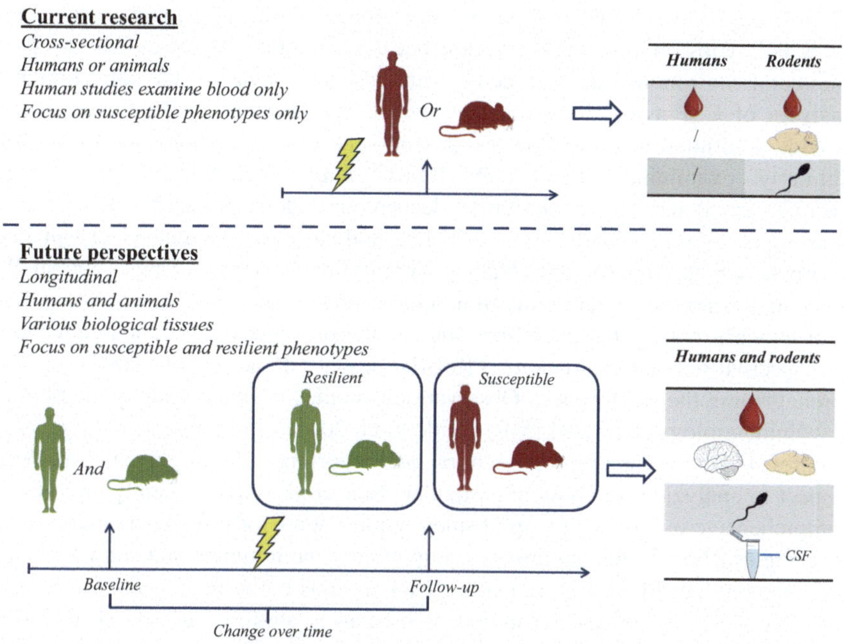

Fig. 2 Current research and future perspectives with regard to miRNA analyses in relation to PTSD. Green and red silhouettes represent mental health or illness, respectively. The lightning bolt represents a stressful event (e.g., CSDS for rodents, combat trauma in humans) and the blood drop represents PBMCs as well as serum and plasma analyses. The Eppendorf tube represents a CSF sample

could not only yield additional information with regard to location and quantity of miRNAs but also shed light on the extent to which blood-based miRNA results are informative for the CNS. In this context, it is becoming clear that focusing on exosome-associated biomarkers provides interesting insights into the brain. Exosomes are secreted membrane vesicles, derived from intracellular endosomes that are generated by the endocytic pathway. The exosomal process traffics damaged or excess proteins and miRNAs as cargo from the cytosol of neurons to the extracellular space where the exosomes can be transported from the CNS to the peripheral circulation. Since exosomes are capable of crossing the blood-brain barrier, when secreted from neural cells, they can be accessed through the bloodstream and further isolated and enriched for neural origin using neural-specific membrane markers (Kapogiannis et al. 2015; Goetzl et al. 2015). Recent studies have shown that $A\beta_{42}$ levels in blood exosomes, presumably derived from neurons, were abnormally higher in subjects with mild cognitive impairment (MCI), MCI that progressed to dementia, and AD (Winston et al. 2016). In another study, blood exosomal levels of $A\beta_{42}$ and tau phosphorylated at Thr^{181} and Ser^{396} predicted development of AD 10 years before clinical onset (Fiandaca et al. 2015). Exosomal plasma $A\beta_{42}$ also correlated with CSF levels of phosphorylated tau (Winston et al. 2016). Therefore

CNS-derived blood-based exosomes are extremely interesting biomarker candidates. Following on from this suggestion, one could imagine the use of CSF to reflect the neural environment more directly. Although more invasive, the collection and analyses of CSF-associated exosomes, which are currently understudied, could provide additional and valuable insights into the brain's pathological processes. Similarly, examining several body fluids jointly, including plasma, serum, PBMCs, sperm and CSF, could further deepen our understanding of miRNA distribution and overlap. Finally, the use of longitudinal designs could yield valuable information regarding dynamic changes over time and how these changes potentially relate to differential susceptibility to traumatic stress.

It is worth noting that guidelines such as the prospective-specimen-collection, retrospective-blinded-evaluation (PRoBE) design (Pepe et al. 2008) or the Strengthening the Reporting of Observational studies in Epidemiology for Molecular Epidemiology (STROBE-ME) (Gallo et al. 2011) offer valuable overviews to help researchers in the design, execution, and reporting of biomarker studies. With respect to analyzing miRNAs in particular, Nair et al. (2014) recently provided a comprehensive overview of helpful study requirements for researchers involved in studying miRNAs in human diseases. Importantly, both human and animal studies have shown that differences in genetic backgrounds between subjects can have a considerable effect on the resolution of biomarker studies (Ahanda et al. 2014; Zhao et al. 2010). Therefore, it is critical for future research to take variations in genetic backgrounds into account and correct for additional factors such as current or previous smoking habits, alcohol abuse or medication use of patients. Finally, when comparing the obtained results, one should keep in mind the heterogeneity of miRNA expression in different tissues. Indeed, PBMC or whole blood-derived miRNA profiles will most likely differ from those obtained through serum or plasma.

Taken together, current preclinical and preliminary clinical evidence show great potential for the use of miRNAs as biomarkers of PTSD, which would enable us to detect at-risk individuals and provide specific preventive strategies and early interventions on an individual basis. This approach is especially relevant because currently no true treatment exists for PTSD. Therefore, the presented findings build an emerging foundation for future research to further examine the exact roles of miRNAs in PTSD using appropriate study designs.

References

Absalon S, Kochanek DM, Raghavan V, Krichevsky AM (2013) MiR-26b, upregulated in Alzheimer's disease, activates cell cycle entry, tau-phosphorylation, and apoptosis in postmitotic neurons. J Neurosci 33(37):14645–14659. https://doi.org/10.1523/jneurosci.1327-13.2013

Ahanda M-LE, Zerjal T, Dhorne-Pollet S, Rau A, Cooksey A, Giuffra E (2014) Impact of the genetic background on the composition of the chicken plasma MiRNome in response to a stress. PLoS One 9(12):e114598. https://doi.org/10.1371/journal.pone.0114598

Aksoy-Aksel A, Zampa F, Schratt G (2014) MicroRNAs and synaptic plasticity—a mutual relationship. Philos Trans R Soc Lond B Biol Sci 369(1652):20130515. https://doi.org/10.1098/rstb.2013.0515

American Psychiatric Association (2013) Diagnostic and statistical manual of mental disorders, 5th edn. American Psychiatric Association, Washington, DC

Aravin AA, Hannon GJ, Brennecke J (2007) The Piwi-piRNA pathway provides an adaptive defense in the transposon arms race. Science 318(5851):761–764. https://doi.org/10.1126/science.1146484

Bai M, Zhu X, Zhang Y, Zhang S, Zhang L, Xue L, Yi J, Yao S, Zhang X (2012) Abnormal hippocampal BDNF and miR-16 expression is associated with depression-like behaviors induced by stress during early life. PLoS One 7(10):e46921. https://doi.org/10.1371/journal.pone.0046921

Baker DG, West SA, Nicholson WE, Ekhator NN, Kasckow JW, Hill KK, Bruce AB, Orth DN, Geracioti TD Jr (1999) Serial CSF corticotropin-releasing hormone levels and adrenocortical activity in combat veterans with posttraumatic stress disorder. Am J Psychiatry 156 (4):585–588. https://doi.org/10.1176/ajp.156.4.585

Balakathiresan NS, Chandran R, Bhomia M, Jia M, Li H, Maheshwari RK (2014) Serum and amygdala microRNA signatures of posttraumatic stress: fear correlation and biomarker potential. J Psychiatr Res 57:65–73. https://doi.org/10.1016/j.jpsychires.2014.05.020

Bam M, Yang X, Zhou J, Ginsberg JP, Leyden Q, Nagarkatti PS, Nagarkatti M (2016a) Evidence for epigenetic regulation of pro-inflammatory cytokines, interleukin-12 and interferon gamma, in peripheral blood mononuclear cells from PTSD patients. J Neuroimmune Pharmacol 11 (1):168–181. https://doi.org/10.1007/s11481-015-9643-8

Bam M, Yang X, Zumbrun EE, Zhong Y, Zhou J, Ginsberg JP, Leyden Q, Zhang J, Nagarkatti PS, Nagarkatti M (2016b) Dysregulated immune system networks in war veterans with PTSD is an outcome of altered miRNA expression and DNA methylation. Sci Rep 6:31209. https://doi.org/10.1038/srep31209

Bian S, TL X, Sun T (2013) Tuning the cell fate of neurons and glia by microRNAs. Curr Opin Neurobiol 23(6):928–934. https://doi.org/10.1016/j.conb.2013.08.002

Binder EB, Salyakina D, Lichtner P, Wochnik GM, Ising M, Putz B, Papiol S, Seaman S, Lucae S, Kohli MA, Nickel T, Kunzel HE, Fuchs B, Majer M, Pfennig A, Kern N, Brunner J, Modell S, Baghai T, Deiml T, Zill P, Bondy B, Rupprecht R, Messer T, Kohnlein O, Dabitz H, Bruckl T, Muller N, Pfister H, Lieb R, Mueller JC, Lohmussaar E, Strom TM, Bettecken T, Meitinger T, Uhr M, Rein T, Holsboer F, Muller-Myhsok B (2004) Polymorphisms in FKBP5 are associated with increased recurrence of depressive episodes and rapid response to antidepressant treatment. Nat Genet 36(12):1319–1325. https://doi.org/10.1038/ng1479

Binder EB, Bradley RG, Liu W, Epstein MP, Deveau TC, Mercer KB, Tang Y, Gillespie CF, Heim CM, Nemeroff CB, Schwartz AC, Cubells JF, Ressler KJ (2008) Association of FKBP5 polymorphisms and childhood abuse with risk of posttraumatic stress disorder symptoms in adults. JAMA 299(11):1291–1305. https://doi.org/10.1001/jama.299.11.1291

Blechert J, Michael T, Vriends N, Margraf J, Wilhelm FH (2007) Fear conditioning in posttraumatic stress disorder: evidence for delayed extinction of autonomic, experiential, and behavioural responses. Behav Res Ther 45(9):2019–2033. https://doi.org/10.1016/j.brat.2007.02.012

Bocchio-Chiavetto L, Maffioletti E, Bettinsoli P, Giovannini C, Bignotti S, Tardito D, Corrada D, Milanesi L, Gennarelli M (2013) Blood microRNA changes in depressed patients during antidepressant treatment. Eur Neuropsychopharmacol 23(7):602–611. https://doi.org/10.1016/j.euroneuro.2012.06.013

Borchert GM, Lanier W, Davidson BL (2006) RNA polymerase III transcribes human microRNAs. Nat Struct Mol Biol 13(12):1097–1101. http://www.nature.com/nsmb/journal/v13/n12/suppinfo/nsmb1167_S1.html

Camussi G, Deregibus M-C, Bruno S, Grange C, Fonsato V, Tetta C (2011) Exosome/microvesicle-mediated epigenetic reprogramming of cells. Am J Cancer Res 1(1):98–110

Chen Y-J, Luo J, Yang G-Y, Yang K, Wen S-Q, Zou S-Q (2012) Mutual regulation between microRNA-373 and methyl-CpG-binding domain protein 2 in hilar cholangiocarcinoma. World J Gastroenterol 18(29):3849–3861. https://doi.org/10.3748/wjg.v18.i29.3849

Chen RJ, Kelly G, Sengupta A, Heydendael W, Nicholas B, Beltrami S, Luz S, Peixoto L, Abel T, Bhatnagar S (2015) MicroRNAs as biomarkers of resilience or vulnerability to stress. Neuroscience 305:36–48. https://doi.org/10.1016/j.neuroscience.2015.07.045

Cho J-H, Lee I, Hammamieh R, Wang K, Baxter D, Scherler K, Etheridge A, Kulchenko A, Gautam A, Muhie S, Chakraborty N, Galas DJ, Jett M, Hood L (2014) Molecular evidence of stress-induced acute heart injury in a mouse model simulating posttraumatic stress disorder. Proc Natl Acad Sci 111(8):3188–3193

Chrousos GP, Gold PW (1992) The concepts of stress and stress system disorders. Overview of physical and behavioral homeostasis. JAMA 267(9):1244–1252

Dajas-Bailador F, Bonev B, Garcez P, Stanley P, Guillemot F, Papalopulu N (2012) microRNA-9 regulates axon extension and branching by targeting Map 1b in mouse cortical neurons. Nat Neurosci 15:697. https://doi.org/10.1038/nn.3082

Daskalakis NP, Cohen H, Nievergelt CM, Baker DG, Buxbaum JD, Russo SJ, Yehuda R (2016) New translational perspectives for blood-based biomarkers of PTSD: from glucocorticoid to immune mediators of stress susceptibility. Exp Neurol 284(Pt B):133–140. https://doi.org/10.1016/j.expneurol.2016.07.024

Davis-Dusenbery BN, Hata A (2010) Mechanisms of control of microRNA biogenesis. J Biochem 148(4):381–392. https://doi.org/10.1093/jb/mvq096

Dias BG, Goodman JV, Ahluwalia R, Easton AE, Andero R, Ressler KJ (2014) Amygdala-dependent fear memory consolidation via miR-34a and Notch signaling. Neuron 83 (4):906–918. https://doi.org/10.1016/j.neuron.2014.07.019

Doxakis E (2010) Post-transcriptional regulation of alpha-synuclein expression by mir-7 and mir-153. J Biol Chem 285(17):12726–12734. https://doi.org/10.1074/jbc.M109.086827

El Khoury-Malhame M, Reynaud E, Soriano A, Michael K, Salgado-Pineda P, Zendjidjian X, Gellato C, Eric F, Lefebvre M-N, Rouby F, Samuelian J-C, Anton J-L, Blin O, Khalfa S (2011) Amygdala activity correlates with attentional bias in PTSD. Neuropsychologia 49 (7):1969–1973. https://doi.org/10.1016/j.neuropsychologia.2011.03.025

Etheridge A, Lee I, Hood L, Galas D, Wang K (2011) Extracellular microRNA: a new source of biomarkers. Mutat Res 717(1–2):85–90. https://doi.org/10.1016/j.mrfmmm.2011.03.004

Etkin A, Wager TD (2007) Functional neuroimaging of anxiety: a meta-analysis of emotional processing in PTSD, social anxiety disorder, and specific phobia. Am J Psychiatry 164 (10):1476–1488. https://doi.org/10.1176/appi.ajp.2007.07030504

Fabbri M, Garzon R, Cimmino A, Liu Z, Zanesi N, Callegari E, Liu S, Alder H, Costinean S, Fernandez-Cymering C, Volinia S, Guler G, Morrison CD, Chan KK, Marcucci G, Calin GA, Huebner K, Croce CM (2007) MicroRNA-29 family reverts aberrant methylation in lung cancer by targeting DNA methyltransferases 3A and 3B. Proc Natl Acad Sci U S A 104 (40):15805–15810. https://doi.org/10.1073/pnas.0707628104

Fabian MR, Sonenberg N (2012) The mechanics of miRNA-mediated gene silencing: a look under the hood of miRISC. Nat Struct Mol Biol 19(6):586–593. https://doi.org/10.1038/nsmb.2296

Fiandaca MS, Kapogiannis D, Mapstone M, Boxer A, Eitan E, Schwartz JB, Abner EL, Petersen RC, Federoff HJ, Miller BL, Goetzl EJ (2015) Identification of preclinical Alzheimer's disease by a profile of pathogenic proteins in neurally derived blood exosomes: a case-control study. Alzheimers Dement 11(6):600–607.e601. https://doi.org/10.1016/j.jalz.2014.06.008

Friedman RC, Farh KK, Burge CB, Bartel DP (2009) Most mammalian mRNAs are conserved targets of microRNAs. Genome Res 19(1):92–105. https://doi.org/10.1101/gr.082701.108

Gallo V, Egger M, McCormack V, Farmer PB, Ioannidis JPA, Kirsch-Volders M, Matullo G, Phillips DH, Schoket B, Stromberg U, Vermeulen R, Wild C, Porta M, Vineis P (2011)

Strengthening the reporting of observational studies in epidemiology-molecular epidemiology STROBE-ME: an extension of the STROBE statement. J Clin Epidemiol 64(12):1350–1363. https://doi.org/10.1016/j.jclinepi.2011.07.010

Gapp K, Jawaid A, Sarkies P, Bohacek J, Pelczar P, Prados J, Farinelli L, Miska E, Mansuy IM (2014) Implication of sperm RNAs in transgenerational inheritance of the effects of early trauma in mice. Nat Neurosci 17(5):667–669. https://doi.org/10.1038/nn.3695

Geracioti TD Jr, Baker DG, Kasckow JW, Strawn JR, Jeffrey Mulchahey J, Dashevsky BA, Horn PS, Ekhator NN (2008) Effects of trauma-related audiovisual stimulation on cerebrospinal fluid norepinephrine and corticotropin-releasing hormone concentrations in post-traumatic stress disorder. Psychoneuroendocrinology 33(4):416–424. https://doi.org/10.1016/j.psyneuen.2007.12.012

Gibbings DJ, Ciaudo C, Erhardt M, Voinnet O (2009) Multivesicular bodies associate with components of miRNA effector complexes and modulate miRNA activity. Nat Cell Biol 11 (9):1143–1149. https://doi.org/10.1038/ncb1929

Gill JM, Saligan L, Woods S, Page G (2009) PTSD is associated with an excess of inflammatory immune activities. Perspect Psychiatr Care 45(4):262–277. https://doi.org/10.1111/j.1744-6163.2009.00229.x

Goetzl EJ, Boxer A, Schwartz JB, Abner EL, Petersen RC, Miller BL, Kapogiannis D (2015) Altered lysosomal proteins in neural-derived plasma exosomes in preclinical Alzheimer disease. Neurology 85(1):40–47. https://doi.org/10.1212/wnl.0000000000001702

Griggs EM, Young EJ, Rumbaugh G, Miller CA (2013) MicroRNA-182 regulates amygdala-dependent memory formation. J Neurosci 33(4):1734

Hanin G, Shenhar-Tsarfaty S, Yayon N, Hoe YY, Bennett ER, Sklan EH, Rao DC, Rankinen T, Bouchard C, Geifman-Shochat S, Shifman S, Greenberg DS, Soreq H (2014) Competing targets of microRNA-608 affect anxiety and hypertension. Hum Mol Genet 23(17):4569–4580

Hanke M, Hoefig K, Merz H, Feller AC, Kausch I, Jocham D, Warnecke JM, Sczakiel G (2010) A robust methodology to study urine microRNA as tumor marker: microRNA-126 and microRNA-182 are related to urinary bladder cancer. Urol Oncol 28(6):655–661. https://doi.org/10.1016/j.urolonc.2009.01.027

Hu Z, Li Z (2017) miRNAs in synapse development and synaptic plasticity. Curr Opin Neurobiol 45:24–31. https://doi.org/10.1016/j.conb.2017.02.014

Hu YK, Wang X, Li L, YH D, Ye HT, Li CY (2013) MicroRNA-98 induces an Alzheimer's disease-like disturbance by targeting insulin-like growth factor 1. Neurosci Bull 29 (6):745–751. https://doi.org/10.1007/s12264-013-1348-5

Issler O, Chen A (2015) Determining the role of microRNAs in psychiatric disorders. Nat Rev Neurosci 16(4):201–212. https://doi.org/10.1038/nrn3879

Jin Y, Lee CG (2013) Single nucleotide polymorphisms associated with microRNA regulation. Biomol Ther 3(2):287–302. https://doi.org/10.3390/biom3020287

Jovasevic V, Corcoran KA, Leaderbrand K, Yamawaki N, Guedea AL, Chen HJ, Shepherd GM, Radulovic J (2015) GABAergic mechanisms regulated by miR-33 encode state-dependent fear. Nat Neurosci 18(9):1265–1271. https://doi.org/10.1038/nn.4084

Kapogiannis D, Boxer A, Schwartz JB, Abner EL, Biragyn A, Masharani U, Frassetto L, Petersen RC, Miller BL, Goetzl EJ (2015) Dysfunctionally phosphorylated type 1 insulin receptor substrate in neural-derived blood exosomes of preclinical Alzheimer's disease. FASEB J 29 (2):589–596. https://doi.org/10.1096/fj.14-262048

Karl A, Schaefer M, Malta LS, Dörfel D, Rohleder N, Werner A (2006) A meta-analysis of structural brain abnormalities in PTSD. Neurosci Biobehav Rev 30(7):1004–1031. https://doi.org/10.1016/j.neubiorev.2006.03.004

Klengel T, Mehta D, Anacker C, Rex-Haffner M, Pruessner JC, Pariante CM, Pace TW, Mercer KB, Mayberg HS, Bradley B, Nemeroff CB, Holsboer F, Heim CM, Ressler KJ, Rein T, Binder

EB (2013) Allele-specific FKBP5 DNA demethylation mediates gene-childhood trauma interactions. Nat Neurosci 16(1):33–41. https://doi.org/10.1038/nn.3275

Konopka W, Kiryk A, Novak M, Herwerth M, Parkitna JR, Wawrzyniak M, Kowarsch A, Michaluk P, Dzwonek J, Arnsperger T, Wilczynski G, Merkenschlager M, Theis FJ, Kohr G, Kaczmarek L, Schutz G (2010) MicroRNA loss enhances learning and memory in mice. J Neurosci 30(44):14835–14842. https://doi.org/10.1523/JNEUROSCI.3030-10.2010

Kouzarides T (2007) Chromatin modifications and their function. Cell 128(4):693–705. https://doi.org/10.1016/j.cell.2007.02.005

Lai CY, SL Y, Hsieh MH, Chen CH, Chen HY, Wen CC, Huang YH, Hsiao PC, Hsiao CK, Liu CM, Yang PC, Hwu HG, Chen WJ (2011) MicroRNA expression aberration as potential peripheral blood biomarkers for schizophrenia. PLoS One 6(6):e21635. https://doi.org/10.1371/journal.pone.0021635

Lee Y, Ahn C, Han J, Choi H, Kim J, Yim J, Lee J, Provost P, Radmark O, Kim S, Kim VN (2003) The nuclear RNase III Drosha initiates microRNA processing. Nature 425(6956):415–419. https://doi.org/10.1038/nature01957

Lemieux AM, Coe CL (1995) Abuse-related posttraumatic stress disorder: evidence for chronic neuroendocrine activation in women. Psychosom Med 57(2):105–115

Lewis BP, Burge CB, Bartel DP (2005) Conserved seed pairing, often flanked by adenosines, indicates that thousands of human genes are microRNA targets. Cell 120(1):15–20. https://doi.org/10.1016/j.cell.2004.12.035

Li C, Liu Y, Liu D, Jiang H, Pan F (2016) Dynamic alterations of miR-34c expression in the hypothalamus of male rats after early adolescent traumatic stress. Neural Plast 2016:5249893. https://doi.org/10.1155/2016/5249893

Lin Q, Wei W, Coelho CM, Li X, Baker-Andresen D, Dudley K, Ratnu VS, Boskovic Z, Kobor MS, Sun YE, Bredy TW (2011) The brain-specific microRNA miR-128b regulates the formation of fear-extinction memory. Nat Neurosci 14(9):1115–1117. https://doi.org/10.1038/nn.2891

Liu C, Teng Z-Q, Santistevan NJ, Szulwach KE, Guo W, Jin P, Zhao X (2010) Epigenetic regulation of miR-184 by MBD1 governs neural stem cell proliferation and differentiation. Cell Stem Cell 6(5):433–444. https://doi.org/10.1016/j.stem.2010.02.017

Liu W-M, Pang RTK, Chiu PCN, Wong BPC, Lao K, Lee K-F, Yeung WSB (2012) Sperm-borne microRNA-34c is required for the first cleavage division in mouse. Proc Natl Acad Sci U S A 109(2):490–494. https://doi.org/10.1073/pnas.1110368109

Magill ST, Cambronne XA, Luikart BW, Lioy DT, Leighton BH, Westbrook GL, Mandel G, Goodman RH (2010) microRNA-132 regulates dendritic growth and arborization of newborn neurons in the adult hippocampus. Proc Natl Acad Sci U S A 107(47):20382–20387. https://doi.org/10.1073/pnas.1015691107

Martin CG, Kim H, Yun S, Livingston W, Fetta J, Mysliwiec V, Baxter T, Gill JM (2017) Circulating miRNA associated with posttraumatic stress disorder in a cohort of military combat veterans. Psychiatry Res 251:261–265. https://doi.org/10.1016/j.psychres.2017.01.081

Meewisse ML, Reitsma JB, de Vries GJ, Gersons BP, Olff M (2007) Cortisol and post-traumatic stress disorder in adults: systematic review and meta-analysis. Br J Psychiatry 191:387–392. https://doi.org/10.1192/bjp.bp.106.024877

Mehta D, Gonik M, Klengel T, Rex-Haffner M, Menke A, Rubel J, Mercer KB, Putz B, Bradley B, Holsboer F, Ressler KJ, Muller-Myhsok B, Binder EB (2011) Using polymorphisms in FKBP5 to define biologically distinct subtypes of posttraumatic stress disorder: evidence from endocrine and gene expression studies. Arch Gen Psychiatry 68(9):901–910. https://doi.org/10.1001/archgenpsychiatry.2011.50

Memczak S, Jens M, Elefsinioti A, Torti F, Krueger J, Rybak A, Maier L, Mackowiak SD, Gregersen LH, Munschauer M, Loewer A, Ziebold U, Landthaler M, Kocks C, le Noble F, Rajewsky N (2013) Circular RNAs are a large class of animal RNAs with regulatory potency. Nature 495(7441):333–338. http://www.nature.com/nature/journal/v495/n7441/abs/nature11928.html#supplementary-information

Michopoulos V, Powers A, Gillespie CF, Ressler KJ, Jovanovic T (2017) Inflammation in fear- and anxiety-based disorders: PTSD, GAD, and beyond. Neuropsychopharmacology 42 (1):254–270. https://doi.org/10.1038/npp.2016.146

Mitchell PS, Parkin RK, Kroh EM, Fritz BR, Wyman SK, Pogosova-Agadjanyan EL, Peterson A, Noteboom J, O'Briant KC, Allen A, Lin DW, Urban N, Drescher CW, Knudsen BS, Stirewalt DL, Gentleman R, Vessella RL, Nelson PS, Martin DB, Tewari M (2008) Circulating microRNAs as stable blood-based markers for cancer detection. Proc Natl Acad Sci U S A 105(30):10513–10518. https://doi.org/10.1073/pnas.0804549105

Nair VS, Pritchard CC, Tewari M, Ioannidis JP (2014) Design and analysis for studying microRNAs in human disease: a primer on -Omic technologies. Am J Epidemiol 180 (2):140–152. https://doi.org/10.1093/aje/kwu135

Neigh GN, Ali FF (2016) Co-morbidity of PTSD and immune system dysfunction: opportunities for treatment. Curr Opin Pharmacol 29:104–110. https://doi.org/10.1016/j.coph.2016.07.011

Norberg MM, Krystal JH, Tolin DF (2008) A meta-analysis of D-cycloserine and the facilitation of fear extinction and exposure therapy. Biol Psychiatry 63(12):1118–1126. https://doi.org/10.1016/j.biopsych.2008.01.012

Park NJ, Zhou H, Elashoff D, Henson BS, Kastratovic DA, Abemayor E, Wong DT (2009) Salivary microRNA: discovery, characterization, and clinical utility for oral cancer detection. Clin Cancer Res 15(17):5473–5477. https://doi.org/10.1158/1078-0432.CCR-09-0736

Pepe MS, Feng Z, Janes H, Bossuyt PM, Potter JD (2008) Pivotal evaluation of the accuracy of a biomarker used for classification or prediction: standards for study design. J Natl Canc Inst 100 (20):1432–1438. https://doi.org/10.1093/jnci/djn326

Peschansky VJ, Wahlestedt C (2014) Non-coding RNAs as direct and indirect modulators of epigenetic regulation. Epigenetics 9(1):3–12. https://doi.org/10.4161/epi.27473

Pritchard CC, Cheng HH, Tewari M (2012) MicroRNA profiling: approaches and considerations. Nat Rev Genet 13(5):358–369. https://doi.org/10.1038/nrg3198

Raabe FJ, Spengler D (2013) Epigenetic risk factors in PTSD and depression. Front Psych 4:80. https://doi.org/10.3389/fpsyt.2013.00080

Ramalingam P, Palanichamy JK, Singh A, Das P, Bhagat M, Kassab MA, Sinha S, Chattopadhyay P (2014) Biogenesis of intronic miRNAs located in clusters by independent transcription and alternative splicing. RNA 20(1):76–87. https://doi.org/10.1261/rna.041814.113

Rassoulzadegan M, Grandjean V, Gounon P, Vincent S, Gillot I, Cuzin F (2006) RNA-mediated non-mendelian inheritance of an epigenetic change in the mouse. Nature 441(7092):469–474. https://doi.org/10.1038/nature04674

Rodgers AB, Morgan CP, Bronson SL, Revello S, Bale TL (2013) Paternal stress exposure alters sperm microRNA content and reprograms offspring HPA stress axis regulation. J Neurosci 33 (21):9003–9012. https://doi.org/10.1523/JNEUROSCI.0914-13.2013

Rong H, Liu TB, Yang KJ, Yang HC, Wu DH, Liao CP, Hong F, Yang HZ, Wan F, Ye XY, Xu D, Zhang X, Chao CA, Shen QJ (n.d.) MicroRNA-134 plasma levels before and after treatment for bipolar mania. J Psychiatr Res 45(1):92–95. https://doi.org/10.1016/jjpsychires201004.028

Rose NR, Klose RJ (2014) Understanding the relationship between DNA methylation and histone lysine methylation. Biochim Biophys Acta 1839(12):1362–1372. https://doi.org/10.1016/j.bbagrm.2014.02.007

Sabbagh JJ, O'Leary JC, Blair LJ, Klengel T, Nordhues BA, Fontaine SN, Binder EB, Dickey CA (2014) Age-associated epigenetic Upregulation of the FKBP5 gene selectively impairs stress resiliency. PLoS One 9(9):e107241. https://doi.org/10.1371/journal.pone.0107241

Sato F, Tsuchiya S, Meltzer SJ, Shimizu K (2011) MicroRNAs and epigenetics. FEBS J 278 (10):1598–1609. https://doi.org/10.1111/j.1742-4658.2011.08089.x

Satrom P, Snove O, Rossi JJ (2007) Epigenetics and microRNAs. Pediatr Res 61(5 Part 2):17R–23R

Schmidt U, Holsboer F, Rein T (2011) Epigenetic aspects of posttraumatic stress disorder. Dis Markers 30(2–3):77–87. https://doi.org/10.3233/DMA-2011-0749

Schmidt MV, Paez-Pereda M, Holsboer F, Hausch F (2012) The prospect of FKBP51 as a drug target. ChemMedChem 7(8):1351–1359. https://doi.org/10.1002/cmdc.201200137

Schmidt U, Herrmann L, Hagl K, Novak B, Huber C, Holsboer F, Wotjak CT, Buell DR (2013) Therapeutic action of fluoxetine is associated with a reduction in prefrontal cortical miR-1971 expression levels in a mouse model of posttraumatic stress disorder. Front Psych 4:66. https://doi.org/10.3389/fpsyt.2013.00066

Schouten M, Aschrafi A, Bielefeld P, Doxakis E, Fitzsimons CP (2013) MicroRNAs and the regulation of neuronal plasticity under stress conditions. Neuroscience 241:188–205. https://doi.org/10.1016/j.neuroscience.2013.02.065

Scott GK, Mattie MD, Berger CE, Benz SC, Benz CC (2006) Rapid alteration of microRNA levels by histone deacetylase inhibition. Cancer Res 66(3):1277–1281. https://doi.org/10.1158/0008-5472.can-05-3632

Sherin JEN, Charles B (2011) Post-traumatic stress disorder: the neurobiological impact of psychological trauma. Dialogues Clin Neurosci 13(3):263

Sherin JE, Nemeroff CB (2011) Post-traumatic stress disorder: the neurobiological impact of psychological trauma. Dialogues Clin Neurosci 13(3):263–278

Shin LM, Rauch SL, Pitman RK (2006) Amygdala, medial prefrontal cortex, and hippocampal function in PTSD. Ann N Y Acad Sci 1071:67–79. https://doi.org/10.1196/annals.1364.007

Sinkkonen L, Hugenschmidt T, Berninger P, Gaidatzis D, Mohn F, Artus-Revel CG, Zavolan M, Svoboda P, Filipowicz W (2008) MicroRNAs control de novo DNA methylation through regulation of transcriptional repressors in mouse embryonic stem cells. Nat Struct Mol Biol 15(3):259–267. http://www.nature.com/nsmb/journal/v15/n3/suppinfo/nsmb.1391_S1.html

Smalheiser NR, Lugli G, Rizavi HS, Torvik VI, Turecki G, Dwivedi Y (2012) MicroRNA expression is down-regulated and reorganized in prefrontal cortex of depressed suicide subjects. PLoS One 7(3):e33201. https://doi.org/10.1371/journal.pone.0033201

Smith ME (2005) Bilateral hippocampal volume reduction in adults with post-traumatic stress disorder: a meta-analysis of structural MRI studies. Hippocampus 15(6):798–807. https://doi.org/10.1002/hipo.20102

Smith B, Treadwell J, Zhang D, Ly D, McKinnell I, Walker PR, Sikorska M (2010) Large-scale expression analysis reveals distinct microRNA profiles at different stages of human neurodevelopment. PLoS One 5(6):e11109. https://doi.org/10.1371/journal.pone.0011109

Taylor D, Gercel-Taylor C (2013) The origin, function, and diagnostic potential of RNA within extracellular vesicles present in human biological fluids. Front Genet 4(142). https://doi.org/10.3389/fgene.2013.00142

Turchinovich A, Samatov T, Tonevitsky A, Burwinkel B (2013) Circulating miRNAs: cell-cell communication function? Front Genet 4(119). https://doi.org/10.3389/fgene.2013.00119

Valadi H, Ekstrom K, Bossios A, Sjostrand M, Lee JJ, Lotvall JO (2007) Exosome-mediated transfer of mRNAs and microRNAs is a novel mechanism of genetic exchange between cells. Nat Cell Biol 9(6):654–659. http://www.nature.com/ncb/journal/v9/n6/suppinfo/ncb1596_S1.html

Van den Hove DL, Kompotis K, Lardenoije R, Kenis G, Mill J, Steinbusch HW, Lesch KP, Fitzsimons CP, De Strooper B, Rutten BP (2014) Epigenetically regulated microRNAs in Alzheimer's disease. Neurobiol Aging 35(4):731–745. https://doi.org/10.1016/j.neurobiolaging.2013.10.082

Venkatesh S, Workman JL (2015) Histone exchange, chromatin structure and the regulation of transcription. Nat Rev Mol Cell Biol 16(3):178–189. https://doi.org/10.1038/nrm3941

Vetere G, Barbato C, Pezzola S, Frisone P, Aceti M, Ciotti M, Cogoni C, Ammassari-Teule M, Ruberti F (2014) Selective inhibition of miR-92 in hippocampal neurons alters contextual fear memory. Hippocampus 24(12):1458–1465. https://doi.org/10.1002/hipo.22326

Vickers KC, Palmisano BT, Shoucri BM, Shamburek RD, Remaley AT (2011) MicroRNAs are transported in plasma and delivered to recipient cells by high-density lipoproteins. Nat Cell Biol 13(4):423–433. https://doi.org/10.1038/ncb2210

Volk N, Paul ED, Haramati S, Eitan C, Fields BK, Zwang R, Gil S, Lowry CA, Chen A (2014) MicroRNA-19b associates with Ago2 in the amygdala following chronic stress and regulates the adrenergic receptor beta 1. J Neurosci 34(45):15070–15082. https://doi.org/10.1523/JNEUROSCI.0855-14.2014

Volk N, Pape JC, Engel M, Zannas AS, Cattane N, Cattaneo A, Binder EB, Chen A (2016) Amygdalar MicroRNA-15a is essential for coping with chronic stress. Cell Rep 17 (7):1882–1891. https://doi.org/10.1016/j.celrep.2016.10.038

Wagner J, Riwanto M, Besler C, Knau A, Fichtlscherer S, Roxe T, Zeiher AM, Landmesser U, Dimmeler S (2013) Characterization of levels and cellular transfer of circulating lipoprotein-bound microRNAs. Arterioscler Thromb Vasc Biol 33(6):1392–1400. https://doi.org/10.1161/ATVBAHA.112.300741

Wang G, van der Walt JM, Mayhew G, Li YJ, Zuchner S, Scott WK, Martin ER, Vance JM (2008) Variation in the miRNA-433 binding site of FGF20 confers risk for Parkinson disease by overexpression of alpha-synuclein. Am J Hum Genet 82(2):283–289. https://doi.org/10.1016/j.ajhg.2007.09.021

Wang K, Zhang S, Weber J, Baxter D, Galas DJ (2010) Export of microRNAs and microRNA-protective protein by mammalian cells. Nucleic Acids Res 38:7248. https://doi.org/10.1093/nar/gkq601

Wang RY, Phang RZ, Hsu PH, Wang WH, Huang HT, Liu IY (2013) In vivo knockdown of hippocampal miR-132 expression impairs memory acquisition of trace fear conditioning. Hippocampus 23(7):625–633. https://doi.org/10.1002/hipo.22123

Weber JA, Baxter DH, Zhang S, Huang DY, Huang KH, Lee MJ, Galas DJ, Wang K (2010) The microRNA spectrum in 12 body fluids. Clin Chem 56(11):1733–1741. https://doi.org/10.1373/clinchem.2010.147405

Wingo AP, Almli LM, Stevens JS, Klengel T, Uddin M, Li Y, Bustamante AC, Lori A, Koen N, Stein DJ, Smith AK, Aiello AE, Koenen KC, Wildman DE, Galea S, Bradley B, Binder EB, Jin P, Gibson G, Ressler KJ (2015) DICER1 and microRNA regulation in post-traumatic stress disorder with comorbid depression. Nat Commun 6:10106. https://doi.org/10.1038/ncomms10106

Winston CN, Goetzl EJ, Akers JC, Carter BS, Rockenstein EM, Galasko D, Masliah E, Rissman RA (2016) Prediction of conversion from mild cognitive impairment to dementia with neuronally derived blood exosome protein profile. Alzheimers Dement (Amst) 3:63–72. https://doi.org/10.1016/j.dadm.2016.04.001

Yang X, WW D, Li H, Liu F, Khorshidi A, Rutnam ZJ, Yang BB (2013) Both mature miR-17-5p and passenger strand miR-17-3p target TIMP3 and induce prostate tumor growth and invasion. Nucleic Acids Res 41(21):9688–9704. https://doi.org/10.1093/nar/gkt680

Yehuda R (2001) Biology of posttraumatic stress disorder. J Clin Psychiatry 62(Suppl 17):41–46

Yehuda R (2006) Advances in understanding neuroendocrine alterations in PTSD and their therapeutic implications. Ann N Y Acad Sci 1071:137–166. https://doi.org/10.1196/annals.1364.012

Yehuda R, Bierer LM, Schmeidler J, Aferiat DH, Breslau I, Dolan S (2000) Low cortisol and risk for PTSD in adult offspring of holocaust survivors. Am J Psychiatry 157(8):1252–1259. https://doi.org/10.1176/appi.ajp.157.8.1252

Yehuda R, Daskalakis NP, Bierer LM, Bader HN, Klengel T, Holsboer F, Binder EB (2016) Holocaust exposure induced intergenerational effects on FKBP5 methylation. Biol Psychiatry 80(5):372–380. https://doi.org/10.1016/j.biopsych.2015.08.005

Young EA, Breslau N (2004) Cortisol and catecholamines in posttraumatic stress disorder: an epidemiologic community study. Arch Gen Psychiatry 61(4):394–401

Zamore PD (2002) Ancient pathways programmed by small RNAs. Science 296 (5571):1265–1269. https://doi.org/10.1126/science.1072457

Zannas AS, Provencal N, Binder EB (2015) Epigenetics of posttraumatic stress disorder: current evidence, challenges, and future directions. Biol Psychiatry 78(5):327–335. https://doi.org/10.1016/j.biopsych.2015.04.003

Zhao H, Shen J, Medico L, Wang D, Ambrosone CB, Liu S (2010) A pilot study of circulating miRNAs as potential biomarkers of early stage breast cancer. PLoS One 5(10):e13735. https://doi.org/10.1371/journal.pone.0013735

Zhao H, Yao R, Cao X, Wu G (2011) Neuroimmune modulation following traumatic stress in rats: evidence for an immunoregulatory cascade mediated by c-Src, miRNA222 and PAK1. J Neuroinflammation 8(1):159. https://doi.org/10.1186/1742-2094-8-159

Zheng D, Sabbagh JJ, Blair LJ, Darling AL, Wen X, Dickey CA (2016) MicroRNA-511 binds to FKBP5 mRNA, which encodes a chaperone protein, and regulates neuronal differentiation. J Biol Chem 291(34):17897–17906. https://doi.org/10.1074/jbc.M116.727941

Zhou J, Nagarkatti P, Zhong Y, Ginsberg JP, Singh NP, Zhang J, Nagarkatti M (2014) Dysregulation in microRNA expression is associated with alterations in immune functions in combat veterans with post-traumatic stress disorder. PLoS One 9(4):e94075. https://doi.org/10.1371/journal.pone.0094075

Zimmermann P, Bruckl T, Nocon A, Pfister H, Binder EB, Uhr M, Lieb R, Moffitt TE, Caspi A, Holsboer F, Ising M (2011) Interaction of FKBP5 gene variants and adverse life events in predicting depression onset: results from a 10-year prospective community study. Am J Psychiatry 168(10):1107–1116. https://doi.org/10.1176/appi.ajp.2011.10111577

Animal Models of PTSD: A Critical Review

Elizabeth I. Flandreau and Mate Toth

Abstract The goals of animal research in post-traumatic stress disorder (PTSD) include better understanding the neurophysiological etiology of PTSD, identifying potential targets for novel pharmacotherapies, and screening drugs for their potential use as PTSD treatment in humans. Diagnosis of PTSD relies on a patient interview and, as evidenced by changes to the diagnostic criteria in the DSM-5, an adequate description of this disorder in humans is a moving target. Therefore, it may seem insurmountable to model the construct of PTSD in animals such as rodents. Fortunately, the neural circuitry involved in fear and anxiety, thought to be essential to the etiology of PTSD in humans, is highly conserved throughout evolution. Furthermore, many symptoms can be modeled using behavioral tests that have face, construct, and predictive validity. Because PTSD is precipitated by a definite traumatic experience, animal models can simulate the induction of PTSD, and test causal factors with longitudinal designs. Accordingly, several animal models of physical and psychological trauma have been established. This review discusses the widely used animal models of PTSD in rodents, and overviews their strengths and weaknesses in terms of face, construct, and predictive validity.

Keywords Animal model • Anxiety • Fear learning • Predator stress • PTSD • Social defeat

E.I. Flandreau (✉) and M. Toth
Grand Valley State University, 1 Campus Drive, Allendale, MI 49401, USA
Department of Behavioral Neurobiology, Hungarian Academy of Sciences, Institute of Experimental Medicine, 43 Szigony Street, Budapest 1083, Hungary
e-mail: flandree@gvsu.edu; toth.mate@koki.mta.hu

Curr Topics Behav Neurosci (2018) 38: 47–68
DOI 10.1007/7854_2016_65
Published Online: 9 April 2016

Contents

1 Animal Models of Traumatic Stress

Diagnostic criteria for PTSD according to DSM-5 include expression – for over 1 month – of debilitating symptoms from four clusters following exposure to traumatic events such as threat of death, or actual or threatened serious injury or violence:

- Cluster B: Intrusion (e.g., nightmares, flashbacks, intrusive thoughts, and physiological reactions to trauma reminders).
- Cluster C: Avoidance (intentionally avoiding trauma-related people, places, or activities).
- Cluster D: Negative alterations in cognitions and mood (e.g., dissociative amnesia, negative perception of self and world, anhedonia, social withdrawal).
- Cluster E: Alterations in arousal and reactivity (e.g., irritability, aggression, problems concentrating, sleep disturbances, and hypervigilance).

The goal of this chapter is to critically review current animal models of trauma exposure and how the behavioral and physiological outcomes of these models translate to these symptom clusters.

For models of PTSD to reach high translational value, essential requirements have been defined (Siegmund and Wotjak 2006): (1) the trauma must be severe, (2) a relatively short duration should be sufficient to provoke PTSD-like symptoms, (3) the intensity of the trauma should predict the severity of outcome, and (4) the stressor should induce persistent or progressive PTSD-like alterations with (5) significant interindividual variability in outcomes. Existing models apply stressors of a physical, psychological, social, or combined nature.

1.1 Physical Stressors

Restraint stress, underwater holding, and electric shock are all stressors of a physical nature. However, it is generally accepted that these lead to a combined physical and psychological stress from the rodents' perspective. Application of these stressors varies greatly between laboratories in terms of duration and severity. Although most of these can also be applied as chronic stressors, this discussion focuses on acute stressors in keeping with the above second criteria for an animal model of PTSD.

1.1.1 Restraint/Immobilization Stress

Both restraint and immobilization stress involve placing rodents in an enclosed chamber allowing for minimal or no movement. Total immobilization can be considered the most severe of the restraint methods. Two hours of complete immobilization has been shown to increase anxiety-like behavior in the elevated plus maze and open-field tests (Andero et al. 2013; Mitra et al. 2005), increase compulsive-like behavior in the marble-burying test (Kedia and Chattarji 2014), reduce declarative memory performance in the water maze task, increase fear learning (Andero et al. 2013), and increase REM sleep (Meerlo et al. 2001). Importantly, avoidance and morphological changes mostly manifested after a significant delay (i.e., 10 days) (Andero et al. 2013; Mitra et al. 2005; Kedia and Chattarji 2014). In line with endocrine data from PTSD patients (Yehuda et al. 2006), when immobilization stress was followed 1, 7, or 13 days later by a 20-min reminder restraint, significant hypoactivity of the hypothalamic pituitary adrenal (HPA) axis was induced, as measured by decreased concentrations of ACTH and corticosterone (Harvey et al. 2006).

1.1.2 Underwater Holding

In rats, underwater trauma typically involves 40 s of forced swimming followed by a 20-s forced submersion (Richter-Levin 1998). Rats exposed to this stress exhibit increased startle reactivity and anxiety-like behavior in the elevated plus maze both immediately, 7, and 30 days later with reduced corticosterone levels 7 days after stress (Richter-Levin 1998; Cohen et al. 2004; Moore et al. 2012). Additionally, contextual fear with altered limbic activity persists for over a month in this model (Ritov et al. 2016).

1.1.3 Single Prolonged Stress

Single prolonged stress (SPS) paradigms typically involve three stressors: 2-h restraint followed by forced swim, followed by ether anesthesia. These stressors induce psychological, physiological, and endocrine stress, respectively. Rats exposed to this SPS procedure exhibit increased anxiety-like behavior in the elevated plus maze, enhanced fear acquisition, and reduced fear extinction learning (Imanaka et al. 2006; Knox et al. 2012; Takahashi et al. 2006; Wang et al. 2008; Yamamoto et al. 2009). Halothane-combined SPS exposure also increased anxiety-like behavior; corticosterone concentration was elevated 1 day following SPS but normalized by day 7 (Harvey et al. 2006). Interestingly, when rats were exposed to a reminder restraint, increased anxiety-like behavior was associated with a decreased corticosterone response, suggesting that the initial trauma is insufficient to produce PTSD-like symptoms in the absence of a reminder (Harvey et al. 2006). Increased anxiety-like behavior in the open field and increased immobility in the forced swim test 1 and 7 (but not 4) days were also apparent following SPS (Wu et al. 2016). The authors suggest that relevant compensatory changes occur in the first week following SPS exposure explain the timing of these effects (Wu et al. 2016).

Similar effects have been observed in mice: restraint, group swim, rat bedding (as a predator exposure) followed by ether exposure until unconsciousness led to enhanced cue-induced freezing, reduced extinction, and increased HPA axis negative feedback in the dexamethasone-suppression test, presumably due to elevated hippocampal expression of the glucocorticoid receptor (Perrine et al. 2016).

PTSD-like disruptions in sleep have also been observed in rats exposed to SPS. EEG recordings revealed that on the day of SPS exposure, rats exhibited increased REM sleep and decreased wakefulness (Vanderheyden et al. 2015). While these changes were not sustained over time, a decrease in non-REM sleep was observed during the active phase up to 7 days following trauma (Vanderheyden et al. 2015).

Morphological changes were also observed in SPS models including apoptotic volume loss in the hippocampus and the dorsal raphe nucleus (Han et al. 2013; Liu et al. 2012a) that may correspond with human findings reporting volumetric decreases in these brain regions in PTSD (Carrion et al. 2009; Gilbertson et al. 2002).

Selective serotonin reuptake inhibitors (SSRIs), a common treatment in PTSD patients (Steckler and Risbrough 2012), have been shown to reduce or prevent PTSD-like behavioral and endocrine symptoms in some SPS paradigms (Wang et al. 2008; Perrine et al. 2016) but not others (Takahashi et al. 2006). This difference could be due to laboratory differences in SPS design, SSRI administration, or behavioral testing. Of note, SSRIs are also ineffective in a portion of human PTSD patients (Stein et al. 2002); therefore successful symptom alleviation with SSRIs may not represent a meaningful concern from a validity perspective. Another possibility is that while SSRIs successfully treat comorbid anxiety and depression in PTSD patients, they may not truly address the core symptoms. As evidence for this hypothesis, a recent study demonstrated that the SSRI escitalopram reduced depressive-like and avoidance symptoms, but not fear extinction deficits in mice exposed to SPS (Lin et al. 2016).

1.1.4 Electric Shock

Uncontrollable and unpredictable foot shock is the most common method of stress exposure and even a single exposure can lead to long-lasting behavioral changes (reviewed in Bali and Jaggi 2015). Significant advantages of electric shock are (1) reliable delivery and precise control of the number and amperage of the shocks, length of the session, and inter-shock intervals; (2) well-defined and reproducible context and environmental cues by using standard boxes with lights and speakers; and (3) adjustable changes in contextual modalities, cues, and reminders. Another potential advantage of using electric shock is that, compared to restraint stress, there is less habituation (Siegmund and Wotjak 2007a). A potential limitation of electric shock as a model of PTSD-like trauma is that compared to human subjects and other animal trauma models, there is relatively little individual variability in terms of sensitivity and resiliency; the vast majority of subjects exposed to electric shock will go on to display behavioral consequences to stress (reviewed in Bali and Jaggi 2015).

The underlying vulnerability, progressive development, and persistence of PTSD symptoms can only be modeled using long-term testing covering multiple symptom domains. Consistent with clinical criteria, several rodent studies reveal progressive development and persistence (3–5 weeks following single or multiple shocks, with or without subsequent reminders) of PTSD-like symptoms, including social withdrawal or avoidance, defensive behavior, hypervigilance, sleep disturbances, and generalization of fear (Siegmund and Wotjak 2007a; Cullen et al. 2015; Louvart et al. 2005; Mikics et al. 2008a, b; Philbert et al. 2011; Pynoos et al. 1996). Importantly, these studies applied somewhat higher amperage (0.8–3.0 mA) compared to most conditioned fear paradigms focusing on fear learning characteristics on a short term (~0.5–1.5 mA). Increased fear response over time in vulnerable individuals has been conceptualized as "incubation of fear" (Siegmund and Wotjak 2007a; Elharrar et al. 2013; Eysenck 1968; Milad et al. 2009) that refers to recurrent recalls leading to sensitization of fear and stress responses, and reconsolidation of traumatic memories, in contrast with the progressive habituation and extinction of traumatic memories in well-adjusting (aka resistant) individuals. Because exposure therapies in clinical settings correspond well with reexposures of animals to shock context or cues, animal models will likely provide important mechanistic information about how cue-based, out-of-context (i.e., reminders in therapy office), and "in-the-context" (virtual reality) exposure therapies are mediated and may be enhanced effectively.

It is important to note some confusion between paradigms that use electric shock as a *single excessive traumatic stressor* to induce lasting symptoms of PTSD and those that induce *conditioned fear* to understand the elements of fear learning. Electric shock as a trauma exposure, as described above, focuses on *long-term* outcomes on multiple PTSD domains (e.g., freezing, startle, avoidance, endocrine, and neuromorphological outcomes), and as such, it is more specific to PTSD. In contrast, fear conditioning, described below, investigates neurobiological mechanisms underlying learning processes relevant to fear and anxiety disorders, uses *short-term* protocols, and measures conditioned fear response (i.e., freezing) or enhanced unconditioned fear response (i.e., fear-potentiated startle).

Shock-Induced Conditioned Fear

Conditioned fear models are based on the classical findings of Pavlov, who originally described that an association is formed between a neutral "conditioned" stimulus and an unconditioned stimulus following repeated co-presentations. Fear conditioning investigates elements of fear learning: fear acquisition, recall, extinction, and reinstatement and renewal of fear memories (Quirk et al. 2010; VanElzakker et al. 2014).

Most conditioned fear studies deliver scrambled electric shocks via metal grid floor in acoustically isolated plastic boxes with no escape. Major parameters (e.g., conditioned stimulus intensity, number of acquisition trials, intensity of shock) can be modified to optimize phenotypic outcome to avoid ceiling/floor effects. These modifications are essential given the significant differences across species (mice vs. rats) and strains in fear learning, presumably due to differences in nociception, exploratory activity, or other background characteristics (Bolivar et al. 2001; Keeley et al. 2015; Keum et al. 2016; Schaap et al. 2013; Stiedl et al. 1999). Shock exposure typically ranges from one to ten 0.3–3 mA shocks lasting between 1 and 2 s each, administered during a 5- to 30-min session (Mikics et al. 2008b; Aliczki and Haller 2015; Davis and Gould 2007; Lehner et al. 2010; Roozendaal et al. 2006; Sanford et al. 2010; Shoji et al. 2014; Siegmund and Wotjak 2007b). For cued associations, shocks are commonly co-presented with novel and salient cues such as sound, light, or odor. These specific cues can be used to dissect context- and cue-dependent fear recall, which are mediated in part by unique neural circuits (Marschner et al. 2008; Phillips and LeDoux 1992). Importantly, these neural circuits are highly conserved across species such that morphological and functional changes in animal models are analogous to those in PTSD patients (Acheson et al. 2012). An overview of conditioned fear literature is beyond the limits of this review; for more detailed reviews see Herry and Johansen (2014); Milad and Quirk (2012); and VanElzakker et al. (2014).

There are two major arguments in support of simple conditioned fear models in PTSD research. First, its face validity is high given that a single shock session as traumatic experience induces fear response with strong associative memory. Second, several reports show alterations of fear conditioning characteristics in PTSD including enhanced fear acquisition, blunted extinction, and relapse (Acheson et al. 2015; Jovanovic et al. 2012; Norrholm et al. 2011). Indeed, conditioned fear paradigms are translatable between animal and human laboratories by using similar procedural protocols (Parsons and Ressler 2013; Risbrough et al. 2016; Singewald et al. 2015). Accordingly, fruitful interactions between animal and human conditioning models have been reported, particularly in endeavors to develop effective pharmacological enhancers of extinction learning (Bowers and Ressler 2015; Johnson et al. 2012). Animal studies also advanced our understanding of how associative components of fear memories are acquired, consolidated, and extinguished spontaneously or following repeated exposures to trauma-related stimuli. There exists a growing body of evidence mapping the specific prefrontal cortex, hippocampal, and amygdala neurocircuits that modulate fear learning and inhibition via distinct neurotransmitter systems (Haubensak et al. 2010; Liu et al. 2012b; Rozeske et al. 2015). In support of translational efforts, these studies suggest that fear circuitry is comparable between rodents and humans (Mahan and Ressler 2012).

Despite the advantages listed above, there exist major concerns about short-term conditioned fear paradigms as models of PTSD. Namely, most models ignore the *temporal and differential diagnostic characteristics* of PTSD. According to the 5th edition of Diagnostic and Statistical Manual of Mental Disorders (DSM-5; American Psychiatric Association 2013), acute symptoms must be separated from long-term, chronic symptomatology. The former is defined as "acute stress disorder" and may be an adaptive stress-coping response in well-adjusting individuals, whereas in vulnerable individuals, PTSD symptoms are progressive and persistent and develop after a delay (DSM-5). Moreover, PTSD is no longer categorized as an "anxiety disorder" in DSM-5; rather, it is defined as "trauma- and stressor-related disorder." Accordingly, differential diagnostic criteria of PTSD require multiple-domain testing, especially given the heterogeneity of PTSD.

Fear-based reexperiencing (or intrusive memories) is a core symptom and is thought to be best modeled by fear conditioning in animals and humans. Acute and chronic phases of fear learning may have differential mechanisms and potential treatment targets; there is differential neuronal activation in the prefrontal cortex, amygdala, hippocampus, and monoaminergic systems during acute (24 h post-trauma) vs. remote (3–5 weeks) cued fear recall (Cullen et al. 2015; Mikics et al. 2008b; Frankland et al. 2004; Goshen et al. 2011; Tulogdi et al. 2012). Models focusing exclusively on acute phenotypic changes, short-term fear learning characteristics, and associative memory components of PTSD may have questionable construct validity and specificity. Considering the long-lasting behavioral consequences following a single shock session in several studies (Siegmund and Wotjak 2007a), conditioned fear models could potentially detect long-term PTSD-like outcomes if testing were extended.

1.2 Social and Psychological Stressors

1.2.1 Social Defeat Stress (SDS)

Social defeat stress (SDS) has been used as a robust stressor in both rat (Koolhaas et al. 2013) and mouse models (Golden et al. 2011) to induce depression (reviewed in Krishnan and Nestler 2011) and PTSD-like symptoms (reviewed in Daskalakis and Yehuda 2014). Evidence for SDS as a valid model for PTSD includes the following: a single defeat can lead to anxiety-like behavior, and behavioral abnormalities continue long after initial defeat (Pulliam et al. 2010).

As with other stress models, the intensity and duration of SDS can be modified. Most mouse SDS models follow a protocol in which experimental subjects as "intruders" are exposed to aggressive resident mice for 5–10 min, followed by 24 h of sensory (but not physical) exposure (Golden et al. 2011). This procedure is repeated for 5–10 days during which the intruder is exposed to different aggressor (Golden et al. 2011). PTSD-like symptoms following this social defeat exposure include decreased social exploration, anhedonia, and increased anxiety-like behavior, which can be reduced with chronic but not acute treatment with antidepressant drugs (Berton et al. 2006; Der-Avakian et al. 2014). Importantly, the PTSD-like symptoms are expressed by only 60–70% of exposed subjects, providing an inherent model of sensitivity and resiliency (Golden et al. 2011). Similar rates of sensitivity and resiliency have been achieved by 10 days of "witnessed social defeat stress." In this modification, the test mouse witnesses a peer exposed to an aggressive resident and then spends 24 h in sensory contact with that aggressor (Warren et al. 2013).

Another modified SDS model employs 6 h of indirect exposure of a partially restrained intruder mouse to an aggressive resident; within those 6 h, researchers randomly allow 1–3 direct exposures of 1 min each (Hammamieh et al. 2012). The authors argue that this model increases the unpredictability and uncontrollability of the stress exposure; this hypothesis is supported by data showing increased freezing behavior 28 days following a 10-day exposure, and increased grooming behavior 42 days following a 10-day exposure (Hammamieh et al. 2012; Muhie et al. 2015).

Although most SDS models violate the criteria of a single traumatic exposure (Yehuda and Antelman 1993), SDS may be relevant for socially induced or combat-related PTSD, which typically do involve multiple exposures. SDS may also be particularly relevant for comorbidity with depression. An important limitation of the SDS model is the lack of protocol for social stress in females. It has been suggested that ovariectomized female mice from an aggressive strain could be primed with male steroids and used as an appropriate resident for an SDS model in females (Daskalakis and Yehuda 2014), indicating that future studies are needed to establish models in females.

1.2.2 Predator Stress

In predator stress models, rodents are exposed to their natural predators or to their odor. In live predator exposure, the predator is well fed and habituated to rodents; therefore no physical injuries occur. In contrast to the physical stressors, which apply well-defined and precise traumatic experience, predator stress models focus on ecological validity to increase translatability.

Predator exposure immediately elicits unconditioned fear and marked stress responses in rodents (Adamec et al. 2006; Adamec and Shallow 1993; Blanchard and Blanchard 1989; Blanchard et al. 1998; Cohen et al. 2003; Dielenberg and McGregor 2001; Zoladz et al. 2008). A single predator exposure (1) increases avoidance behavior of both novel and predator-related stimuli (cluster C in DSM-5; Cohen et al. 2004; Toth et al. 2016), (2) induces cognitive impairments and negative mood-like states as measured by spatial learning in the Morris water maze (cluster D; Diamond et al. 2006), and (3) increases hyperarousal and hypervigilance as measured by startle reactivity (cluster E; Adamec et al. 2010). Depending on the specific predator stress model employed, enduring behavioral changes can last up to a month or more after exposure, replicating the chronic nature of PTSD (Adamec and Shallow 1993; Zoladz et al. 2012). These behavioral changes coincide with long-term structural changes of the hippocampus and the amygdala that are in line with circuit changes observed in PTSD patients (Diamond et al. 2006; Adamec et al. 2012; Mitra et al. 2009).

In a modified predator exposure, rodents are in a porous container to allow sensory but not physical exposure to live cats (Diamond et al. 1999; Mesches et al. 1999). The authors argue that besides offering physical protection, the container can enhance fear by increasing uncontrollability and helplessness. The exposure in this model is 30–75 min long. To maximize the predator interest across the session, food is placed on top of the container in which the rodent is held. This single exposure can induce long-term memory and cognitive deficits in rats (Diamond et al. 1999; Mesches et al. 1999; Woodson et al. 2003). In a version of this model, rodents are also reexposed to the same procedure 10 days later. Because the reexposure is identical to the traumatic event itself, it is not clear if this modification simulates "flashback-like memory retrieval" as the authors indicate or simply increases the "dose" of trauma. According to a recent study, a single reexposure instigated PTSD-like symptoms while a second reexposure was not able to further increase predator-induced effects and actually caused a slight decrease in PTSD-like symptoms (Zoladz et al. 2015), potentially due to habituation.

Social instability after the initial trauma has been used as a chronic stressor to exacerbate PTSD-like behaviors (Zoladz et al. 2008). Three weeks after this procedure, increased anxiety-like behavior, startle hyperreactivity, and spatial memory deficits are observed (Zoladz et al. 2008, 2012). This combined paradigm also results in reduced thymus and adrenal weights, reduced basal glucocorticoid levels, and increased dexamethasone suppression of the HPA axis (Zoladz et al. 2008, 2012), physiological and endocrine changes corresponding with HPA abnormalities in PTSD (Yehuda et al. 1993). This paradigm was also shown to impair spatial memory, presumably secondary to structural and functional plasticity changes in the hippocampus (Diamond et al. 2006; Mesches et al. 1999; Sandi et al. 2005). These symptoms and structural changes are similar to observations in PTSD patients (O'Doherty et al. 2015).

Exposure only to olfactory cues from a predator can also induce behavioral signs of fear in rats and mice, and activate brain regions mediating defensive, fear, and anxiety-like responses, such as the lateral septum, extended amygdala, paraventricular nucleus of the hypothalamus, hypothalamic circuits mediating defensive reaction, and periaqueductal gray (Dielenberg and McGregor 2001; Canteras et al. 2015; Takahashi 2014). Predator odor methods use litter or bedding from a predator, fur and/or urine, a cat-worn collar, or trimethylthiazoline, a synthetic component of fox feces. Based on data from field and laboratory studies, the fear-eliciting efficacy of predator odor exposure is highly variable; fur and used litter appear to be the most effective odor stimuli (Takahashi 2014; Apfelbach et al. 2005).

In support of the predictive validity of predator stress paradigms, benzodiazepines, SSRIs, and the tricyclic antidepressant tianeptine have all been shown to efficiently reduce some PTSD-like symptoms of predator exposure (Dielenberg and McGregor 2001; Vouimba et al. 2006; Zoladz et al. 2013).

2 Operationalizing PTSD-Like Symptoms in Animal Models

The DSM-5 divides PTSD symptoms into four major symptom clusters: intrusion, avoidance, negative alterations in cognitions and mood, and alterations in arousal and reactivity. There exist numerous validated behavioral tests in rodents designed to operationalize these constructs. These behavioral tests can be used to assess sensitivity or resiliency to trauma exposure and to screen novel pharmaceutical treatments.

Assessing intrusive (i.e., reexperiencing) symptoms (Cluster B) is not possible in animal models; however measuring physiological reactions to trauma-related cues can operationalize fear memory processes thought to underlie reexperiencing symptoms. Approach-avoidance conflict tests such as the elevated plus maze, elevated zero maze, open-field arena, and light-dark box are designed to measure avoidance behavior (Cluster C) and generalized-like anxiety in different contexts, and they are widely used in preclinical research with reasonable predictive validity, although there are limitations in their use (Adamec 1997; Bailey and Crawley 2009; Belzung and Griebel 2001; File and Seth 2003; McCormick and Green 2013; Olson et al. 2011). Trauma-specific avoidance can also be measured via introduction of trauma-related cues, e.g., predator scent in the case of predator stress, novel conspecific in the case of social defeat stress, and training context in the case of active and passive avoidance.

Negative alterations in mood (depression-like symptoms; Cluster D) are commonly measured by tests of "behavioral despair," i.e., reduced activity in forced swim and tail suspension tests or decreased preference of sucrose as a sign of anhedonia, or reduced intracranial self-stimulation. Predominantly, these tests are based on their predictive validity for antidepressant medications which are also somewhat effective in PTSD (Cryan et al. 2005). Developing additional measures of negative mood in rodents should be an active area of continued research to expand reliability and validity and minimize locomotion and memory confounds.

Modeling negative alterations in cognition (Cluster D) is highly limited in animals as these symptoms are manifested in distorted, negatively tuned self- and environment interpretations. However, more general cognitive deficits are commonly measured, i.e., attention, spatial learning, and social recognition. For spatial learning and memory, Morris water maze became a golden standard test (Vorhees and Williams 2006), although less stressful versions, i.e., Barnes maze, T-maze, or radial maze (Sharma et al. 2010), are also widely used and may be advantageous in certain paradigms where testing with minimal stress is crucial. Attention and more complex learning characteristics can be reliably assessed in rodents using the five-choice paradigm (Chudasama and Robbins 2004); there is no data available on this specific symptom domain in the most common models of PTSD (see Table 1). As specific neurocircuits are involved in these tasks with documented changes in PTSD (Gilbertson et al. 2002; Moser and Moser 1998), identification of neurobiological changes including structural and functional alterations of the prefrontal and hippocampal networks can increase the construct validity of these models.

Trauma-induced hyperarousal and hyperreactivity can be assessed by startle amplitude, its habituation, and prepulse inhibition (Adamec et al. 2010). Hypervigilance can be measured using object-burying paradigms or locomotor changes (Mikics et al. 2008a; Philbert et al. 2011). Irritability-impulsivity can be approximated using the resident-intruder test, and delayed discounting paradigm (Fodor et al. 2014; Mar and Robbins 2007); however, none of the described models have been shown to result in such externalizing-like symptoms. Sleep disturbances (altered structure or quantitative changes in stages) can be measured using EEG signals (Pawlyk et al. 2008).

Table 1 Summary of symptoms induced in animal models of PTSD

	Cluster B: intrusive symptoms	Cluster C: avoidance symptoms	Cluster D: negative alterations in mood and cognition		Cluster E: hyperarousal symptoms					Criterion F: lasting symptoms
	"Signs of" intrusive memories	Increased avoidance	Anhedonia/ decreased activity	Cognitive dysfunctions	Irritability/ aggression	Hyper-vigilance	Exaggerated startle	Concentration problems	Sleep disturbance	Symptoms present
Restraint/ immobilization stress	Freezing (Andero et al. 2013)	Plus-maze (Andero et al. 2013; Mitra et al. 2005)		Spatial memory (Andero et al. 2013)		Marble burying (Kedia and Chattarji 2014)			Increased REM (Meerlo et al. 2001)	>10 Days
Underwater trauma	Freezing (Richter-Levin 1998)	Plus-maze (Moore et al. 2012)		Spatial memory (Richter-Levin 1998)			Enhanced reactivity (Cohen et al. 2004)			>Month
Single prolonged stress	Freezing (Takahashi et al. 2006; Yamamoto et al. 2009)	Plus-maze (Harvey et al. 2006; Imanaka et al. 2006)	Forced swim (Wu et al. 2016)	Disrupted extinction learning (Knox et al. 2012)			Enhanced reactivity (Khan and Liberzon 2004)		Increased REM (Vanderheyden et al. 2015)	>2 Weeks
Electric shock	Freezing, fear generalization (Siegmund and Wotjak 2007a; Cullen et al. 2015; Louvart et al. 2005; Mikics et al. 2008b)	Plus maze, social interaction, novelty- and social avoidance (Siegmund and Wotjak 2007a; Louvart et al. 2005; Mikics et al. 2008a; Philbert et al. 2011)	Forced swim (Siegmund and Wotjak 2007a)			Object burying, hyperlocomotion (Mikics et al. 2008a; Philbert et al. 2011)	Enhanced reactivity, fear-potentiated startle (Pynoos et al. 1996; Cassella and Davis 1985; Davis 1989)		Sleep fragmentation (Philbert et al. 2011)	>Month
Social defeat	Fear generalization (Rygula et al. 2006)	Plus maze, social avoidance (Berton et al. 2006; Golden et al. 2011)	Self-stimulation and sucrose preference (Der-Avakian et al. 2014; Rygula et al. 2006)				Enhanced reactivity (Pulliam et al. 2010)		Sleep fragmentation	>Month
Predator stress	Freezing (Zoladz et al. 2015)	Plus maze (Zoladz et al. 2015)		Spatial and working memory (Diamond et al. 1999; Woodson et al. 2003; Zoladz et al. 2015)			Enhanced reactivity (Zoladz et al. 2008; Adamec et al. 2010)			>Month

Examples/references are from representative studies showing replicated findings in validated tests

For instance, sleep disturbances in trauma-exposed rats correlated with other behavioral symptoms; rats with the greatest increase in REM sleep following SPS exposure also had the largest percent of freezing behavior in a fear conditioning paradigm (Vanderheyden et al. 2015).

To outline a current status of preclinical models of PTSD, Table 1 summarizes which clusters and symptoms have been induced in particular models, also indicating symptoms, which need to be addressed as data are not available. Noteworthy, we did not include models and symptoms, where repeated stress exposure is required to induce phenotypic changes, e.g., anhedonia or exaggerated startle following repeated exposures, to be more specific to PTSD, and eliminate mixed (e.g., depression-like) models.

3 The "Cutoff Behavioral Criteria" Model of PTSD: Focus on Vulnerability and Resiliency Factors

Large epidemiological studies estimated the incidence of PTSD following traumatic events around 10–20% with significant sex differences (i.e., 8–13% for men and 20–30% for women), although these rates are highly dependent on the type of the trauma (Breslau et al. 1991, 1997; Kessler et al. 1995). Accordingly, animal models have sought to use individual variability as a means to identify the neurobiological risk factors and underpinnings of PTSD.

The "cutoff behavioral criteria" established by Cohen and colleagues (2006a, 2012) used predator stress as the trauma; according to predefined behavioral "cutoff" criteria, traumatized animals are assigned to "extreme behavioral response" (EBR; aka vulnerable) and "minimal behavioral response" (MBR; aka resilient) groups based on their post-stress startle reactivity and performance on the elevated plus maze. Although potentially other behavioral, endocrine, or physiological measurements could be used as cutoff criteria, startle reactivity and avoidance on the elevated plus maze detect changes in measures of hyper-alertness and overall behavioral avoidance/anxiety, modeling two main symptom clusters of PTSD. To maximize effect size and interpretability of the data, the group of rodents between EBR and MBR is considered to have a "partial behavioral response," and not investigated (Cohen et al. 2006a).

The cutoff criteria described by Cohen and colleagues are fairly conservative, yet the rate of EBR varies substantially between strains. For rodents to meet EBR criteria they must spend the entire 5-min elevated plus session in the closed arms, and exhibit >800 units startle amplitude without habituation. MBR criteria required rodents to spend less than 1 of 5 min in closed arms of the elevated plus, and to exhibit <700 units startle amplitude with significant habituation over time (Matar et al. 2013). Using these criteria with Sprague-Dawley rats, EBR rates 24 h following predator stress may be as high as 90%. However, similar to humans, acute-anxiety symptoms fade in well-adapted animals; the EBR rate drops sharply in the days immediately following predator stress (Cohen et al. 2004). For Sprague-Dawley rats, EBR incidence settles at around 25% (Cohen et al. 2003), which is comparable with human epidemiological data (Breslau et al. 1991; Kessler et al. 1995). The EBR phenotype can persist for more than 3 weeks in the 25% of experimental subjects classified as EBR (Cohen et al. 2004), which is again compatible with the temporal characteristic of human PTSD (Cohen et al. 2004). However, EBR rates vary significantly between strains in rats (10% and 50% in Lewis and in Fisher rats, respectively) (Cohen et al. 2006a, b), and mice (from 6% to 55% in the most widely used strains, DBA/2J and C57Bl/6J, as two extremes of the spectrum, respectively) (Cohen et al. 2008). Importantly, extreme behavioral response at baseline (without stress exposure) is hardly detectable in experimental populations, i.e., 1.3% (Cohen et al. 2012), confirming that EBR is clearly induced by trauma exposure, and lasts for weeks only in a vulnerable subpopulation (i.e., 25%).

Studies using these behavioral cutoff criteria reported enhanced stress reactivity in EBR subjects as measured by increased heart rate and sympathetic activity, and higher plasma corticosterone and adrenocorticotropic hormone concentrations (Cohen et al. 2003, 2005). EBR subjects also exhibited additional plasticity-related changes in the hippocampus, such as reduced expression of BDNF, synaptophysin, and ERK1/2 pathway elements, and elevated expression of glucocorticoid receptor and postsynaptic density protein-95 mRNA (Matar et al. 2013; Cohen et al. 2014; Kozlovsky et al. 2007). Importantly, acute and 7-day-long treatment with the SSRI sertraline was able to decrease anxiety, reduce startle reactivity, and significantly decrease the number of rodents meeting EBR criteria (Matar et al. 2006).

Importantly, this cutoff criteria approach can be applied to other PTSD models to compare vulnerable and resilient subjects. An underwater trauma model found that EBR rats exhibited lasting anxiety and hyperarousal (Cohen et al. 2004). In another study, mice were exposed to electric shock (14× 1-s-long shocks of 1 mA with variable intervals over 85-min period) followed 24 h later by a shock “trigger” (5× 1-s-long shocks of 0.7 mA with fixed intervals over 5 min) and tested behavioral and endocrine outcomes 7–16 days later. Based on a cumulative 5-test criteria system (upper quintiles in the marble-burying test, acoustic startle, pre-pulse inhibition, light-dark box, and home cage locomotor activity), researchers categorized subjects as “resilient” or “PTSD-like” (20–20% of the population). Compared to resilient subjects, mice with PTSD-like phenotype exhibited blunted corticosterone response to acute restraint stress with significant upregulation of hippocampal glucocorticoid receptors (Lebow et al. 2012). Moreover, PTSD-like phenotype was associated with altered expression of a number of “PTSD candidate” genes (i.e., previously reported as risk polymorphisms in human studies) in the bed nucleus of stria terminalis (Lebow et al. 2012).

In summary, despite low genetic variability, animal models of PTSD using inbred strains provide substantial individual variability for application of cutoff criteria. Models that can reliably and robustly identify resistant and vulnerable populations may be highly valuable to test the causal contribution of developmental risk factors and candidate mechanisms identified by human studies (e.g., single polymorphisms).

4 Special Challenges

While the animal models described above have strong face and predictive validity, reaching full diagnosis (based on human DSM-5 criteria) in a single animal model may not be feasible. Most of the symptoms operationalized in rodent PTSD models such as associative fear, avoidance, anhedonia, or hypervigilance are also described in other anxiety or mood disorders; thus specificity of a given model for PTSD is a challenge. Additionally, some symptoms are difficult (if not impossible) to model in animals, e.g., intrusive memories and flashbacks, feeling emotional numb, or detachment. These challenges might mean that we cannot create a rat or mouse model that is comprehensive and specific to PTSD. However, it is relevant to note that the symptoms of PTSD have significant overlap with mood and anxiety disorders and, furthermore, PTSD is frequently found comorbid with mood, anxiety, and substance-use disorders (Sareen 2014). PTSD patients also show extensive heterogeneity in symptom presentation, with over 20 possible symptoms described in DSM-5. To account for this heterogeneity, preclinical and clinical characterizations have shifted focus from diagnosis to domain-based conception of disorders based on biological mechanisms and their domain-like organization (i.e., Research Domain Criteria-RDoC proposed by NIMH), e.g., distinct and definite mechanisms underlying reexperiencing and hyperarousal in PTSD (Cuthbert 2014; Cuthbert and Insel 2013). This approach may increase translatability of animal findings in the clinics in the future, and benefit for our understanding of the pathogenesis of PTSD.

Although the above-described models have been found to successfully model some aspects of PTSD, lack of data on particular symptoms in particular models (e.g., externalizing or attention symptoms), negative findings (likely significant amount unpublished), and inconsistencies between laboratories still exist, which is difficult to interpret as experimental procedures show high variability between laboratories. Each laboratory has a unique method of trauma exposure and optimized behavioral test battery and specific optimized protocols to assess symptoms. For example "incubation time" periods between trauma and testing (~1–5 weeks), testing time (light or dark cycle), or the order of tests in a battery, and the specific tests included in a battery, all vary between laboratories.

5 Conclusions and Future Perspectives

Development of valid animal models with high translational values is a constant challenge for preclinical research and needs recurrent updates to maximize efficacy. On the one hand, diagnostic criteria are constantly shifting targets as clinical and epidemiological data are accumulating. Additionally, identification and interpretation of measured behaviors in animal models is a complex task as variables or behavioral symptoms must be ecologically valid in the behavioral repertoire of the experimental species and still translatable to humans.

To better understand the complex etiology and mechanisms of PTSD and specific contributions of different symptom domains will require wider screening of phenotypic changes in different models. From a translational standpoint, it would be particularly important to identify pathways linked to vulnerability and resilience against traumatic stress either causally or as robust and measurable predictive markers (e.g., blood-based markers). In this respect, the "cutoff behavioral criteria" strategy and comparative studies contrasting biological characteristics of resilient and vulnerable populations provide an essential approach either to clarify the causal role of candidate mechanisms identified by humans studies (e.g., polymorphisms) or to implicate new targets to test them in human studies.

Although large animal numbers and extensive screening are necessary to produce behavioral extremes with the necessary statistical power for reliable and reproducible results, the highly variable outcomes observed after trauma exposure in humans clearly warrant the application of such responder/nonresponder classification approaches based on the fact that only a portion of trauma-exposed individuals develop PTSD symptoms.

As a final comment, the past decade has seen a substantial increase in PTSD research in both humans and animal models. Building upon this literature is essential to developing tools that can actually improve the human experience. Although sex differences were beyond the scope of this review, to effectively study this topic will require expanding the subject pool more consistently to include females. The NIMH mandate to include female cells, tissue, and subjects in preclinical research will certainly push the field towards addressing this gap.

References

Acheson DT, Gresack JE, Risbrough VB (2012) Hippocampal dysfunction effects on context memory: possible etiology for posttraumatic stress disorder. Neuropharmacology 62:674–685

Acheson DT, Geyer MA, Baker DG, Nievergelt CM, Yurgil K, Risbrough VB, Team M-I (2015) Conditioned fear and extinction learning performance and its association with psychiatric symptoms in active duty Marines. Psychoneuroendocrinology 51:495–505

Adamec R (1997) Transmitter systems involved in neural plasticity underlying increased anxiety and defense—implications for understanding anxiety following traumatic stress. Neurosci Biobehav Rev 21:755–765

Adamec RE, Shallow T (1993) Lasting effects on rodent anxiety of a single exposure to a cat. Physiol Behav 54:101–109

Adamec R, Head D, Blundell J, Burton P, Berton O (2006) Lasting anxiogenic effects of feline predator stress in mice: sex differences in vulnerability to stress and predicting severity of anxiogenic response from the stress experience. Physiol Behav 88:12–29

Adamec R, Fougere D, Risbrough V (2010) CRF receptor blockade prevents initiation and consolidation of stress effects on affect in the predator stress model of PTSD. Int J Neuropsychopharmacol 13:747–757

Adamec R, Hebert M, Blundell J, Mervis RF (2012) Dendritic morphology of amygdala and hippocampal neurons in more and less predator stress responsive rats and more and less spontaneously anxious handled controls. Behav Brain Res 226:133–146

Aliczki M, Haller J (2015) Electric shock as model of post-traumatic stress disorder in rodents. In: Martin CR, Preedy VR, Patel VB (eds) Comprehensive guide to post-traumatic stress disorder. Springer International Publishing, Cham

American Psychiatric Association (2013) Diagnostic and statistical manual of mental disorders, 5th edn. American Psychiatric Publishing, Washington, DC
Andero R, Brothers SP, Jovanovic T, Chen YT, Salah-Uddin H, Cameron M, Bannister TD, Almli L, Stevens JS, Bradley B et al (2013) Amygdala-dependent fear is regulated by Oprl1 in mice and humans with PTSD. Sci Transl Med 5:188ra173
Apfelbach R, Blanchard CD, Blanchard RJ, Hayes RA, McGregor IS (2005) The effects of predator odors in mammalian prey species: a review of field and laboratory studies. Neurosci Biobehav Rev 29:1123–1144
Bailey KR, Crawley JN (2009) Anxiety-related behaviors in mice. In: Buccafusco JJ (ed) Methods of behavior analysis in neuroscience. CRC Press, Boca Raton, FL
Bali A, Jaggi AS (2015) Electric foot shock stress adaptation: does it exist or not? Life Sci 130:97–102
Belzung C, Griebel G (2001) Measuring normal and pathological anxiety-like behaviour in mice: a review. Behav Brain Res 125:141–149
Berton O, McClung CA, Dileone RJ, Krishnan V, Renthal W, Russo SJ, Graham D, Tsankova NM, Bolanos CA, Rios M et al (2006) Essential role of BDNF in the mesolimbic dopamine pathway in social defeat stress. Science 311:864–868
Blanchard RJ, Blanchard DC (1989) Antipredator defensive behaviors in a visible burrow system. J Comp Psychol 103:70–82
Blanchard RJ, Nikulina JN, Sakai RR, McKittrick C, McEwen B, Blanchard DC (1998) Behavioral and endocrine change following chronic predatory stress. Physiol Behav 63:561–569
Bolivar VJ, Pooler O, Flaherty L (2001) Inbred strain variation in contextual and cued fear conditioning behavior. Mamm Genome 12:651–656
Bowers ME, Ressler KJ (2015) An overview of translationally informed treatments for posttraumatic stress disorder: animal models of pavlovian fear conditioning to human clinical trials. Biol Psychiatry 78:E15–E27
Breslau N, Davis GC, Andreski P, Peterson E (1991) Traumatic events and posttraumatic stress disorder in an urban population of young adults. Arch Gen Psychiatry 48:216–222
Breslau N, Davis GC, Andreski P, Peterson EL, Schultz LR (1997) Sex differences in posttraumatic stress disorder. Arch Gen Psychiatry 54:1044–1048
Canteras NS, Pavesi E, Carobrez AP (2015) Olfactory instruction for fear: neural system analysis. Front Neurosci 9:276
Carrion VG, Weems CF, Watson C, Eliez S, Menon V, Reiss AL (2009) Converging evidence for abnormalities of the prefrontal cortex and evaluation of midsagittal structures in pediatric posttraumatic stress disorder: an MRI study. Psychiatry Res 172:226–234
Cassella JV, Davis M (1985) Fear-enhanced acoustic startle is not attenuated by acute or chronic imipramine treatment in rats. Psychopharmacology (Berl) 87:278–282
Chudasama Y, Robbins TW (2004) Psychopharmacological approaches to modulating attention in the five-choice serial reaction time task: implications for schizophrenia. Psychopharmacology (Berl) 174:86–98
Cohen H, Zohar J, Matar M (2003) The relevance of differential response to trauma in an animal model of posttraumatic stress disorder. Biol Psychiatry 53:463–473
Cohen H, Zohar J, Matar MA, Zeev K, Loewenthal U, Richter-Levin G (2004) Setting apart the affected: the use of behavioral criteria in animal models of post traumatic stress disorder. Neuropsychopharmacology 29:1962–1970
Cohen H, Zohar J, Matar MA, Kaplan Z, Geva AB (2005) Unsupervised fuzzy clustering analysis supports behavioral cutoff criteria in an animal model of posttraumatic stress disorder. Biol Psychiatry 58:640–650
Cohen H, Matar MA, Richter-Levin G, Zohar J (2006a) The contribution of an animal model toward uncovering biological risk factors for PTSD. Ann N Y Acad Sci 1071:335–350
Cohen H, Zohar J, Gidron Y, Matar MA, Belkind D, Loewenthal U, Kozlovsky N, Kaplan Z (2006b) Blunted HPA axis response to stress influences susceptibility to posttraumatic stress response in rats. Biol Psychiatry 59:1208–1218

Cohen H, Geva AB, Matar MA, Zohar J, Kaplan Z (2008) Post-traumatic stress behavioural responses in inbred mouse strains: can genetic predisposition explain phenotypic vulnerability? Int J Neuropsychopharmacol 11:331–349

Cohen H, Kozlovsky N, Alona C, Matar MA, Joseph Z (2012) Animal model for PTSD: from clinical concept to translational research. Neuropharmacology 62:715–724

Cohen H, Kozlovsky N, Matar MA, Zohar J, Kaplan Z (2014) Distinctive hippocampal and amygdalar cytoarchitectural changes underlie specific patterns of behavioral disruption following stress exposure in an animal model of PTSD. Eur Neuropsychopharmacol 24:1925–1944

Cryan JF, Mombereau C, Vassout A (2005) The tail suspension test as a model for assessing antidepressant activity: review of pharmacological and genetic studies in mice. Neurosci Biobehav Rev 29:571–625

Cullen PK, Gilman TL, Winiecki P, Riccio DC, Jasnow AM (2015) Activity of the anterior cingulate cortex and ventral hippocampus underlie increases in contextual fear generalization. Neurobiol Learn Mem 124:19–27

Cuthbert BN (2014) The RDoC framework: facilitating transition from ICD/DSM to dimensional approaches that integrate neuroscience and psychopathology. World Psychiatryn 13:28–35

Cuthbert BN, Insel TR (2013) Toward the future of psychiatric diagnosis: the seven pillars of RDoC. BMC Med 11:126

Daskalakis NP, Yehuda R (2014) Principles for developing animal models of military PTSD. Eur J Psychotraumatol 5

Davis M (1989) Sensitization of the acoustic startle reflex by footshock. Behav Neurosci 103:495–503

Davis JA, Gould TJ (2007) beta2 subunit-containing nicotinic receptors mediate the enhancing effect of nicotine on trace cued fear conditioning in C57BL/6 mice. Psychopharmacology (Berl) 190:343–352

Der-Avakian A, Mazei-Robison MS, Kesby JP, Nestler EJ, Markou A (2014) Enduring deficits in brain reward function after chronic social defeat in rats: susceptibility, resilience, and antidepressant response. Biol Psychiatry 76:542–549

Diamond DM, Park CR, Heman KL, Rose GM (1999) Exposing rats to a predator impairs spatial working memory in the radial arm water maze. Hippocampus 9:542–552

Diamond DM, Campbell AM, Park CR, Woodson JC, Conrad CD, Bachstetter AD, Mervis RF (2006) Influence of predator stress on the consolidation versus retrieval of long-term spatial memory and hippocampal spinogenesis. Hippocampus 16:571–576

Dielenberg RA, McGregor IS (2001) Defensive behavior in rats towards predatory odors: a review. Neurosci Biobehav Rev 25:597–609

Elharrar E, Warhaftig G, Issler O, Sztainberg Y, Dikshtein Y, Zahut R, Redlus L, Chen A, Yadid G (2013) Overexpression of corticotropin-releasing factor receptor type 2 in the bed nucleus of stria terminalis improves posttraumatic stress disorder-like symptoms in a model of incubation of fear. Biol Psychiatry 74:827–836

Eysenck HJ (1968) A theory of the incubation of anxiety-fear responses. Behav Res Ther 6:309–321

File SE, Seth P (2003) A review of 25 years of the social interaction test. Eur J Pharmacol 463:35–53

Fodor A, Barsvari B, Aliczki M, Balogh Z, Zelena D, Goldberg SR, Haller J (2014) The effects of vasopressin deficiency on aggression and impulsiveness in male and female rats. Psychoneuroendocrinology 47:141–150

Frankland PW, Bontempi B, Talton LE, Kaczmarek L, Silva AJ (2004) The involvement of the anterior cingulate cortex in remote contextual fear memory. Science 304:881–883

Gilbertson MW, Shenton ME, Ciszewski A, Kasai K, Lasko NB, Orr SP, Pitman RK (2002) Smaller hippocampal volume predicts pathologic vulnerability to psychological trauma. Nat Neurosci 5:1242–1247

Golden SA, Covington HE 3rd, Berton O, Russo SJ (2011) A standardized protocol for repeated social defeat stress in mice. Nat Protoc 6:1183–1191
Goshen I, Brodsky M, Prakash R, Wallace J, Gradinaru V, Ramakrishnan C, Deisseroth K (2011) Dynamics of retrieval strategies for remote memories. Cell 147:678–689
Hammamieh R, Chakraborty N, De Lima TC, Meyerhoff J, Gautam A, Muhie S, D'Arpa P, Lumley L, Carroll E, Jett M (2012) Murine model of repeated exposures to conspecific trained aggressors simulates features of post-traumatic stress disorder. Behav Brain Res 235:55–66
Han F, Yan S, Shi Y (2013) Single-prolonged stress induces endoplasmic reticulum-dependent apoptosis in the hippocampus in a rat model of post-traumatic stress disorder. PLoS One 8: e69340
Harvey BH, Brand L, Jeeva Z, Stein DJ (2006) Cortical/hippocampal monoamines, HPA-axis changes and aversive behavior following stress and restress in an animal model of post-traumatic stress disorder. Physiol Behav 87:881–890
Haubensak W, Kunwar PS, Cai H, Ciocchi S, Wall NR, Ponnusamy R, Biag J, Dong HW, Deisseroth K, Callaway EM et al (2010) Genetic dissection of an amygdala microcircuit that gates conditioned fear. Nature 468:270–276
Herry C, Johansen JP (2014) Encoding of fear learning and memory in distributed neuronal circuits. Nat Neurosci 17:1644–1654
Imanaka A, Morinobu S, Toki S, Yamawaki S (2006) Importance of early environment in the development of post-traumatic stress disorder-like behaviors. Behav Brain Res 173:129–137
Johnson LR, McGuire J, Lazarus R, Palmer AA (2012) Pavlovian fear memory circuits and phenotype models of PTSD. Neuropharmacology 62:638–646
Jovanovic T, Kazama A, Bachevalier J, Davis M (2012) Impaired safety signal learning may be a biomarker of PTSD. Neuropharmacology 62:695–704
Kedia S, Chattarji S (2014) Marble burying as a test of the delayed anxiogenic effects of acute immobilisation stress in mice. J Neurosci Methods 233:150–154
Keeley RJ, Bye C, Trow J, McDonald RJ (2015) Strain and sex differences in brain and behaviour of adult rats: learning and memory, anxiety and volumetric estimates. Behav Brain Res 288:118–131
Kessler RC, Sonnega A, Bromet E, Hughes M, Nelson CB (1995) Posttraumatic stress disorder in the National Comorbidity Survey. Arch Gen Psychiatry 52:1048–1060
Keum S, Park J, Kim A, Park J, Kim KK, Jeong J, Shin HS (2016) Variability in empathic fear response among 11 inbred strains of mice. Genes Brain Behav 15:231–242
Khan S, Liberzon I (2004) Topiramate attenuates exaggerated acoustic startle in an animal model of PTSD. Psychopharmacology (Berl) 172:225–229
Knox D, George SA, Fitzpatrick CJ, Rabinak CA, Maren S, Liberzon I (2012) Single prolonged stress disrupts retention of extinguished fear in rats. Learn Mem 19:43–49
Koolhaas JM, Coppens CM, de Boer SF, Buwalda B, Meerlo P, Timmermans PJ (2013) The resident-intruder paradigm: a standardized test for aggression, violence and social stress. J Vis Exp e4367
Kozlovsky N, Matar MA, Kaplan Z, Kotler M, Zohar J, Cohen H (2007) Long-term down-regulation of BDNF mRNA in rat hippocampal CA1 subregion correlates with PTSD-like behavioural stress response. Int J Neuropsychopharmacol 10:741–758
Krishnan V, Nestler EJ (2011) Animal models of depression: molecular perspectives. Curr Top Behav Neurosci 7:121–147
Lebow M, Neufeld-Cohen A, Kuperman Y, Tsoory M, Gil S, Chen A (2012) Susceptibility to PTSD-like behavior is mediated by corticotropin-releasing factor receptor type 2 levels in the bed nucleus of the stria terminalis. J Neurosci 32:6906–6916
Lehner M, Wislowska-Stanek A, Maciejak P, Szyndler J, Sobolewska A, Krzascik P, Plaznik A (2010) The relationship between pain sensitivity and conditioned fear response in rats. Acta Neurobiol Exp (Wars) 70:56–66

Lin CC, Tung CS, Liu YP (2016) Escitalopram reversed the traumatic stress-induced depressed and anxiety-like symptoms but not the deficits of fear memory. Psychopharmacology (Berl) 233(7):1135–1146

Liu D, Xiao B, Han F, Wang E, Shi Y (2012a) Single-prolonged stress induces apoptosis in dorsal raphe nucleus in the rat model of posttraumatic stress disorder. BMC Psychiatry 12:211

Liu X, Ramirez S, Pang PT, Puryear CB, Govindarajan A, Deisseroth K, Tonegawa S (2012b) Optogenetic stimulation of a hippocampal engram activates fear memory recall. Nature 484:381–385

Louvart H, Maccari S, Ducrocq F, Thomas P, Darnaudery M (2005) Long-term behavioural alterations in female rats after a single intense footshock followed by situational reminders. Psychoneuroendocrinology 30:316–324

Mahan AL, Ressler KJ (2012) Fear conditioning, synaptic plasticity and the amygdala: implications for posttraumatic stress disorder. Trends Neurosci 35:24–35

Mar AC, Robbins TW (2007) Delay discounting and impulsive choice in the rat. Curr Protoc Neurosci Chapter 8, Unit 8 22

Marschner A, Kalisch R, Vervliet B, Vansteenwegen D, Buchel C (2008) Dissociable roles for the hippocampus and the amygdala in human cued versus context fear conditioning. J Neurosci 28:9030–9036

Matar MA, Cohen H, Kaplan Z, Zohar J (2006) The effect of early poststressor intervention with sertraline on behavioral responses in an animal model of post-traumatic stress disorder. Neuropsychopharmacology 31:2610–2618

Matar MA, Zohar J, Cohen H (2013) Translationally relevant modeling of PTSD in rodents. Cell Tissue Res 354:127–139

McCormick CM, Green MR (2013) From the stressed adolescent to the anxious and depressed adult: investigations in rodent models. Neuroscience 249:242–257

Meerlo P, Easton A, Bergmann BM, Turek FW (2001) Restraint increases prolactin and REM sleep in C57BL/6 J mice but not in BALB/cJ mice. Am J Physiol Regul Integr Comp Physiol 281:R846–R854

Mesches MH, Fleshner M, Heman KL, Rose GM, Diamond DM (1999) Exposing rats to a predator blocks primed burst potentiation in the hippocampus in vitro. J Neurosci 19:RC18

Mikics E, Baranyi J, Haller J (2008a) Rats exposed to traumatic stress bury unfamiliar objects—a novel measure of hyper-vigilance in PTSD models? Physiol Behav 94:341–348

Mikics E, Toth M, Varju P, Gereben B, Liposits Z, Ashaber M, Halasz J, Barna I, Farkas I, Haller J (2008b) Lasting changes in social behavior and amygdala function following traumatic experience induced by a single series of foot-shocks. Psychoneuroendocrinology 33:1198–1210

Milad MR, Quirk GJ (2012) Fear extinction as a model for translational neuroscience: ten years of progress. Annu Rev Psychol 63:129–151

Milad MR, Pitman RK, Ellis CB, Gold AL, Shin LM, Lasko NB, Zeidan MA, Handwerger K, Orr SP, Rauch SL (2009) Neurobiological basis of failure to recall extinction memory in posttraumatic stress disorder. Biol Psychiatry 66:1075–1082

Mitra R, Jadhav S, McEwen BS, Vyas A, Chattarji S (2005) Stress duration modulates the spatiotemporal patterns of spine formation in the basolateral amygdala. Proc Natl Acad Sci U S A 102:9371–9376

Mitra R, Adamec R, Sapolsky R (2009) Resilience against predator stress and dendritic morphology of amygdala neurons. Behav Brain Res 205:535–543

Moore NL, Gauchan S, Genovese RF (2012) Differential severity of anxiogenic effects resulting from a brief swim or underwater trauma in adolescent male rats. Pharmacol Biochem Behav 102:264–268

Moser MB, Moser EI (1998) Distributed encoding and retrieval of spatial memory in the hippocampus. J Neurosci 18:7535–7542

Muhie S, Gautam A, Meyerhoff J, Chakraborty N, Hammamieh R, Jett M (2015) Brain transcriptome profiles in mouse model simulating features of post-traumatic stress disorder. Mol Brain 8:14
Norrholm SD, Jovanovic T, Olin IW, Sands LA, Karapanou I, Bradley B, Ressler KJ (2011) Fear extinction in traumatized civilians with posttraumatic stress disorder: relation to symptom severity. Biol Psychiatry 69:556–563
O'Doherty DC, Chitty KM, Saddiqui S, Bennett MR, Lagopoulos J (2015) A systematic review and meta-analysis of magnetic resonance imaging measurement of structural volumes in posttraumatic stress disorder. Psychiatry Res 232:1–33
Olson VG, Rockett HR, Reh RK, Redila VA, Tran PM, Venkov HA, Defino MC, Hague C, Peskind ER, Szot P, Raskind MA (2011) The role of norepinephrine in differential response to stress in an animal model of posttraumatic stress disorder. Biol Psychiatry 70:441–448
Parsons RG, Ressler KJ (2013) Implications of memory modulation for post-traumatic stress and fear disorders. Nat Neurosci 16:146–153
Pawlyk AC, Morrison AR, Ross RJ, Brennan FX (2008) Stress-induced changes in sleep in rodents: models and mechanisms. Neurosci Biobehav Rev 32:99–117
Perrine SA, Eagle AL, George SA, Mulo K, Kohler RJ, Gerard J, Harutyunyan A, Hool SM, Susick LL, Schneider BL et al (2016) Severe, multimodal stress exposure induces PTSD-like characteristics in a mouse model of single prolonged stress. Behav Brain Res 303:228–237
Philbert J, Pichat P, Beeske S, Decobert M, Belzung C, Griebel G (2011) Acute inescapable stress exposure induces long-term sleep disturbances and avoidance behavior: a mouse model of post-traumatic stress disorder (PTSD). Behav Brain Res 221:149–154
Phillips RG, LeDoux JE (1992) Differential contribution of amygdala and hippocampus to cued and contextual fear conditioning. Behav Neurosci 106:274–285
Pulliam JV, Dawaghreh AM, Alema-Mensah E, Plotsky PM (2010) Social defeat stress produces prolonged alterations in acoustic startle and body weight gain in male Long Evans rats. J Psychiatr Res 44:106–111
Pynoos RS, Ritzmann RF, Steinberg AM, Goenjian A, Prisecaru I (1996) A behavioral animal model of posttraumatic stress disorder featuring repeated exposure to situational reminders. Biol Psychiatry 39:129–134
Quirk GJ, Pare D, Richardson R, Herry C, Monfils MH, Schiller D, Vicentic A (2010) Erasing fear memories with extinction training. J Neurosci 30:14993–14997
Richter-Levin G (1998) Acute and long-term behavioral correlates of underwater trauma—potential relevance to stress and post-stress syndromes. Psychiatry Res 79:73–83
Risbrough VB, Glenn DE, Baker DG (2016) On the road to translation for PTSD treatment: theoretical and practical considerations of the use of human models of conditioned fear for drug development. Curr Top Behav Neurosci 28:173–196
Ritov G, Boltyansky B, Richter-Levin G (2016) A novel approach to PTSD modeling in rats reveals alternating patterns of limbic activity in different types of stress reaction. Mol Psychiatry 21(5):630–641. doi:10.1038/mp.2015.169
Roozendaal B, Hui GK, Hui IR, Berlau DJ, McGaugh JL, Weinberger NM (2006) Basolateral amygdala noradrenergic activity mediates corticosterone-induced enhancement of auditory fear conditioning. Neurobiol Learn Mem 86:249–255
Rozeske RR, Valerio S, Chaudun F, Herry C (2015) Prefrontal neuronal circuits of contextual fear conditioning. Genes Brain Behav 14:22–36
Rygula R, Abumaria N, Domenici E, Hiemke C, Fuchs E (2006) Effects of fluoxetine on behavioral deficits evoked by chronic social stress in rats. Behav Brain Res 174:188–192
Sandi C, Woodson JC, Haynes VF, Park CR, Touyarot K, Lopez-Fernandez MA, Venero C, Diamond DM (2005) Acute stress-induced impairment of spatial memory is associated with decreased expression of neural cell adhesion molecule in the hippocampus and prefrontal cortex. Biol Psychiatry 57:856–864
Sanford LD, Yang L, Wellman LL, Liu X, Tang X (2010) Differential effects of controllable and uncontrollable footshock stress on sleep in mice. Sleep 33:621–630

Sareen J (2014) Posttraumatic stress disorder in adults: impact, comorbidity, risk factors, and treatment. Can J Psychiatry 59:460–467

Schaap MW, van Oostrom H, Doornenbal A, van't Klooster J, Baars AM, Arndt SS, Hellebrekers LJ (2013) Nociception and conditioned fear in rats: strains matter. PLoS One 8:e83339

Sharma S, Rakoczy S, Brown-Borg H (2010) Assessment of spatial memory in mice. Life Sci 87:521–536

Shoji H, Takao K, Hattori S, Miyakawa T (2014) Contextual and cued fear conditioning test using a video analyzing system in mice. J Vis Exp (85). doi:10.3791/50871

Siegmund A, Wotjak CT (2006) Toward an animal model of posttraumatic stress disorder. Ann N Y Acad Sci 1071:324–334

Siegmund A, Wotjak CT (2007a) A mouse model of posttraumatic stress disorder that distinguishes between conditioned and sensitised fear. J Psychiatr Res 41:848–860

Siegmund A, Wotjak CT (2007b) Hyperarousal does not depend on trauma-related contextual memory in an animal model of posttraumatic stress disorder. Physiol Behav 90:103–107

Singewald N, Schmuckermair C, Whittle N, Holmes A, Ressler KJ (2015) Pharmacology of cognitive enhancers for exposure-based therapy of fear, anxiety and trauma-related disorders. Pharmacol Ther 149:150–190

Steckler T, Risbrough V (2012) Pharmacological treatment of PTSD—established and new approaches. Neuropharmacology 62:617–627

Stein MB, Kline NA, Matloff JL (2002) Adjunctive olanzapine for SSRI-resistant combat-related PTSD: a double-blind, placebo-controlled study. Am J Psychiatry 159:1777–1779

Stiedl O, Radulovic J, Lohmann R, Birkenfeld K, Palve M, Kammermeier J, Sananbenesi F, Spiess J (1999) Strain and substrain differences in context- and tone-dependent fear conditioning of inbred mice. Behav Brain Res 104:1–12

Takahashi LK (2014) Olfactory systems and neural circuits that modulate predator odor fear. Front Behav Neurosci 8:72

Takahashi T, Morinobu S, Iwamoto Y, Yamawaki S (2006) Effect of paroxetine on enhanced contextual fear induced by single prolonged stress in rats. Psychopharmacology (Berl) 189:165–173

Toth M, Flandreau EI, Deslauriers J, Geyer MA, Mansuy IM, Merlo Pich E, Risbrough VB (2016) Overexpression of forebrain CRH during early life increases trauma susceptibility in adulthood. Neuropsychopharmacology 41:1681–1690

Tulogdi A, Soros P, Toth M, Nagy R, Biro L, Aliczki M, Klausz B, Mikics E, Haller J (2012) Temporal changes in c-Fos activation patterns induced by conditioned fear. Brain Res Bull 88:359–370

Vanderheyden WM, George SA, Urpa L, Kehoe M, Liberzon I, Poe GR (2015) Sleep alterations following exposure to stress predict fear-associated memory impairments in a rodent model of PTSD. Exp Brain Res 233:2335–2346

VanElzakker MB, Dahlgren MK, Davis FC, Dubois S, Shin LM (2014) From Pavlov to PTSD: the extinction of conditioned fear in rodents, humans, and anxiety disorders. Neurobiol Learn Mem 113:3–18

Vorhees CV, Williams MT (2006) Morris water maze: procedures for assessing spatial and related forms of learning and memory. Nat Protoc 1:848–858

Vouimba RM, Munoz C, Diamond DM (2006) Differential effects of predator stress and the antidepressant tianeptine on physiological plasticity in the hippocampus and basolateral amygdala. Stress 9:29–40

Wang W, Liu Y, Zheng H, Wang HN, Jin X, Chen YC, Zheng LN, Luo XX, Tan QR (2008) A modified single-prolonged stress model for post-traumatic stress disorder. Neurosci Lett 441:237–241

Warren BL, Vialou VF, Iniguez SD, Alcantara LF, Wright KN, Feng J, Kennedy PJ, Laplant Q, Shen L, Nestler EJ, Bolanos-Guzman CA (2013) Neurobiological sequelae of witnessing stressful events in adult mice. Biol Psychiatry 73:7–14

Woodson JC, Macintosh D, Fleshner M, Diamond DM (2003) Emotion-induced amnesia in rats: working memory-specific impairment, corticosterone-memory correlation, and fear versus arousal effects on memory. Learn Mem 10:326–336

Wu Z, Tian Q, Li F, Gao J, Liu Y, Mao M, Liu J, Wang S, Li G, Ge D et al (2016) Behavioral changes over time in post-traumatic stress disorder: insights from a rat model of single prolonged stress. Behav Processes 124:123–129

Yamamoto S, Morinobu S, Takei S, Fuchikami M, Matsuki A, Yamawaki S, Liberzon I (2009) Single prolonged stress: toward an animal model of posttraumatic stress disorder. Depress Anxiety 26:1110–1117

Yehuda R, Antelman SM (1993) Criteria for rationally evaluating animal models of posttraumatic stress disorder. Biol Psychiatry 33:479–486

Yehuda R, Southwick SM, Krystal JH, Bremner D, Charney DS, Mason JW (1993) Enhanced suppression of cortisol following dexamethasone administration in posttraumatic stress disorder. Am J Psychiatry 150:83–86

Yehuda R, Yang RK, Buchsbaum MS, Golier JA (2006) Alterations in cortisol negative feedback inhibition as examined using the ACTH response to cortisol administration in PTSD. Psychoneuroendocrinology 31:447–451

Zoladz PR, Conrad CD, Fleshner M, Diamond DM (2008) Acute episodes of predator exposure in conjunction with chronic social instability as an animal model of post-traumatic stress disorder. Stress 11:259–281

Zoladz PR, Fleshner M, Diamond DM (2012) Psychosocial animal model of PTSD produces a long-lasting traumatic memory, an increase in general anxiety and PTSD-like glucocorticoid abnormalities. Psychoneuroendocrinology 37:1531–1545

Zoladz PR, Fleshner M, Diamond DM (2013) Differential effectiveness of tianeptine, clonidine and amitriptyline in blocking traumatic memory expression, anxiety and hypertension in an animal model of PTSD. Prog Neuropsychopharmacol Biol Psychiatry 44:1–16

Zoladz PR, Park CR, Fleshner M, Diamond DM (2015) Psychosocial predator-based animal model of PTSD produces physiological and behavioral sequelae and a traumatic memory four months following stress onset. Physiol Behav 147:183–192

Integrating NIMH Research Domain Criteria (RDoC) into PTSD Research

Ulrike Schmidt and Eric Vermetten

Abstract Three and a half decades of research on posttraumatic stress disorder (PTSD) has produced substantial knowledge on the pathobiology of this frequent and debilitating disease. However, despite all research efforts, so far no drug that has specifically targeted PTSD core symptoms progressed to clinical use. Instead, although not overly efficient, serotonin re-uptake inhibitors continue to be considered the gold standard of PTSD pharmacotherapy. The psychotherapeutic treatment and symptom-oriented drug therapy options available for PTSD treatment today show some efficacy, although not in all PTSD patients, in particular not in a substantial percent of those suffering from the detrimental sequelae of repeated childhood trauma or in veterans with combat related PTSD. PTSD has this in common with other psychiatric disorders – in particular effective treatment for incapacitating conditions such as resistant major depression, chronic schizophrenia, and frequently relapsing obsessive-compulsive disorder as well as dementia has not yet been developed through modern neuropsychiatric research.

In response to this conundrum, the National Institute of Mental Health launched the Research Domain Criteria (RDoC) framework which aims to leave diagnosis-oriented psychiatric research behind and to move on to the use of research domains overarching the traditional diagnosis systems. To the best of our knowledge, the paper at hand is the first that has systematically assessed the utility of the RDoC system for PTSD research. Here, we review core findings in neurobiological PTSD

U. Schmidt
Trauma Outpatient Unit and RG Molecular Psychotraumatology, Clinical Department, Max Planck Institute of Psychiatry, Kraepelinstrasse 10, Munich 80804, Germany

E. Vermetten (✉)
Department Psychiatry, Leiden University Medical Center Utrecht, Albinusdreef 2, Leiden 2333 ZA, The Netherlands

Arq Psychotruama Research Group, Diemen, The Netherlands
e-mail: e.vermetten@lumc.nl

Curr Topics Behav Neurosci (2018) 38: 69–92
DOI 10.1007/7854_2017_1
Published Online: 13 March 2017

research and match them to the RDoC research domains and units of analysis. Our synthesis reveals that several core findings in PTSD such as amygdala overactivity have been linked to all RDoC domains without further specification of their distinct role in the pathophysiological pathways associated with these domains. This circumstance indicates that the elucidation of the cellular and molecular processes ultimately decisive for regulation of psychic processes and for the expression of psychopathological symptoms is still grossly incomplete. All in all, we find the RDoC research domains to be useful but not sufficient for PTSD research. Hence, we suggest adding two novel domains, namely stress and emotional regulation and maintenance of consciousness. As both of these domains play a role in various if not in all psychiatric diseases, we judge them to be useful not only for PTSD research but also for psychiatric research in general.

Keywords Emotion regulation • Neurobiology • PTSD • RDOC • Research • Stress

Contents

1 Introduction

Even if only a minority of trauma-exposed individuals develop posttraumatic stress disorder (PTSD) [1], it is one of the most prevalent psychiatric diseases worldwide [2]. The prevalence of PTSD increases with increasing numbers of lifetime traumatic events [3, 4]. In the US general population, lifetime PTSD prevalence is 6.8% [5], in Germany 2.8% [6] and in The Netherlands 7.4% [7]. Core symptoms of PTSD include aversive re-experiencing of traumatic events, avoidance anxiety, nervous hyperarousal, and emotional numbing [8]. PTSD significantly reduces the patients' quality of life and increases their unemployment, mortality, morbidity, and suicidality rates [9, 10]. The fact that PTSD was not officially recognized as a diagnosis earlier than 1980 [11] probably contributes to the persistent deficit in drugs targeting PTSD-specific symptoms. Undoubtedly, there is an urgent need for new treatments, since only 60% of PTSD patients respond to the first line drug treatment of PTSD [12], i.e. to serotonin reuptake inhibitors (SSRI). Furthermore, a substantial number of patients suffering from PTSD do not benefit from exposure-based interventions [13], the current gold standard for PTSD psychotherapy [14].

Aiming to overcome this unsatisfactory situation in PTSD treatment, large numbers of preclinical and clinical studies have been performed. To date, it is widely accepted that genetic polymorphisms predispose some individuals to trauma-mediated changes in the dynamic epigenome [15, 16] and probably also in the miRNome [17–19]. These trauma-elicited miRNomic and epigenetic changes can in turn alter the expression of distinct proteins and consequently lead to disturbances in organ function, e.g. endocrine dysfunction and amygdala overactivity. Thus, an individual's trauma and stress tolerance level depends on the interplay of environmental and predefined biological factors, in other words on gene–environment interaction (G $\times$ E). Of course, there is no *single* gene or polymorphism that conveys strong risk for PTSD; instead, psychiatric diseases are, in general, considered to be multigenic [20–22].

Besides the function and regulation of various brain regions such as the amygdala, the hippocampus, and the prefrontal cortex (PFC), the two major stress hormone systems, namely the hypothalamic–pituitary–adrenal (HPA) axis and the sympathetic nervous system (SNS) have repeatedly been shown to be altered in PTSD [23–25, 26]. A meta-analysis indicates that the ventromedial PFC (vmPFC) is the most consistently reported hypoactive, and the amygdala the most consistently reported hyperactive brain region in PTSD [27]. The vmPFC putatively fails to constrain the amygdala thereby leading to, among other symptoms, increased fear responses and impaired extinction of trauma reminders in PTSD patients [23]. Dysfunction of the hippocampus, which, like the vmPFC [23], was repeatedly reported to be smaller in various populations of PTSD patients [28, 29], may contribute to the PTSD-associated impairment in the recognition of safe contexts and memory deficits for neutral stimuli [23, 30]. Animal studies suggest that this PTSD-associated hippocampal volume loss might be mediated by a trauma-elicited transient increase in glucocorticoids [31].

Glucocorticoid homeostasis is regulated by the HPA axis which is widely, but not unequivocally [32], accepted to be hypofunctional in PTSD [33]. The inconsistent findings on HPA axis function in PTSD patients may be explained by different HPA axis responder types which were recently identified in a population of female PTSD patients [34]: HPA axis responder and non-responder PTSD patients differed in the prevalence of combined adult and early life trauma (ELT), the intensity of trauma-related dissociative symptoms, as well as in post-stress expression levels of peripheral FK506 binding protein 51 (FKBP51) [34], an inhibitor of glucocorticoid receptor (GR) signaling [35]. FKBP51 and its gene *FKBP5* are well-characterized candidate molecules for affective disorders such as major depression, anxiety disorders and PTSD [24, 36, 37]. Genetic polymorphisms in *FKBP5* [37, 38] but, interestingly, not in the gene of its target GR [39], are associated with PTSD. In mice, null mutation of *Fkbp5* significantly improved endocrine and behavioral stress coping [40] and, in parallel, alleviated the stress-induced loss of hippocampal synapsin [41], thereby probably counteracting stress-induced hippocampal shrinkage [42].

In contrast to HPA axis function, SNS function was found to be elevated in PTSD [43]. Peripheral and cerebrospinal fluid (CSF) levels of norepinephrine, a major effector hormone of the SNS, are increased in PTSD patients [44–46] and correlate positively with PTSD symptom severity [43]. Consistent with these findings, several studies showed that both alpha- and beta-adrenoceptor blockers are effective in the treatment of PTSD-associated psychopathology [47, 48]. Besides the SNS, the HPA axis and various brain regions, a variety of other systems and networks such as neurotransmitters and neuropeptides have been extensively studied for their role in PTSD [49], among them the serotonin and the dopamine systems [49] as well as anxiolytic neuropeptides [50, 51]. In addition to molecular and fMRI studies, a plethora of (neuro)psychological studies has been performed with PTSD patients. These studies revealed that PTSD patients suffer, inter alia, from deficits in attention [52], regulation of the stress response [53], emotional processing [54, 55], and cognition [56, 57].

2 The RDoC System

It is disappointing that, although urgently needed, no novel drug treatment has yet sprung from this increased knowledge of the biological and psychological mechanisms of PTSD. Since this problem is the same with any other psychiatric disorder, it is compellingly logical that there is either a lack of implementation of psychiatric research findings into clinical practice and/or that the design of the studies hitherto performed is not suitable for drug development. The latter was suggested by many authors and resulted in the proposition of novel research concepts for psychiatry, for instance in symptom-based approaches [58–60] as well as in the Research Domain Criteria (RDoC) concept framed by the US National Institute of Mental Health (NIMH) [61].

Symptom-based approaches suggest assembling patient study cohorts according to major complaints rather than according to diagnoses classified in the two leading psychiatric classification manuals, i.e. the fifth edition of the Diagnostic and Statistical Manual of Mental Disorders (DSM-5) [8] and the 10th edition of the International Classification of Diseases (ICD10).[62]. The RDoC project instead proposes to integrate neuroscientific findings with research in psychopathology in order to identify neurobiological and behavioral dimensions across the current disease categories [63]. These dimensions were created to promote the establishment of a novel biologically informed psychiatric nosology [63, 64], thereby addressing the "comorbidity problem" (p.9) of DSM-5 and ICD10 [65].

RDoC defines psychiatric disorders as pathobiological conditions that involve brain circuits implicated in specific domains of behavior, cognition, and emotion. RDoC does not concentrate on pathological conditions, instead its dimensions cover the range from pathological to non-pathological. The two-dimensional matrix of the RDoC framework comprises five *research domains* which are proposed to be analyzed with seven *units of analysis* [66] that are all weighted equally. The following five research domains were defined so far: (1) negative valence systems (fear, anxiety, loss, frustrative nonreward), (2) positive valence systems (reward learning, reward valuation, habits), (3) cognitive systems (attention, perception, declarative memory, working memory, cognitive control), (4) systems for social processes (attachment formation, social communication, perception of self, perception of others), and (5) arousal/modulatory systems (arousal, circadian rhythm, sleep and wakefulness) [66]. The RDoC framework proposes to analyze these research domains by taking the following units of analysis into account: genes, molecules, cells, neural circuits, physiology, behaviors, and self-reports [66].

There are several manuscripts proposing a reconceptualization of a variety of categorical psychiatric diagnoses and of psychiatric symptoms according to the RDoC system, for example of the diagnoses major depression [67], panic disorder [68], and schizophrenia [69] and of the symptoms inhibitory control [70], auditory hallucinations [70], and pediatric disinhibited eating [71]. Other research teams aim to optimize the RDoC framework by suggesting novel research domains such as the domain of social cognition [72].

3 Integration of the RDoC System into PTSD Research

3.1 State of the Literature

We have found three manuscripts on RDoC and PTSD [73–75]. In their narrative review, Montalvo-Ortiz and colleagues welcomed the RDoC approach for psychiatric genetics and epigenetics since genes are linked to distinct phenotypes and behaviors rather than to multifaceted syndromes such as DSM-5 or ICD10 diagnoses [74]. Another excellent narrative review on PTSD pathobiology was provided by Gerald Young. He suggested to enrich the RDoC framework by adding further candidate endophenotypes [75]. Bauer and colleagues proposed the psychophysiological posterior probability score (PPPS), a composite measure of psychophysiological reactivity to script-driven imagery, as a disease marker for PTSD and categorized it as the "sort of quantifiable physiological or biological trait" (p. 1042) defined by the RDoC system [73].

To the best of our knowledge, there is so far no manuscript integrating PTSD research into the RDoC framework. In the following, we provide a narrative review of neurobiological findings and concomitantly analyze the relevance of the RDoC concept to PTSD research. A summary of our synthesis is provided in Table 1. We conclude this paper with suggestions for future directions in the application of the RDoC framework to PTSD research.

Table 1 Integration of findings on PTSD pathopsychobiology in the RDoC framework

Construct	Negative valence systems	Positive valence systems	Cognitive systems	Arousal
Self-reports	Fear to be confronted with trauma-related cues, anxiety	Appetite ↓, interest ↓ libido ↓ Feeling to feel nothing (numbing)	Deficits in: memory, attention, concentration. Attention and memory bias to threat	Nervousness, sleeping problems
Behaviors	Anxious avoidance which can transform into generalized avoidance	Anhedonia (only partial overlap with emotional numbing)	Concentration and memory deficits, reduced school + work performance	Nervousness, sleeping problems, jumpiness
Physiology	Fear conditioning, fear extinction, attention to threat stimuli, impaired neuroendocrine release of hormones	Reward functioning, memory	Explicit memory pathway: temporal and diencephalic neural network. Implicit memory pathway: right occipital cortex	Startle response ↑ SNS activity ↑
Neural circuits	BLA + mPFC promote fear conditioning Fear extinction: HC→+ ILC →+ BLA □ CA PTSD: mPFC fails to inhibit AMY. HC has impaired safety signaling	Anhedonia: mesocorticolimbic reward pathway (VTA →BLA, the mPFC + NA) Motivation: dopamine consumption: opioid + GABA receptors	AMY activity ↑, insular activity ↑, HC ↑+↓ (depending on type of cue), attention bias: AMY, dorsal ACC, insula, (vmPFC)	Locus coeruleus (LC)-norepinephrine arousal system (in particular LC → AMY circuit)
Cells	BLA: inhibitory interneurons CA: output neurons	Dopaminergic, gabaergic + opioidergic neurons	Various cell types. Further studies needed!	Adrenergic neurons
Molecules	D-cycloserine, L-DOPA, yohimbine	SSRIs, oxytocin Numbing: ?	HAT inhibitors HDAC inhibitors	Adrenoreceptor-blockers such as prazosin, propranolol
Genes	BDNF val66val, COMTval158met, 5-HTTLPR	DAT	Picture recognition: methylation of GR promoter, fear ext./cons.: variety of epigenetic mechanisms	Deletion variant of α2B adrenergic receptor, polymorphism in β2-adrenergic receptor (ADRB2)

BLA basolateral amygdala, *CA* central amygdala, *mPFC* medial prefrontal cortex, *ILC* infralimbic cortex, *HC* hippocampus, *SNS* sympathetic nervous system, *BDNF* brain derived neurotrophic factor, *COMT* catechol-O-methyltransferase, *5-HTTLPR* serotonin transporter gene, *SSRIs* serotonin re-uptake inhibitors, *DAT* dopamine transporter, *SNS* sympathetic nervous system, *AMY* amygdala, *ACC* anterior cingulate cortex, *VTA* ventral tegmental area, *NA* nucleus accumbens, *GR* glucocorticoid receptor. Symbols: →, influence; →+, activation; □, inhibition, ↓, reduction; ↑, increase. Note that the domain "social processes" is not reviewed here

3.2 Negative Valence Systems

The relevance of negative valence systems to PTSD is supported by multiple lines of research. PTSD patients report and show fear and anxiety symptoms, in particular anxious avoidance of trauma-related cues [8]. This specific avoidance anxiety tends to generalize as neutral trauma-unrelated environmental cues can turn into trigger cues when the stimuli occur during a flashback or intrusion. The core physiological mechanisms related to avoidance anxiety are fear conditioning and fear extinction. Fear conditioning in PTSD is biased toward stimuli with higher emotional intensity than the original conditioned-fear stimulus [76]. The overgeneralized fear usually becomes harmful even though the learning of fear is an evolutionarily beneficial response mechanism [77]. Extinction deficits and cue generalization appears to be predominantly associated with PTSD symptoms but not general anxiety or depression, suggesting this is an important dimension for understanding PTSD pathology [78]. The neural circuits underlying fear conditioning and extinction and their pathological alterations have been extensively studied. The basolateral amygdala (BLA) is widely accepted as the main neural structure in which information of unconditioned and conditioned stimuli are integrated [79]. Impairment of the function of the BLA [80] and of the prelimbic division of the medial PFC (mPFC) disrupts the learning of fear [79]. Upon presentation of an extinguished cue during extinction training, the hippocampus activates the infralimbic cortex that stimulates inhibitory interneurons in the BLA which, in turn, prevent conditioned responding by inhibiting the output neurons in the central amygdala [79, 81]. In PTSD, the vmPFC is hypoactive and presumably fails to constrain the amygdala, thereby leading to increased fear responses and impaired extinction of trauma reminders [23, 28]. Furthermore, presumably, upon remembering and reconsolidating the trauma, the hippocampus fails to utilize environmental contextual cues to signal safety [23, 82]. Impairments in fear expression and extinction have been widely accepted to be central to PTSD pathobiology and thus have been analyzed in much more detail than given here – for review see [83]. Although the two major stress hormone systems, the HPA axis and the SNS, are of undisputed importance in psychiatric disorders, their roles in fear extinction, and in particular in PTSD-associated fear extinction learning deficits, have not yet been fully investigated [83].

The amygdala, the PFC, and the hippocampus harbor high concentrations of brain derived neurotrophic factor (BDNF), a protein implicated in synaptic plasticity [84]. Carriers of a distinct BDNF polymorphism, the BDNF val66met-allele, were shown to exhibit impaired extinction learning together with an elevated activity of the vmPFC and the amygdala during extinction trials [83, 84]. Accordingly, in comparison to val66val carriers, PTSD patients carrying the val66met-allele showed a poorer response to (fear extinction based) exposure therapy [85]. Besides BDNF val66met, polymorphisms, genes encoding for the catechol-O-methyltransferase (COMT) and for the serotonin transporter have also been associated with impaired fear extinction learning in PTSD patients [83].

A number of animal and clinical studies aimed at establishing a pharmacotherapy for PTSD-associated deficits in memory reconsolidation and fear extinction learning and suggested D-cycloserine as a potential therapeutic option [86, 87]. However, reports of the efficacy of D-cycloserine augmentation of fear extinction on PTSD psychotherapy revealed inconsistent results [88–90]. Other drugs or potential drugs, such as neuropeptide S, which has been shown to enhance the effects of D-cycloserine [87], L-DOPA and the indole alkaloid yohimbine [91], and the neuropeptide oxytocin [92, 93] have been suggested as augmentative treatment for PTSD psychotherapy. Since the extinction-optimizing drugs hitherto tested are far from being recommendable therapeutic options, future research should place particular emphasis on bridging this development gap. Although the neural circuits and molecular pathways of memory reconsolidation, fear learning, and extinction deficits associated with pathological anxiety are already quite well understood, this level of understanding does not yet appear sufficient for systematic drug development. One potential challenge to current study designs for pharmacotherapeutic enhancement of exposure therapy is that cognitive enhancers may strengthen contextual learning during extinction, limiting generalization of recall, thus treatments may need to be administered in multiple settings or during in vivo exposure to support general symptom reduction [94, 95].

3.3 Positive Valence Systems

In relation to negative valence systems, positive valence systems such as reward learning and reward valuation are understudied topics in PTSD research. The first systematic review on reward processing in PTSD was published in 2015 [96]. Reward processing is thought to underlie anhedonia [96], a symptom which is not pathognomonic for PTSD since it can be observed in a variety of other psychiatric disorders as well, predominantly in major depression [97]. Anhedonia is related to the symptom of emotional numbing [97] which is more characteristic for PTSD patients and thus belongs to the DSM-5 core symptoms of PTSD [8]. Both anhedonia and emotional numbing appear to reflect emotional inexpressiveness and insensitivity to emotional stimulation [97]. However, in accordance with the hypothesis of a bivariate regulation of aversion and appetition [98], emotional numbing reflects a diminished goal-oriented behavior "in response to incentives and positive stimuli at the cognitive-experiential level" (p.464) while anhedonia reflects a loss of pleasure and interest in previously pleasurable activities [97]. As, to the best of our knowledge and to our surprise, the neurobiology of emotional numbing is still elusive, we concentrate here on integrating anhedonia into the RDoC framework.

Reward processing comprises two steps, namely reward motivation ("wanting") and reward consumption ("liking") [96]. The anticipation of a reward stimulates a feeling of desire ("wanting") which motivates a behavior directed at consumption of the reward. In detail, reward cues stimulate the meso-corticolimbic reward pathway leading finally, through dopaminergic activation of the motor cortices and the dorsal striatum, to an approach behavior. This mesolimbic reward pathway comprises a bundle of dopaminergic fibers originating from the ventral tegmental area (VTA). These fibers innervate limbic structures including the BLA, the mPFC, and the nucleus accumbens [99]. This mesolimbic pathway is also, at least partly, involved in the sense of pleasure in response to a reward ("liking"). However, this reward consumption-associated feeling of joy is mainly mediated by GABA and opioid receptors [96]. Both motivational and consummatory anhedonia have been reported in PTSD patients [96]. PTSD-associated reward deficits were reported more often in female PTSD patients and in studies analyzing social stimuli [96].

The molecular pathology of PTSD-associated impairments in reward processing is still elusive. Studies on this topic are rare. However, one of the few reported an increase in striatal dopamine transporter (DAT) availability in PTSD patients [100]. Elevation of DAT availability leads to a reduction in dopaminergic transmission and was hence suggested to underlie deficits in motivational reward processing [96]. Accordingly, dopamine agonists were proposed to overcome PTSD-associated impairments in reward processing [101]. Selective serotonin reuptake inhibitors (SSRIs) are known to enhance striatal function [96] and lead to full remission in about a third of PTSD patients [12]. Thus, in SSRI-sensitive PTSD patients, SSRIs may exert their therapeutic effects, at least in part, through enhancement of motivational reward processing [96, 102]. Besides SSRIs, the anxiolytic neuropeptide oxytocin has been recently shown to augment the sensitivity of the reward pathway during reward anticipation in PTSD patients versus trauma-exposed controls [103]. This finding is in accordance with the fact that oxytocin is proposed as a potential cognitive enhancer in PTSD treatment [91, 104].

3.4 Cognitive Systems

Cognitive systems are undoubtedly affected in PTSD. PTSD patients were repeatedly reported to exhibit deficits in attention [52] and planning [105] as well as in declarative (explicit) [106] and working memory [107]. Moreover, memory deficits, in particular impairments in fear extinction memory, also play a role in the above-discussed pathobiology of avoidance anxiety (see section "Negative Valence Systems"). PTSD patients are overengaged in scanning for potential environmental threats. They have an attentional bias [52] and a memory bias [108, 109] towards threat and negative stimuli at the expense of other cognitive processes [105]. The amygdala, the dorsal anterior cingulate cortex (ACC), the insula, and possibly also the vmPFC (mixed findings) were reported to be active in PTSD patients during tasks of negative attention [105].

PTSD-associated cognition deficits are promoted by negative emotionality through interactions between the amygdala and the hippocampus [105]. Most studies agree on the presence of amygdala overactivity in PTSD [27] whereas reports on hippocampal activity are mixed, possibly due to the fact that some studies employed general negative stimuli while others used trauma-specific cues [105]. The latter can induce false memories and a reduction in the activity of the hippocampus [105].

In contrast to explicit memory, which mediates the encoding and recall of facts [110], implicit or nondeclarative memory refers to the unconscious recall of encoded items. One facet of implicit memory is repetition priming – it refers to a bias or facilitation in retrieval of an encoded stimulus due to prior processing of a related or the same stimulus. “Repetition priming is perceptual when it reflects prior processing of stimulus form” and “conceptual when it reflects prior processing of stimulus meaning” (p. 494, [110]). Many, but not all, studies on this topic have reported that PTSD is associated with an increase in perceptional priming [111]. Together with diminished fear extinction and enhanced conditioning (see section “Negative Valence Systems”), the increase in perceptual priming of threat cues might be a powerful etiological combination in maintenance and pathogenesis of PTSD. Cognitive deficits in PTSD patients were reported to improve through successful PTSD treatment [105].

In comparison to fMRI studies, studies on the molecular basis of memory deficits in PTSD are scarce. One of the few studies on this topic found that increased methylation of the promoter of the gene encoding for the GR was linked to PTSD risk in genocide survivors as well as to reduced picture recognition in healthy men [112]. In fear consolidation, the most intensely studied epigenetic mechanism is histone acetylation. Animal models revealed that drugs that block histone acetylation (histone acetyltransferase (HAT) inhibitors) disrupt fear consolidation whereas the prevention of histone deacetylation by histone deacetylase (HDAC) inhibitors was found to increase it [113]. Epigenetic modifications in fear consolidation and extinction have been excellently reviewed by [113] and seem to be promising drug targets for PTSD, at least for PTSD-associated avoidance and possibly also for the aversive recall of traumatic memories.

3.5 Arousal Systems

Nervous hyperarousal belongs to the PTSD core symptoms [8] and shows up inter alia in enhanced nervousness, sleeping problems including nightmares and enhanced jumpiness. It is broadly accepted that SNS overdrive plays a core etiological role in PTSD, in particular in PTSD-associated hyperarousal [23, 24] and may be a risk factor for developing PTSD [114]. Both animal and clinical studies strongly suggest that the major effector hormones of the SNS, adrenaline and noradrenaline, enhance memory storage and that, consequently, excessive SNS activity at the time of trauma exposure might foster the consolidation of traumatic memory thereby promoting it to become intrusive [115]. A deletion variant of the gene encoding the α2B adrenergic receptor (*ADRA2B*) was reported to be linked to enhanced emotional memory both in survivors of the Rwandan genocide survivors and in healthy Swiss control cohort [116]. A polymorphism of another adrenoceptor, β2-adrenergic receptor (*ADRB2*), was found to be associated with PTSD both in male European Americans and in female African Americans [117]. In the latter study, the polymorphism in *ADRB2* interacted with childhood adversity to predict adult PTSD symptoms. In a recent review article, PTSD patients were reported to have elevated peripheral and cerebrospinal fluid (CSF) noradrenaline and adrenaline levels [48]. Accordingly, adrenoreceptor blockers such as the α-1 adrenoreceptor blocker prazosin [118] and the beta-blocker propranolol [47] have been found by several authors to be effective in PTSD treatment. However, promoting a systemic attenuation of SNS activity can bring various side effects such as arterial hypotonia and loss in motivational drive.

In the brain, adrenergic transmission is regulated by the locus coeruleus (LC)-noradrenaline arousal system [119]. The LC is the principal site for brain synthesis of norepinephrine [120]; it mediates arousal and primes neurons to stimulus activation in widespread central regions such as the cerebellum, the hypothalamus, the thalamic relay nuclei, and the amygdala [120]. The LC was shown to mediate cognition through arousal [120]. Enhanced noradrenergic postsynaptic responsiveness, in particular in the circuit spanning from the LC to the BLA, was suggested as a major factor in the pathophysiology of PTSD and other stress-related disorders [121].

3.6 *Systems for Social Processes*

Social processes such as attachment formation, social communication and perception of self and others are clearly affected in PTSD patients, especially in patients with complex PTSD and in those having suffered an interpersonal trauma. However, as the body of literature on this topic comprises mainly psychological experiments and data, we did not review or summarize it here in this paper, which is focusing on neurobiological findings. However, we suggest putting particular emphasis on the study of the concepts of shame, guilt, and paranoid distrust since all of them are particularly frequent in interpersonally traumatized PTSD patients.

4 Proposed Novel Domains for PTSD Research

Table 1 demonstrates that many core findings on PTSD vulnerability and pathogenesis can be easily integrated into the RDoC framework. Hence, the RDoC system is unquestionably useful for PTSD research. However, there are some facets of PTSD pathobiology that do not easily fit into the proposed RDoC research domains, in particular PTSD-associated impairments in emotion processing and dissociative symptoms. For this reason, we propose two novel RDoC domains for PTSD research, the "emotional processes" and the "maintenance of consciousness" domains that we describe in the following and outline in Table 2.

Table 2 Suggested novel domains for PTSD research

Construct	Stress and emotion regulation	Maintenance of consciousness
Self-reports	Emotional numbing, instable mood	Spectrum from feeling briefly disconnected from reality to losing consciousness, amnesic spells/gaps in memory
Behaviors	Loss of interest and emotionality, emotional instability, sometimes with self-injury	Spectrum reaching from brief periods of absentmindedness to seizure-like attacks (spectrum from intrusions to PNES)
Physiology	Central nervous emotion regulation, key structure: PFC dorsal pathway: dorsolateral prefrontal cortex (dlPFC) + lateral parietal cortex ventral pathway: AMY, ventrolateral PFC (vlPFC) + medial prefrontal cortex (mPFC)	Various brain networks, i.e. networks mediating emotion regulation, awareness, executive/cognitive control, attention, self-referential processing, and motor functions. Shift in default and salience network
		Corticolimbic pathway
Neural circuits	Prefrontal/parietal connectivity with subgenual cingulate ↑ in PTSD, AMY activity ↑, impaired top-down-attentional control of emotions	Emotional overmodulation (PFC hyperactivity with consecutive limbic hypersuppression) in dissociative PTSD
	Further studies needed	Corticolimbic model of dissociation
		Various functional and structural alterations in various brain networks mediating: emotion regulation and awareness (anterior cingulate, orbitofrontal cortex, insula), executive/cognitive control (inferior frontal gyrus, ACC), attention (posterior parietal cortex), self-referential processing (precuneus), motor symptoms (suppl. Motor area, precentral gyrus, cerebellum)
		Further studies needed
Cells	Studies needed	Studies needed
Molecules	Sertraline, paroxetine were found to ameliorate stress and emotion regulation dysfunction	Limited evidence for partial opioid-antagonists such as naltrexone, further studies, in particular RCTs, clearly needed!
Genes	Studies needed	First evidence for FKBP51 and MR

dlPFC dorsolateral prefrontal cortex, *vlPFC* ventrolateral prefrontal cortex, *mPFC* medial prefrontal cortex, *AMY* amygdala, *PNES* psychogenic non-epileptic seizures, *FKBP5* FK506 binding protein 51, *MR* mineralocorticoid receptor, *ACC* anterior cingulate cortex. Symbols: →, influence; →+, activation; □, inhibition, ↓, reduction; ↑, increase

4.1 Stress and Emotion Regulation

In our eyes, stress and emotion regulation neither fits properly in the RDoC domain "negative valence systems" nor in any other of the RDoC research domains suggested so far. Accordingly, del Rio-Casanova and colleagues also stated that "emotion regulation should be considered a core domain when constructing clinical phenotypes in trauma spectrum disorders" [122].

Gross and colleagues defined stress and emotion regulation as "processes by which individuals influence which emotions they have, when they have them, and how they experience and express these emotions" [123]. However, definitions vary due to inadequate distinctions of character traits versus cognitive distortions and automatic versus voluntary processes [122].

A substantial body of literature documents the role of stress and emotion regulation in PTSD, complex PTSD, and borderline personality disorder (BPS) [124, 125]. BPS is seen as a trauma spectrum disorder by many researchers due to its comorbidity with PTSD and the high prevalence of traumatic events in the early life of BPS patients [126]. Moreover, BPS has a significant overlap with complex PTSD, but is, nevertheless, considered a separate diagnostic entity [127]. Here we will therefore concentrate on emotion regulation in PTSD which was repeatedly found to be dysfunctional in various facets [125].

The PFC plays a key role in the central nervous emotion regulation network [128]. This network was repeatedly described to comprise a dorsal and ventral subpathway. The dorsal pathway includes brain regions such as the lateral parietal cortex and the dorsolateral prefrontal cortex (dlPFC), and is involved in executive control of emotions. The ventral pathway includes the mPFC, the amygdala, and the ventrolateral PFC (vlPFC) and is mainly involved in in processing of emotions [122]. Both subpathways are key for the cognitive control of emotions and "compete for attentional resources" [122]. Upon evaluation of the emotional component of a stimulus, the activity of the orbitofrontal cortex and the amygdala increases, whereas the activity of the right vlPFC decreases. Conversely, upon linguistic labeling of the same stimulus, limbic area activity decreases [122]. Hence, emotion regulation processing can be understood as the result of an opposition between limbic and prefrontal areas.

One study demonstrated that PTSD patients, in relation to trauma-exposed controls, exhibit increased connectivity of the prefrontal/parietal region with the subgenual cingulate [122]. Findings on mPFC activity in PTSD are mixed. However, studies agree on the presence of amygdala hyperactivity in PTSD [23] and suggest PTSD to be associated with an impaired top-down-attentional control of emotions [122]. In general, in comparison to BPS, studies researching functional and molecular underpinnings of emotion regulation deficits in PTSD are scarce. To the best of our knowledge, there are so far no studies on the genetic or epigenetic basis of emotional regulation dysfunction in PTSD. However, according to the RDoC system, pathways should no longer be studied solely in relation to traditional diagnoses such as BPS and PTSD – it might well be that the molecular alterations associated with emotional dysregulation in BPS also underlie emotional dysfunction in PTSD. Interestingly, in a cohort of patients suffering from chronic PTSD, impairments in the regulation of emotions improved in response to prolonged exposure-based psychotherapy as well as to treatment with the SSRI sertraline [129].

As stress and emotion regulation not only play a role in PTSD, complex PTSD and BPS but also in a variety of other psychiatric diseases, for instance in major depression, eating disorders [77], and bipolar disorder, in concurrence with Fernandez and colleagues [130], we suggest to add this novel domain to the RDoC framework not only for PTSD research, but also for psychiatric research in general.

4.2 Maintenance of Consciousness

The aversive recall of traumatic memories is pathognomonic for PTSD and always goes along with dissociation and usually also with a change in the state of consciousness. An intrusion is a brief recollection of traumatic memories while a flashback can last for hours and flow into various forms of dissociation, for instance into psychogenic non-epileptic seizures (PNES). However, PTSD is not the only psychiatric disorder associated with dissociative symptoms. Derealization and depersonalization, which belong also to the spectrum of dissociative symptoms [131], occur for instance during panic attacks. Hence, the novel domain "maintenance of consciousness" might not be relevant to PTSD research only, but for various psychiatric disorders. During the experience of dissociation there is a change in consciousness [132] which is the result of, but cannot be fully explained by, cognitive processes. Thus, in our eyes, the research domain "cognitive processes" covers only parts of the phenomenon of psychological dissociation.

There are various definitions for psychological dissociation. According to DSM-IV, it is "a disruption in the usually integrated functions of consciousness, memory, identity, or perception of the environment" [133]. Among other reasons, the fact that both frequency and intensity of dissociative symptoms differ significantly among PTSD patients has motivated the definition of the dissociative subtype of PTSD [134]. The assessment of the functional and molecular underpinnings of dissociative phenomena such as PNES is still in its early stages [135] – defining maintenance of consciousness as a novel research domain in the RDoC system might hopefully remedy this gap of knowledge and promote substantial neurobiological research on this important topic, in particular on dissociative symptoms. Studies on the drug treatment of dissociative symptoms are also scarce. Naltrexone, a partial opiate antagonist, is one of the few drugs tested regarding its efficacy in the treatment of dissociative symptoms [136]. Currently, a clinical study tests cognitive behavioral therapy versus standardized medical care in adults suffering from PNES in a multicenter randomized controlled trial (RCT) protocol [137]. To the best of our knowledge, there is so far no RCT that has assessed the effect of any drug treatment on dissociative symptoms. With regard to the suffering of patients experiencing pathological dissociation this is a significant omission.

Thirty percent of PTSD patients have a blunted heart rate response to trauma narrative exposure [138]. Accordingly, in a Trier Social Stress Test (TSST) paradigm, one group of PTSD patients exhibited a blunted HPA reactivity together with an increased prevalence of trauma-related dissociative symptoms (HPA non-responder

PTSD patients) [34]. These non-responder patients showed alterations in the peripheral expression levels on the mineralocorticoid receptor (MR) and of *FKBP5* [34]. These studies suggest that a blunted reactivity of the HPA axis and of the SNS might be associated with the propensity for dissociative symptoms in PTSD patients. Lanius and colleagues state that only patients suffering from the dissociative subtype of PTSD show a frequent and severe overactivity of the prefrontal area with consecutive hypersuppression of limbic regions upon exposure to trauma-narrative scripts (emotional overmodulation) [138]. Emotional under- and overmodulation is present in all PTSD patients at different time-points [138]. A study by Felmingham and colleagues [139] and studies on dissociative amnesia support this corticolimbic model of dissociation [138].

In Table 2, the various brain networks and regions that have been implicated in PNES pathobiology and hence probably play a role in other dissociative symptoms also are summarized from the review by Perez and colleagues [135]. The fact that the networks and regions mentioned in that review comprise the majority of brain areas might hint at a lack of specificity of the results and again stresses the urgent need for further studies on the neurobiology of the maintenance of consciousness, in particular of psychological dissociation.

5 Summary and Conclusions

In summary, decades of PTSD research have considerably advanced our understanding of PTSD pathobiology (see, e.g., [140, 141]). However, none of the potential biomarkers and none of the proposed drugs, except for prazosin, or drug targets have yet progressed to clinical use [141]. Our synthesis reveals that several core findings in PTSD such as amygdala/BLA overactivity can be linked to all RDoC domains for PTSD research but lack further specification of their exact role in the pathways associated with these domains (Tables 1 and 2). This circumstance indicates that the cellular and molecular processes finally decisive for regulation of psychic processes and hence for the expression of psychopathological symptoms have not yet been identified.

The RDoC framework was conceptualized to overcome the translational gap in psychiatric research by detaching it from the concept of traditional psychiatric diagnoses. The currently described integration of important neurobiological findings of PTSD research into the RDoC system put fear processing, reward functioning, explicit/implicit memory pathways and the SNS in the spotlight of PTSD vulnerability and pathogenesis (Table 1) and, furthermore, revealed that PTSD-associated emotional instability and dissociative symptoms are not adequately represented in current RDoC domains. For this reason, we suggest two novel domains, i.e. the domains "maintenance of consciousness" and "stress and emotion regulation" (Table 2) – the latter has recently also been suggested by others to be indispensable for general psychiatric research [130]. Integrating PTSD research findings into these two novel domains revealed large gaps of knowledge in the associated units of analysis "cells," "molecules," and "genes" (Table 2), i.e. in the molecular and cellular processes underlying the regulation of stress and emotion and in the maintenance of consciousness. We hope that the gaps of knowledge in PTSD pathobiology identified here stimulate studies aiming to close them – such studies will certainly profit from the adoption of the RDoC principle.

References

1. Forresia B, Caffoa E, Battagliab M (2015) Gene environment interplays: why PTSD makes a good case for gene–environment interaction studies and how adding a developmental approach can help. In: Martin CR, Preedy VR, Patel VB (eds) Comprehensive guide to post-traumatic stress disorder. Springer, pp 1–3. doi:10.1007/978-3-319-08613-2_84-1. ISBN 978-3-319-08613-2
2. McFall M, Saxon AJ, Thompson CE et al (2005) Improving the rates of quitting smoking for veterans with posttraumatic stress disorder. Am J Psychiatry 162:1311–1319. doi:10.1176/appi.ajp.162.7.1311
3. Dorrington S, Zavos H, Ball H et al (2014) Trauma, post-traumatic stress disorder and psychiatric disorders in a middle-income setting: prevalence and comorbidity. Br J Psychiatry. bjp. bp. 113.141796
4. Ringdal GI, Ringdal K, Simkus A (2007) War-related distress among Kosovar Albanians. J Loss Trauma 13:59–71
5. Kessler RC (1995) Posttraumatic stress disorder in the National Comorbidity Survey. Arch Gen Psychiatry 52:1048. doi:10.1001/archpsyc.1995.03950240066012
6. Hauffa R, Rief W, Brähler E et al (2011) Lifetime traumatic experiences and posttraumatic stress disorder in the German population: results of a representative population survey. J Nerv Ment Dis 199:934–939. doi:10.1097/NMD.0b013e3182392c0d
7. de Vries G-J, Olff M (2009) The lifetime prevalence of traumatic events and posttraumatic stress disorder in the Netherlands. J Trauma Stress 22:259–267
8. American Psychiatric Association (2013) Diagnostic and statistical manual of mental disorders: DSM-5, 5th edn. American Psychiatric Association, Washington, DC
9. Pompili M, Sher L, Serafini G et al (2013) Posttraumatic stress disorder and suicide risk among veterans: a literature review. J Nerv Ment Dis 201:802–812. doi:10.1097/NMD.0b013e3182a21458
10. Rodgers CS, Allard CB, Wansley P (2010) Post-traumatic stress disorder and quality of life in women. In: Handbook of disease burdens and quality of life measures. Springer, pp 3439–3456
11. Scott WJ (1990) PTSD in DSM-III: a case in the politics of diagnosis and disease. Soc Probl 37:294–310. doi:10.2307/800744
12. Berger W, Mendlowicz MV, Marques-Portella C et al (2009) Pharmacologic alternatives to antidepressants in posttraumatic stress disorder: a systematic review. Prog Neuro-Psychopharmacol Biol Psychiatry 33:169–180. doi:10.1016/j.pnpbp.2008.12.004
13. Yehuda R, Bierer LM, Sarapas C et al (2009) Cortisol metabolic predictors of response to psychotherapy for symptoms of PTSD in survivors of the World Trade Center attacks on September 11, 2001. Psychoneuroendocrinology 34:1304–1313. doi:10.1016/j.psyneuen.2009.03.018
14. Cusack K, Jonas DE, Forneris CA et al (2016) Psychological treatments for adults with posttraumatic stress disorder: a systematic review and meta-analysis. Clin Psychol Rev 43:128–141. doi:10.1016/j.cpr.2015.10.003
15. Almli LM, Stevens JS, Smith AK et al (2015) A genome-wide identified risk variant for PTSD is a methylation quantitative trait locus and confers decreased cortical activation to

fearful faces. Am J Med Genet B Neuropsychiatr Genet 168:327–336. doi:10.1002/ajmg.b.32315

16. Klengel T, Mehta D, Anacker C et al (2013) Allele-specific FKBP5 DNA demethylation mediates gene-childhood trauma interactions. Nat Neurosci 16:33–41. doi:10.1038/nn.3275
17. Bam M, Yang X, Zumbrun EE et al (2016) Dysregulated immune system networks in war veterans with PTSD is an outcome of altered miRNA expression and DNA methylation. Sci Rep 6:31209. doi:10.1038/srep31209
18. Schmidt U, Herrmann L, Hagl K et al (2013) Therapeutic action of fluoxetine is associated with a reduction in prefrontal cortical miR-1971 expression levels in a mouse model of posttraumatic stress disorder. Front Psychiatry 4:66. doi:10.3389/fpsyt.2013.00066
19. Schmidt U, Keck ME, Buell DR (2015) miRNAs and other non-coding RNAs in posttraumatic stress disorder: a systematic review of clinical and animal studies. J Psychiatr Res 65:1–8. doi:10.1016/j.jpsychires.2015.03.014
20. Freedman R, Leonard S, Olincy A et al (2001) Evidence for the multigenic inheritance of schizophrenia. Am J Med Genet 105:794–800
21. Martins-de-Souza D (2013) Biomarkers for psychiatric disorders: where are we standing. Markers 35:1–2
22. Pinsonneault J, Sadée W (2003) Pharmacogenomics of multigenic diseases: sex-specific differences in disease and treatment outcome. AAPS Pharm Sci 5:49–61
23. Pitman RK, Rasmusson AM, Koenen KC et al (2012) Biological studies of post-traumatic stress disorder. Nat Rev Neurosci 13:769–787. doi:10.1038/nrn3339
24. Schmidt U, Kaltwasser SF, Wotjak CT (2013) Biomarkers in posttraumatic stress disorder: overview and implications for future research. Dis Markers 35:43–54. doi:10.1155/2013/835876
25. Vermetten E, Bremner JD (2002) Circuits and systems in stress. II. Applications to neurobiology and treatment in posttraumatic stress disorder. Depress Anxiety 16(1):14–38
26. Vermetten E, Bremner JD (2002) Circuits and systems in stress. I. Preclinical studies. Depress Anxiety 15(3):126–147
27. Hayes JP, Hayes SM, Mikedis AM (2012) Quantitative meta-analysis of neural activity in posttraumatic stress disorder. Biol Mood Anxiety Disord 2:9. doi:10.1186/2045-5380-2-9
28. Bremner JD, Randall P, Vermetten E et al (1997) Magnetic resonance imaging-based measurement of hippocampal volume in posttraumatic stress disorder related to childhood physical and sexual abuse – a preliminary report. Biol Psychiatry 41:23–32. doi:10.1016/S0006-3223(96)00162-X
29. Vermetten E, Vythilingam M, Southwick SM et al (2003) Long-term treatment with paroxetine increases verbal declarative memory and hippocampal volume in posttraumatic stress disorder. Biol Psychiatry 54:693–702. doi:10.1016/S0006-3223(03)00634-6
30. Acheson DT, Gresack JE, Risbrough VB (2012) Hippocampal dysfunction effects on context memory: possible etiology for posttraumatic stress disorder. Neuropharmacology 62(2):674–685
31. Crochemore C, Lu J, Wu Y et al (2005) Direct targeting of hippocampal neurons for apoptosis by glucocorticoids is reversible by mineralocorticoid receptor activation. Mol Psychiatry 10:790–798. doi:10.1038/sj.mp.4001679
32. Klaassens ER, Giltay EJ, Cuijpers P et al (2012) Adulthood trauma and HPA-axis functioning in healthy subjects and PTSD patients: a meta-analysis. Psychoneuroendocrinology 37:317–331. doi:10.1016/j.psyneuen.2011.07.003
33. Naughton M, Dinan TG, Scott LV (2014) Corticotropin-releasing hormone and the hypothalamic-pituitary-adrenal axis in psychiatric disease. Handb Clin Neurol 124:69–91. doi:10.1016/B978-0-444-59602-4.00005-8
34. Zaba M, Kirmeier T, Ionescu IA et al (2015) Identification and characterization of HPA-axis reactivity endophenotypes in a cohort of female PTSD patients. Psychoneuroendocrinology 55:102–115. doi:10.1016/j.psyneuen.2015.02.005
35. Fries GR, Gassen NC, Schmidt U, Rein T (2015) The FKBP51-glucocorticoid receptor balance in stress-related mental disorders. Curr Mol Pharmacol 9:126–140

36. Binder EB (2009) The role of FKBP5, a co-chaperone of the glucocorticoid receptor in the pathogenesis and therapy of affective and anxiety disorders. Psychoneuroendocrinology 34: S186–S195
37. van Zuiden M, Kavelaars A, Geuze E et al (2013) Predicting PTSD: pre-existing vulnerabilities in glucocorticoid-signaling and implications for preventive interventions. Brain Behav Immun 30:12–21
38. Binder EB, Bradley RG, Liu W et al (2008) Association of FKBP5 polymorphisms and childhood abuse with risk of posttraumatic stress disorder symptoms in adults. JAMA 299:1291–1305
39. Bachmann AW, Sedgley TL, Jackson RV et al (2005) Glucocorticoid receptor polymorphisms and post-traumatic stress disorder. Psychoneuroendocrinology 30:297–306. doi:10.1016/j.psyneuen.2004.08.006
40. Touma C, Gassen NC, Herrmann L et al (2011) FK506 binding protein 5 shapes stress responsiveness: modulation of neuroendocrine reactivity and coping behavior. Biol Psychiatry 70:928–936. doi:10.1016/j.biopsych.2011.07.023
41. Schmidt U, Buell DR, Ionescu IA et al (2015) A role for synapsin in FKBP51 modulation of stress responsiveness: convergent evidence from animal and human studies. Psychoneuroendocrinology 52:43–58. doi:10.1016/j.psyneuen.2014.11.005
42. Golub Y, Kaltwasser SF, Mauch CP et al (2011) Reduced hippocampus volume in the mouse model of posttraumatic stress disorder. J Psychiatr Res 45:650–659. doi:10.1016/j.jpsychires.2010.10.014
43. Morris MC, Rao U (2013) Psychobiology of PTSD in the acute aftermath of trauma: integrating research on coping, HPA function and sympathetic nervous system activity. Asian J Psychiatry 6:3–21. doi:10.1016/j.ajp.2012.07.012
44. Blanchard EB, Kolb LC, Prins A et al (1991) Changes in plasma norepinephrine to combat-related stimuli among Vietnam veterans with posttraumatic stress disorder. J Nerv Ment Dis 179:371–373
45. Geracioti TD, Baker DG, Ekhator NN et al (2001) CSF norepinephrine concentrations in posttraumatic stress disorder. Am J Psychiatry 158:1227–1230. doi:10.1176/appi.ajp.158.8.1227
46. Yehuda R, Southwick S, Giller EL et al (1992) Urinary catecholamine excretion and severity of PTSD symptoms in Vietnam combat veterans. J Nerv Ment Dis 180:321–325
47. Giustino TF, Fitzgerald PJ, Maren S (2016) Revisiting propranolol and PTSD: memory erasure or extinction enhancement? Neurobiol Learn Mem 130:26–33
48. Hendrickson RC, Raskind MA (2016) Noradrenergic dysregulation in the pathophysiology of PTSD. Exp Neurol. doi: 10.1016/j.expneurol.2016.05.014
49. Bandelow B, Baldwin D, Abelli M et al (2016) Biological markers for anxiety disorders, OCD and PTSD: a consensus statement. Part II. Neurochemistry, neurophysiology and neurocognition. World J Biol Psychiatry 1–53. doi: 10.1080/15622975.2016.1190867
50. Koch SBJ, van Zuiden M, Nawijn L et al (2016) Intranasal oxytocin normalizes amygdala functional connectivity in posttraumatic stress disorder. Neuropsychopharmacol Off Publ Am Coll Neuropsychopharmacol 41:2041–2051. doi:10.1038/npp.2016.1
51. Schmeltzer SN, Herman JP, Sah R (2016) Neuropeptide Y (NPY) and posttraumatic stress disorder (PTSD): a translational update. Exp Neurol. doi: 10.1016/j.expneurol.2016.06.020
52. Block SR, Liberzon I (2016) Attentional processes in posttraumatic stress disorder and the associated changes in neural functioning. Exp Neurol doi: 10.1016/j.expneurol.2016.05.009
53. Herman JP, Ostrander MM, Mueller NK, Figueiredo H (2005) Limbic system mechanisms of stress regulation: hypothalamo-pituitary-adrenocortical axis. Prog Neuro-Psychopharmacol Biol Psychiatry 29(8):1201–1213
54. Litz BT, Orsillo SM, Kaloupek D, Weathers F (2000) Emotional processing in posttraumatic stress disorder. J Abnorm Psychol 109:26

55. Phan KL, Britton JC, Taylor SF et al (2006) Corticolimbic blood flow during Nontraumatic emotional processing in posttraumatic stress disorder. Arch Gen Psychiatry 63:184. doi:10.1001/archpsyc.63.2.184
56. Schuitevoerder S, Rosen JW, Twamley EW et al (2013) A meta-analysis of cognitive functioning in older adults with PTSD. J Anxiety Disord 27:550–558. doi:10.1016/j.janxdis.2013.01.001
57. Vasterling JJ, Duke LM, Brailey K et al (2002) Attention, learning, and memory performances and intellectual resources in Vietnam veterans: PTSD and no disorder comparisons. Neuropsychology 16:5–14. doi:10.1037//0894-4105.16.1.5
58. Fleeson W, Furr RM, Arnold EM (2010) An agenda for symptom-based research. Behav Brain Sci 33:157. doi:10.1017/S0140525X10000750
59. Schmidt U (2015) A plea for symptom-based research in psychiatry. Eur J Psychotraumatol. doi: 10.3402/ejpt.v6.27660
60. Sharpe M, Walker J (2009) Symptoms: a new approach. Psychiatry 8:146–148. doi:10.1016/j.mppsy.2009.03.016
61. Insel T, Cuthbert B, Garvey M et al (2010) Research domain criteria (RDoC): toward a new classification framework for research on mental disorders. Am J Psychiatry 167:748–751. doi:10.1176/appi.ajp.2010.09091379
62. World Health Organization (1992) The ICD-10 classification of mental and behavioural disorders: clinical descriptions and diagnostic guidelines. World Health Organization, Geneva
63. Wildes JE, Marcus MD (2015) Application of the research domain criteria (RDoC) framework to eating disorders: emerging concepts and research. Curr Psychiatry Rep. doi: 10.1007/s11920-015-0572-2
64. Cuthbert BN (2014) The RDoC framework: facilitating transition from ICD/DSM to dimensional approaches that integrate neuroscience and psychopathology: forum – the research domain criteria project. World Psychiatry 13:28–35. doi:10.1002/wps.20087
65. Unoka Z, Richman MJ (2016) Neuropsychological deficits in BPD patients and the moderator effects of co-occurring mental disorders: a meta-analysis. Clin Psychol Rev 44:1–12
66. Morris SE, Cuthbert BN (2012) Research domain criteria: cognitive systems, neural circuits, and dimensions of behavior. Dialogues Clin Neurosci 14:29–37
67. Woody ML, Gibb BE (2015) Integrating NIMH research domain criteria (RDoC) into depression research. Curr Opin Psychol 4:6–12. doi:10.1016/j.copsyc.2015.01.004
68. Hamm AO, Richter J, Pané-Farré C et al (2016) Panic disorder with agoraphobia from a behavioral neuroscience perspective: applying the research principles formulated by the research domain criteria (RDoC) initiative: panic disorder with agoraphobia. Psychophysiology 53:312–322. doi:10.1111/psyp.12553
69. Morris SE, Vaidyanathan U, Cuthbert BN (2016) Changing the diagnostic concept of schizophrenia: the NIMH research domain criteria initiative. In: Li M, Spaulding WD (eds) The neuropsychopathology of schizophrenia. Springer International Publishing, Cham, pp 225–252
70. Badcock JC, Hugdahl K (2014) A synthesis of evidence on inhibitory control and auditory hallucinations based on the research domain criteria (RDoC) framework. Front Hum Neurosci 8:180
71. Tanofsky-Kraff M, Engel S, Yanovski JA et al (2013) Pediatric disinhibited eating: toward a research domain criteria framework. Int J Eat Disord 46:451–455. doi:10.1002/eat.22101
72. Gur RC, Gur RE (2016) Social cognition as an RDoC domain. Am J Med Genet B Neuropsychiatr Genet 171:132–141. doi:10.1002/ajmg.b.32394
73. Bauer MR, Ruef AM, Pineles SL et al (2013) Psychophysiological assessment of PTSD: a potential research domain criteria construct. Psychol Assess 25:1037–1043. doi:10.1037/a0033432

74. Montalvo-Ortiz JL, Gelernter J, Hudziak J, Kaufman J (2016) RDoC and translational perspectives on the genetics of trauma-related psychiatric disorders. Am J Med Genet B Neuropsychiatr Genet 171:81–91. doi:10.1002/ajmg.b.32395
75. Young G (2014) PTSD, endophenotypes, the RDoC, and the DSM-5. Psychol Inj Law 7:75–91. doi:10.1007/s12207-014-9187-x
76. Morey RA, Dunsmoor JE, Haswell CC, Brown VM, Vora A, Weiner J, Stjepanovic D, Wagner HR, Brancu M, Marx CE, Naylor JC (2015) Fear learning circuitry is biased toward generalization of fear associations in posttraumatic stress disorder. Transl Psychiatry 5(12): e700
77. Mahan AL, Ressler KJ (2012) Fear conditioning, synaptic plasticity and the amygdala: implications for posttraumatic stress disorder. Trends Neurosci 35:24–35. doi:10.1016/j.tins.2011.06.007
78. Acheson DT, Geyer MA, Baker DG, Nievergelt CM, Yurgil K, Risbrough VB, Team MI (2015) Conditioned fear and extinction learning performance and its association with psychiatric symptoms in active duty marines. Psychoneuroendocrinology 51:495–505
79. Graham BM, Milad MR (2011) The study of fear extinction: implications for anxiety disorders. Am J Psychiatry 168:1255–1265. doi:10.1176/appi.ajp.2011.11040557
80. LeDoux J (2007) The amygdala. Curr Biol CB 17:R868–R874. doi:10.1016/j.cub.2007.08.005
81. Quirk GJ, Mueller D (2008) Neural mechanisms of extinction learning and retrieval. Neuropsychopharmacology 33:56–72. doi:10.1038/sj.npp.1301555
82. Liberzon I, Abelson JL (2016) Context processing and the neurobiology of post-traumatic stress disorder. Neuron 92(1):14–30
83. Zuj DV, Palmer MA, Lommen MJJ, Felmingham KL (2016) The centrality of fear extinction in linking risk factors to PTSD: a narrative review. Neurosci Biobehav Rev 69:15–35. doi:10.1016/j.neubiorev.2016.07.014
84. Middlemas DS, Bylund DB (2015) Brain-derived neurotrophic factor. In: Stolerman IP, Price LH (eds) Encyclopedia of psychopharmacology. Springer, Heidelberg, pp 310–313
85. Felmingham KL, Dobson-Stone C, Schofield PR et al (2013) The brain-derived neurotrophic factor Val66Met polymorphism predicts response to exposure therapy in posttraumatic stress disorder. Biol Psychiatry 73:1059–1063. doi:10.1016/j.biopsych.2012.10.033
86. Byrne SP, Rapee RM, Richardson R et al (2015) D-cycloserine enhances generalization of fear extinction in children: research article: DCS enhances fear extinction in children. Depress Anxiety 32:408–414. doi:10.1002/da.22356
87. Sartori SB, Maurer V, Murphy C et al (2016) Combined neuropeptide S and D-cycloserine augmentation prevents the return of fear in extinction-impaired rodents: advantage of dual versus single drug approaches. Int J Neuropsychopharmacol Off Sci J Coll Int Neuropsychopharmacol CINP. doi: 10.1093/ijnp/pyv128
88. de Kleine RA, Smits JAJ, Hendriks G-J, et al (2015) Extinction learning as a moderator of d-cycloserine efficacy for enhancing exposure therapy in posttraumatic stress disorder. J Anxiety Disord 34:63–67. doi:10.1016/j.janxdis.2015.06.005
89. Difede J, Cukor J, Wyka K et al (2014) D-cycloserine augmentation of exposure therapy for post-traumatic stress disorder: a pilot randomized clinical trial. Neuropsychopharmacol Off Publ Am Coll Neuropsychopharmacol 39:1052–1058. doi:10.1038/npp.2013.317
90. Scheeringa MS, Weems CF (2014) Randomized placebo-controlled D-cycloserine with cognitive behavior therapy for pediatric posttraumatic stress. J Child Adolesc Psychopharmacol 24:69–77. doi:10.1089/cap.2013.0106
91. Singewald N, Schmuckermair C, Whittle N et al (2015) Pharmacology of cognitive enhancers for exposure-based therapy of fear, anxiety and trauma-related disorders. Pharmacol Ther 149:150–190. doi:10.1016/j.pharmthera.2014.12.004
92. Acheson D, Feifel D, de Wilde S, Mckinney R, Lohr J, Risbrough V (2013) The effect of intranasal oxytocin treatment on conditioned fear extinction and recall in a healthy human sample. Psychopharmacology 229(1):199–208

93. Acheson DT, Risbrough VB (2015) Oxytocin enhancement of fear extinction: a new target for facilitating exposure-based treatments? Biol Psychiatry 78(3):154
94. Risbrough VB, Glenn DE, Baker DG (2016) On the road to translation for PTSD treatment: theoretical and practical considerations of the use of human models of conditioned fear for drug development. Transl Neuropsychopharmacol 28:173–196
95. Vervliet B, Craske MG, Hermans D (2013) Fear extinction and relapse: state of the art. Annu Rev Clin Psychol 9:215–248
96. Nawijn L, van Zuiden M, Frijling JL et al (2015) Reward functioning in PTSD: a systematic review exploring the mechanisms underlying anhedonia. Neurosci Biobehav Rev 51:189–204. doi:10.1016/j.neubiorev.2015.01.019
97. Kashdan TB, Elhai JD, Frueh BC (2006) Anhedonia and emotional numbing in combat veterans with PTSD. Behav Res Ther 44:457–467. doi:10.1016/j.brat.2005.03.001
98. Gable SL, Reis HT, Elliot AJ (2003) Evidence for bivariate systems: an empirical test of appetition and aversion across domains. J Res Personal 37:349–372. doi:10.1016/S0092-6566(02)00580-9
99. Sharf R, Sarhan M, DiLeone RJ (2010) Role of orexin/hypocretin in dependence and addiction. Brain Res 1314:130–138. doi:10.1016/j.brainres.2009.08.028
100. Hoexter MQ, Fadel G, Felício AC et al (2012) Higher striatal dopamine transporter density in PTSD: an in vivo SPECT study with [99mTc]TRODAT-1. Psychopharmacology 224:337–345. doi:10.1007/s00213-012-2755-4
101. Blum K, Giordano J, Oscar-Berman M et al (2012) Diagnosis and healing in veterans suspected of suffering from post-traumatic stress disorder (PTSD) using reward Gene testing and reward circuitry natural dopaminergic activation. J Genet Syndr Gene Ther 3:1000116. doi:10.4172/2157-7412.1000116
102. Heller AS, Johnstone T, Light SN et al (2013) Relationships between changes in sustained Fronto-striatal connectivity and positive affect in major depression resulting from antidepressant treatment. Am J Psychiatry 170:197–206. doi:10.1176/appi.ajp.2012.12010014
103. Nawijn L, van Zuiden M, Koch SBJ et al (2016) Intranasal oxytocin enhances neural processing of monetary reward and loss in post-traumatic stress disorder and traumatized controls. Psychoneuroendocrinology 66:228–237. doi:10.1016/j.psyneuen.2016.01.020
104. Koch SBJ, van Zuiden M, Nawijn L et al (2014) Intranasal oxytocin as strategy for medication-enhanced psychotherapy of PTSD: salience processing and fear inhibition processes. Psychoneuroendocrinology 40:242–256. doi:10.1016/j.psyneuen.2013.11.018
105. Hayes JP, VanElzakker MB, Shin LM (2012) Emotion and cognition interactions in PTSD: a review of neurocognitive and neuroimaging studies. Front Integr Neurosci. doi: 10.3389/fnint.2012.00089
106. Campanella C, Bremner JD (2016) Neuroimaging of PTSD. In: Bremner JD (ed) Posttraumatic stress disorder: from neurobiology to treatment. Wiley, Hoboken, p. 291
107. Schweizer S, Dalgleish T (2016) The impact of affective contexts on working memory capacity in healthy populations and in individuals with PTSD. Emotion 16:16–23. doi:10.1037/emo0000072
108. Golier JA, Yehuda R, Lupien SJ et al (2002) Memory performance in holocaust survivors with posttraumatic stress disorder. Am J Psychiatry 159:1682–1688. doi:10.1176/appi.ajp.159.10.1682
109. Paunovi N, Lundh LG, Ost LG (2002) Attentional and memory bias for emotional information in crime victims with acute posttraumatic stress disorder (PTSD). J Anxiety Disord 16:675–692
110. Wagner AD, Stebbins GT, Masciari F et al (1998) Neuropsychological dissociation between recognition familiarity and perceptual priming in visual long-term memory. Cortex 34:493–511
111. Ehlers A, Michael T, Chen YP, Payne E, Shan S (2006) Enhanced perceptual priming for neutral stimuli in a traumatic context: a pathway to intrusive memories? Memory 14 (3):316–328

112. Vukojevic V, Kolassa I-T, Fastenrath M et al (2014) Epigenetic modification of the glucocorticoid receptor Gene is linked to traumatic memory and post-traumatic stress disorder risk in genocide survivors. J Neurosci 34:10274–10284. doi:10.1523/JNEUROSCI.1526-14.2014
113. Kwapis JL, Wood MA (2014) Epigenetic mechanisms in fear conditioning: implications for treating post-traumatic stress disorder. Trends Neurosci 37:706–720. doi:10.1016/j.tins.2014.08.005
114. Minassian A, Maihofer AX, Baker DG, Nievergelt CM, Geyer MA, Risbrough VB (2015) Association of predeployment heart rate variability with risk of postdeployment posttraumatic stress disorder in active-duty marines. JAMA Psychiat 72(10):979–986
115. Zoladz PR, Diamond DM (2013) Current status on behavioral and biological markers of PTSD: a search for clarity in a conflicting literature. Neurosci Biobehav Rev 37:860–895. doi:10.1016/j.neubiorev.2013.03.024
116. de Quervain DJ-F, Kolassa I-T, Ertl V et al (2007) A deletion variant of the alpha2b-adrenoceptor is related to emotional memory in Europeans and Africans. Nat Neurosci 10:1137–1139. doi:10.1038/nn1945
117. Liberzon I, King AP, Ressler KJ et al (2014) Interaction of the *ADRB2* Gene polymorphism with childhood trauma in predicting adult symptoms of posttraumatic stress disorder. JAMA Psychiat 71:1174. doi:10.1001/jamapsychiatry.2014.999
118. De Berardis D, Marini S, Serroni N et al (2015) Targeting the noradrenergic system in posttraumatic stress disorder: a systematic review and meta-analysis of prazosin trials. Curr Drug Targets 16:1094–1106
119. Bangasser DA, Wiersielis KR, Khantsis S (2016) Sex differences in the locus coeruleus-norepinephrine system and its regulation by stress. Brain Res 1641:177–188. doi:10.1016/j.brainres.2015.11.021
120. Sara SJ, Bouret S (2012) Orienting and reorienting: the locus coeruleus mediates cognition through arousal. Neuron 76:130–141. doi:10.1016/j.neuron.2012.09.011
121. Bangasser DA, Valentino RJ (2014) Sex differences in stress-related psychiatric disorders: neurobiological perspectives. Front Neuroendocrinol 35:303–319. doi:10.1016/j.yfrne.2014.03.008
122. del Río-Casanova L, González A, Páramo M et al (2016) Emotion regulation strategies in trauma-related disorders: pathways linking neurobiology and clinical manifestations. Rev Neurosci. doi: 10.1515/revneuro-2015-0045
123. Gross JJ (1998) The emerging field of emotion regulation: an integrative review. Rev Gen Psychol 2:271–299. doi:10.1037/1089-2680.2.3.271
124. Laddis A (2015) The pathogenesis and treatment of emotion dysregulation in borderline personality disorder. ScientificWorldJournal 2015:179276. doi:10.1155/2015/179276
125. Seligowski AV, Lee DJ, Bardeen JR, Orcutt HK (2015) Emotion regulation and posttraumatic stress symptoms: a meta-analysis. Cogn Behav Ther 44:87–102. doi:10.1080/16506073.2014.980753
126. Vermetten E, Spiegel D (2014) Trauma and dissociation: implications for borderline personality disorder. Curr Psychiatry Rep 16:434. doi:10.1007/s11920-013-0434-8
127. Ford JD, Courtois CA (2014) Complex PTSD, affect dysregulation, and borderline personality disorder. Borderline Personal Disord Emot Dysregulation 1:9. doi:10.1186/2051-6673-1-9
128. Cohen Kadosh K, Luo Q, de Burca C et al (2016) Using real-time fMRI to influence effective connectivity in the developing emotion regulation network. NeuroImage 125:616–626. doi:10.1016/j.neuroimage.2015.09.070
129. Jerud AB, Pruitt LD, Zoellner LA, Feeny NC (2016) The effects of prolonged exposure and sertraline on emotion regulation in individuals with posttraumatic stress disorder. Behav Res Ther 77:62–67. doi:10.1016/j.brat.2015.12.002
130. Fernandez KC, Jazaieri H, Gross JJ (2016) Emotion regulation: a transdiagnostic perspective on a new RDoC domain. Cogn Ther Res 40:426–440. doi:10.1007/s10608-016-9772-2

131. Holmes E, Brown R, Mansell W et al (2005) Are there two qualitatively distinct forms of dissociation? A review and some clinical implications. Clin Psychol Rev 25:1–23. doi:10.1016/j.cpr.2004.08.006
132. Ginzburg K, Koopman C, Butler LD et al (2006) Evidence for a dissociative subtype of post-traumatic stress disorder among help-seeking childhood sexual abuse survivors. J Trauma Dissociation 7:7–27
133. American Psychiatric Association (2000) Diagnostic and statistical manual of mental disorders: DSM-IV-TR, 4th edn. American Psychiatric Association, Washington, DC. Text revision
134. Lanius RA, Brand B, Vermetten E et al (2012) The dissociative subtype of posttraumatic stress disorder: rationale, clinical and neurobiological evidence, and implications. Depress Anxiety 29:701–708. doi:10.1002/da.21889
135. Perez DL, Dworetzky BA, Dickerson BC et al (2015) An integrative Neurocircuit perspective on psychogenic Nonepileptic seizures and functional movement disorders: neural functional unawareness. Clin EEG Neurosci 46:4–15. doi:10.1177/1550059414555905
136. Pape W, Wöller W (2015) Low dose naltrexone in the treatment of dissociative symptoms. Nervenarzt 86:346–351. doi:10.1007/s00115-014-4015-9
137. Goldstein LH, Mellers JDC, Landau S et al (2015) COgnitive behavioural therapy vs standardised medical care for adults with dissociative non-epileptic seizures (CODES): a multicentre randomised controlled trial protocol. BMC Neurol 15:98. doi:10.1186/s12883-015-0350-0
138. Lanius RA, Vermetten E, Loewenstein RJ et al (2010) Emotion modulation in PTSD: clinical and neurobiological evidence for a dissociative subtype. Am J Psychiatry 167:640–647. doi:10.1176/appi.ajp.2009.09081168
139. Felmingham K, Kemp AH, Williams L et al (2008) Dissociative responses to conscious and non-conscious fear impact underlying brain function in post-traumatic stress disorder. Psychol Med 38:1771–1780. doi:10.1017/S0033291708002742
140. Vermetten E, Lanius RA (2012) Biological and clinical framework for posttraumatic stress disorder. Handb Clin Neurol 106:291–342
141. Yehuda R, Hoge CW, McFarlane AC, Vermetten E, Lanius RA, Nievergelt CM, Hobfoll SE, Koenen KC, Neylan TC, Hyman SE (2015) Posttraumatic stress disorder. Nat Rev Primer. doi: 10.1038/nrdp.2015.57

Neurocognition in PTSD: Treatment Insights and Implications

Amy J. Jak, Laura D. Crocker, Robin L. Aupperle, Ashley Clausen, and Jessica Bomyea

Abstract Post-traumatic stress disorder (PTSD) is classified as a traumatic stress-related condition and is most often discussed in terms of emotional dysfunction. However, given that cognitive and emotional processes are intricately intertwined, implemented by overlapping brain networks, and effectively integrated in at least some of the same regions (e.g., prefrontal cortex, for a review, see Crocker et al. 2013), an abundance of literature now highlights the key role that *cognitive* functioning plays in both the development and maintenance (or exacerbation) of PTSD symptoms (Aupperle et al. 2012a; Verfaellie et al. 2012). Findings from this body of work detail objective impairment in neuropsychological function in those with PTSD (Brandes et al. 2002; Hayes et al. 2012a; Koenen et al. 2001). Yet despite the impact of neurocognition on PTSD treatment engagement and success (e.g., Haaland et al. 2016; Nijdam et al. 2015) and conversely, the role of PTSD treatment in normalizing cognitive dysfunction, a much smaller literature exists on neurocognitive changes following treatment for PTSD. Even aside from its role in treatment, cognitive functioning in PTSD has significant implications for daily functioning for individuals with this disorder, as cognition is predictive of school achievement, obtaining and

A.J. Jak (✉)
VA San Diego Healthcare System, San Diego, CA, USA

University of California, San Diego, La Jolla, CA, USA
e-mail: ajak@ucsd.edu

L.D. Crocker and J. Bomyea
VA San Diego Healthcare System, San Diego, CA, USA
e-mail: laura.crocker@va.gov; jessica.bomyea@va.gov

R.L. Aupperle and A. Clausen
Laureate Institute for Brain Research, Tulsa, OK, USA

University of Tulsa, Tulsa, OK, USA
e-mail: raupperle@laureateinstitute.org; aclausen@laureateinstitute.org

Curr Topics Behav Neurosci (2018) 38: 93–116
DOI 10.1007/7854_2016_62
Published Online: 4 April 2016

maintaining employment, job advancement, maintaining relationships, greater wealth, and better health and quality of life (e.g., Diamond and Ling 2016).

Keywords Neuroimaging • Neuropsychology • PTSD • Treatment

Contents

1 Introduction

Post-traumatic stress disorder (PTSD) is classified as a traumatic stress-related condition and is most often discussed in terms of emotional dysfunction. However, given that cognitive and emotional processes are intricately intertwined, implemented by overlapping brain networks, and effectively integrated in at least some of the same regions (e.g., prefrontal cortex, for a review, see Crocker et al. 2013), an abundance of literature now highlights the key role that *cognitive* functioning plays in both the development and maintenance (or exacerbation) of PTSD symptoms (Aupperle et al. 2012a; Verfaellie et al. 2012). Findings from this body of work detail objective impairment in neuropsychological function in those with PTSD (Brandes et al. 2002; Hayes et al. 2012a; Koenen et al. 2001). Yet despite the impact of neurocognition on PTSD treatment engagement and success (e.g., Haaland et al. 2016; Nijdam et al. 2015) and conversely, the role of PTSD treatment in normalizing cognitive dysfunction, a much smaller literature exists on neurocognitive changes following treatment for PTSD. Even aside from its role in treatment, cognitive functioning in PTSD has significant implications for daily functioning for individuals with this disorder, as cognition is predictive of school achievement, obtaining and maintaining employment, job advancement, maintaining relationships, greater wealth, and better health and quality of life (e.g., Diamond and Ling 2016).

Therefore, given the pervasiveness of cognitive symptoms in PTSD, understanding the potential impact of treatment for PTSD on neuropsychological functioning is also essential to understanding neurocognition in PTSD. This review will briefly summarize primary findings across cognitive domains as well as the proposed neural mechanisms for performance differences observed in PTSD. We will also review neuropsychological and neuroimaging outcomes following psychotherapy and psychopharmacology interventions for PTSD. Better understanding of neurocognition in PTSD has the potential to advance our understanding of treatment mechanisms, inform efforts to improve treatments, and better match individuals to specific treatments; these and other future directions for research on neurocognition in PTSD will be discussed.

2 Neurocognition in PTSD

Several reviews and recent meta-analyses have summarized the literature regarding how PTSD relates to neuropsychological function, providing evidence for decreased performance within domains of learning and memory, executive functions, processing

Table 1 Summary of neuropsychological functioning in PTSD and implicated brain regions

Cognitive domain	Summary of primary neuropsychological findings	Implicated brain regions	Exemplar reviews or meta-analyses
Learning and memory	More robust dysfunction for verbal rather than visual information and for initial encoding and retrieval rather than delayed recall	Decreased hippocampal volume, dysfunction in hippocampal activation though directionality of findings mixed, dysfunction in prefrontal cortex (PFC) may also play a role	Brewin et al. (2007), Samuelson (2011) and Scott et al. (2015)
Attention	Decreased performance for basic and sustained attention, particularly for verbal rather than visual information	PFC dysfunction, though the specific PFC regions (anterior cingulate cortex, orbitofrontal cortex, dorsolateral PFC, inferior frontal cortex, etc.) and directionality of findings differ across studies, particularly depending on specific task demands	Aupperle et al. (2012a, b) and Polak et al. (2012)
Working memory	Decreased working memory performance, particularly for verbal rather than visual information		
Executive function	Deficits in response inhibition, some evidence of impairment in speeded attentional switching; however, rule-learning and untimed strategy switching is mostly spared		
Processing speed	Impaired information processing speed (PS), one of the strongest effect sizes in meta-analysis; deficits in PS impact performance in other domains of function	Inferior frontal gyrus	Scott et al. (2015)

speed, attention, and working memory (Aupperle et al. 2012a; Brewin 2011; Polak et al. 2012; Samuelson 2011; Scott et al. 2015). Below we review some of the primary findings for each of these domains, as well as the proposed neural mechanisms for performance differences observed in PTSD (see Table 1 for a summary).

2.1 Memory

There is strong evidence of memory deficits in PTSD. The disorder involves persistent re-experiencing of traumatic memories, and the animal literature shows that prolonged or severe stress influences hippocampal structure and function (Brewin 2011; McEwen and Sapolsky 1995). Individual studies and meta-analyses support the existence of memory dysfunction in PTSD, which is most robust for verbal (rather than visual) information and for initial encoding and retrieval information (rather than delayed recall; see (Scott et al. 2015)).

There is also consistent evidence that PTSD is related to decreased hippocampal volume. Bremmer and colleagues were among the first to show reduced hippocampal volume in individuals with PTSD (Bremner et al. 1995), with evidence of smaller hippocampi along the continuum from trauma exposed individuals (Woon et al. 2010) to chronic PTSD (Kitayama et al. 2005). There is also suggestion that smaller hippocampal volume is a risk factor for persistence of PTSD (Apfel et al. 2011; van Rooij et al. 2015a, b). There has been mixed evidence, however, as to whether hippocampal volume is related to observed decreases in memory performance for PTSD patients, with some research findings smaller hippocampi related to poorer memory performances (Bremner et al. 1995; Tischler et al. 2006; Vythilingam et al. 2005), while others found no association between hippocampal volume and memory performance (Lindauer et al. 2006; Neylan et al. 2004; Stein et al. 1997; Woodward et al. 2009). In addition, there is evidence that hippocampal or medial temporal lobe activation may relate to PTSD, but the directionality of findings are mixed (Geuze et al. 2008; Shin et al. 2004).

Given the profile of findings within the memory domain (i.e., more robust deficits in initial learning), impairment may also be related to prefrontal cortex (PFC; or connections between limbic regions and PFC) dysfunction (Scott et al. 2015), as well as volumetric differences in PFC regions, particularly within the anterior cingulate cortex (ACC; for review, see (Samuelson 2011; Stillman and Aupperle 2015)). Though the possibility that PFC volume or activation differences relate to memory performance has not been thoroughly investigated, PTSD was associated with decreased PFC activation (as well as greater medial temporal activation) during a paired-associates learning task (Geuze et al. 2008). In addition, a recent study found that medial PFC and ACC activation during interference processing, and connectivity of the medial and lateral PFC during resting state, was related to overall neuropsychological performance for combat Veterans (including verbal memory; Aupperle et al. 2014).

2.2 *Attention, Working Memory, and Executive Function*

Similar to findings associated with memory, decreased performance on basic attention, sustained attention, and working memory tasks have also been reported in PTSD (for review, see Polak et al. 2012; Aupperle et al. 2012b), particularly for verbal (versus visual) stimuli (Jenkins et al. 2000; Samuelson et al. 2006). Many studies assessing sustained attention in PTSD have reported increased errors of commission (i.e., incorrectly responding to distractor stimuli), suggesting difficulty with response inhibition (Vasterling et al. 1998; Wu et al. 2010). Response inhibition deficits in PTSD are supported by studies utilizing tasks that directly assess for inhibitory function, such as the Go/No-Go, Stop signal, and color-word Stroop (Wu et al. 2010; Falconer et al. 2008; Leskin and White 2007; Shucard et al. 2008). In addition, PTSD is associated with increased intrusions during verbal memory tests, providing further evidence of inhibitory dysfunction (Lindauer et al. 2006; Vasterling et al. 1998). Neuropsychological research provides some (albeit inconsistent) support for impairment in speed-reliant, attentional switching (Jenkins et al. 2000; Leskin and White 2007; Lagarde et al. 2010; Twamley et al. 2009); however, planning, rule-learning, and untimed strategy switching are mostly spared in PTSD (Aupperle et al. 2012a; Polak et al. 2012; Vasterling et al. 1998; Lagarde et al. 2010; Twamley et al. 2009).

A few neuroimaging studies have been conducted using tasks related to attention or executive functioning. These studies utilized tasks of sustained attention (e.g., auditory oddball task), response inhibition or interference (e.g., Go/No-Go, Stop Signal, and multisource interference task), and working memory (e.g., N-back like tasks), and consistently report that PTSD is related to dysfunction in PFC recruitment (Shin et al. 2001; Falconer et al. 2008; Aupperle et al. 2016; Bryant et al. 2005; Moores et al. 2008). However, the specific PFC regions (ACC, orbitofrontal cortex [OFC], dorsolateral PFC, inferior frontal cortex, etc.) and directionality of findings reported differ across studies. These discrepancies may be because PTSD is associated with difficulty appropriately up- and down-regulating PFC regions given specific task demands (Aupperle et al. 2016), with some indication for increased PFC recruitment during easier tasks, but decreased recruitment during more difficult tasks (Aupperle et al. 2016; Moores et al. 2008). These findings may reflect successful compensatory efforts that break down with more difficult tasks, or perhaps difficulties down-regulating medial PFC regions involved in default mode, self-referential, or emotional processing (Aupperle et al. 2016).

2.3 *Processing Speed*

Information processing speed is consistently impaired in other mental health conditions, notably depression (den Hartog et al. 2003; Marazziti et al. 2010; Tsourtos et al. 2002), and has one of the strongest effect sizes in a recent meta-analysis of neurocognitive functioning in PTSD (Scott et al. 2015); however, it has not consistently been a focus of neuropsychological studies in PTSD. Processing speed is related to prefrontal (i.e., inferior frontal gyrus) activation (Usui et al. 2009) though specific mechanisms for potential processing speed deficits in PTSD remain unclear. Reductions in cognitive resources more generally (e.g., due to concurrent emotional processing, hypervigilance, or decreased sleep) could reduce attention to relevant stimuli and the task at hand, thus influencing processing speed (Scott et al. 2015; Shucard et al. 2008; Morey et al. 2009). In support of this, one functional magnetic resonance imaging (fMRI) study reported that greater medial PFC and amygdala activation during emotional processing in PTSD related to reduced performance on the symbol digit modalities task (SDMT; (Aupperle et al. 2012b)). Aberrations in reward processing could also contribute to decreased motivation, thus influencing processing speed across tasks. While there is evidence for deficits in reward learning and decreased striatal responses to reward in PTSD (Sailer et al. 2008), it has yet to be investigated whether these effects relate to neuropsychological performance. As processing speed deficits would impact performance across all other domains of function, clarifying the cognitive and neural mechanisms for these deficits could be an important foundation of knowledge for understanding neuropsychological function in PTSD.

3 Neurocognition and Treatment for PTSD

3.1 Neuropsychological Function and PTSD Psychotherapy

Empirically supported treatments for PTSD such as Cognitive Processing Therapy (CPT; Resick et al. 2008) and Prolonged Exposure (PE; Foa et al. 2008) have reasonable efficacy for reducing the hallmark symptoms of PTSD such as avoidance and re-experiencing (Cusack et al. 2016). Despite a large literature documenting the neuropsychological impairments associated with PTSD and an equally large literature detailing good efficacy of trauma-focused therapy for treating PTSD overall, very little data exists on the neurocognitive outcomes following treatment for PTSD.

Resilience-oriented treatment for PTSD focuses on re-engaging the approach motivational system (as opposed to extinguishing fear conditioning) and enhancing positive emotional engagement (Kent et al. 2011). A small randomized controlled trial (RCT) comparing group resilience-oriented treatment to a wait-list control condition revealed small-to-medium effect size improvements in memory and executive functioning in the intervention group over the control group (Kent et al. 2011). Although not an RCT, a small study of 15 women assessed pre- and post-individual trauma-focused therapy revealed improvements in executive functioning with medium effect sizes in addition to psychological symptom improvement (Walter et al. 2010). However, other cognitive domains were not measured in this study. Preliminary data from an ongoing trial comparing CPT to a hybrid treatment that combines CPT with elements of cognitive rehabilitation (SMART-CPT) in Iraq and Afghanistan war era Veterans with PTSD and a history of mild to moderate traumatic brain injury (TBI; Jak et al. 2015) shows clinically significant decreases in PTSD symptoms and improvements in processing speed, executive functioning, and memory (Crocker et al. 2015). A recent single case study also documented cognitive improvements following CPT with an Iraq/Afghanistan Veteran with PTSD in the domains of processing speed and visual attention, but not memory or executive functioning, in addition to a clinically significant decline in self-reported PTSD symptoms (Boyd et al. 2016).

There is evidence that poorer cognitive functioning pre-trauma may serve as a risk factor for the development of PTSD, but that neuropsychological function may also be further exacerbated by the experience of trauma (Aupperle et al. 2012a). Because most trauma-focused therapies rely on imaginal exposure, cognitive restructuring, and ostensibly retrieval and storage of new information, it is reasonable to suspect that cognitive weaknesses or impairments may also negatively impact one's ability to benefit from PTSD treatment. At least one target of trauma-focused psychotherapy is reducing fear reactions by strengthening the ventromedial PFC inhibition of the amygdala. When there are pre-treatment memory or inhibitory impairments, the response to trauma-focused therapy may be attenuated, at least in part due to compromised functioning of the brain regions involved in the fear network. Therefore, recent work has also centered on understanding how mild impairments in memory, attention, and/or executive functioning at baseline might impact treatment outcomes. If verbal memory is dysfunctional at treatment outset, it is hypothesized that traumatic memories cannot be as effectively reconsolidated and re-experiencing symptoms may not decrease sufficiently (Nijdam et al. 2015). Similarly, if executive function deficits exist, it may be difficult for patients to inhibit their responses or modify their thought processes as part of therapy. Therefore, improving verbal memory or executive functions could potentially improve treatment response to trauma-focused therapy.

In an RCT comparing eye movement desensitization and reprocessing (EMDR) to brief eclectic psychotherapy (BEP), lower verbal memory scores at baseline were significantly associated with poorer PTSD treatment outcomes for both treatment modalities when controlling for initial PTSD symptom severity (Nijdam et al. 2015). Another small study of 23 individuals with PTSD who received cognitive behavioral therapy (CBT) for PTSD underwent memory and attention assessment both before and after treatment. Memory performances (particularly verbal encoding), even when controlling for pre-treatment depression, PTSD severity, and attention, predicted improvement in psychological symptoms and PTSD recovery (Wild and Gur 2008). Pre-treatment/post-treatment comparisons of neuropsychological performances were not made, however. Consistent with these findings, a recent non-randomized study of female Veterans who completed group psychotherapy involving cognitive restructuring, exposure therapy, and skills training found that better pre-treatment learning and memory predicted successful treatment outcomes, whereas pre-treatment inhibition/switching did not (Haaland et al. 2016). However, treatment-related improvements in inhibition/switching, but not learning/memory and working memory, were related to greater reductions in PTSD symptoms, even when controlling for changes in depression. A large epidemiological study suggested that PTSD symptoms of re-experiencing and arousal negatively impacted verbal recall. These core symptoms of PTSD were also inversely related to pre-trauma attention, processing speed, and verbal intelligence. That is, lower cognitive functioning prior to trauma seemed to place one at higher risk for increased symptoms/poorer outcomes after trauma but also that the trauma itself negatively impacted cognition (Parslow and Jorm 2007).

3.2 Neuroimaging and PTSD Psychotherapy

There has been an abundance of studies examining the neural correlates of PTSD using a variety of paradigms. Several reviews and meta-analyses indicate that PSTD is associated with dysfunction in lateral PFC, ACC, ventromedial PFC, insula, amygdala, and hippocampus (e.g., Etkin and Wager 2007; Garfinkel and Liberzon 2009; Hayes et al. 2012b; Simmons and Matthews 2012). These regions are part of networks that implement key processes relevant for PTSD, including fear conditioning, threat processing, top-down control of emotional responses, and processing of contextual information (Hayes et al. 2012a; Yehuda and LeDoux 2007). However, very few studies have examined structural or functional brain changes associated with PTSD treatment.

Findings are mixed across the few published studies of structural brain changes after psychotherapy. Levy-Gigi et al. (2013) found that prior to CBT, individuals diagnosed with PTSD exhibited smaller hippocampi than control participants; however, treatment appeared to normalize hippocampal volume in PTSD patients such that there were no group differences after treatment. Further, changes in PTSD symptoms were correlated with changes in hippocampal volume. These changes were specific to the hippocampus, as there was no evidence that treatment was associated with alterations in medial OFC, amygdala, or total brain volume. Lindauer et al. (Lindauer et al. 2006) also found that PTSD patients had smaller hippocampi at baseline relative to control individuals; however, BEP was not associated with hippocampal volume changes, nor changes in parahippocampal gyrus and amygdala. Similarly, van Rooij et al. (van Rooij et al. 2015a, b) found that patients considered non-remitters after trauma-focused therapy exhibited smaller left hippocampal volume relative to remitted patients and combat controls both pre- and post-treatment, but there were no treatment-related changes observed in the hippocampus.

Several fMRI studies have shown treatment-related alterations in activation in the regions most commonly identified as exhibiting dysfunction in PTSD. For example, CBT, one of the most effective psychological treatments for PTSD that targets maladaptive thoughts and behaviors, has been associated with functional changes in the ACC (Aupperle et al. 2013; Felmingham et al. 2007; Roy et al. 2010; Thomaes et al. 2012), dorsolateral (Aupperle et al. 2013; Roy et al. 2010) and ventrolateral (Felmingham et al. 2007) PFC, insula (Aupperle et al. 2013; Thomaes et al. 2012), amygdala (Aupperle et al. 2013; Felmingham et al. 2007; Roy et al. 2010; Peres et al. 2011), and hippocampus (Felmingham et al. 2007). Changes have also been observed in other brain areas as a function of treatment, including specific regions in temporal and parietal cortices (e.g., van Rooij et al. 2015a; Felmingham et al. 2007; Farrow et al. 2005), though less consistently so. Similar regions have shown treatment-related changes as a function of remission status after therapy completion. For example, amygdala activation decreased from pre- to post-treatment for non-remitters but remained consistent across time for remitters (van Rooij et al. 2016). In addition, non-remitters exhibited greater activation in insula and dorsal ACC at both baseline and post-treatment in response to negative pictures relative to those who remitted. Consistent with these findings, Simmons et al. (2013) found that remitters showed decreased activation in the insula from pre- to post-treatment in anticipation of negative images, whereas insula activation increased across time in non-remitters when anticipating positive images. Importantly, Simmons et al. (2013) found that functional connectivity between the left ventral anterior insula and other regions, including the right cingulate and mid-posterior insula, increased from pre- to post-treatment in remitters.

3.3 Neuropsychological Function and PTSD Pharmacotherapy

The pathophysiology of PTSD is associated with dysfunction in biological systems that regulate stress responses including norepinephrine, hypothalamic-pituitary adrenal, noradrenergic, serotonergic, and glutamatergic systems (Bailey et al. 2013; Puetz et al. 2015; Ravindran and Stein 2009; Steckler and Risbrough 2012). Medications targeting these systems, specifically selective serotonin reuptake inhibitors (SSRIs) including sertraline and paroxetine, which act on the serotonergic system, are effective pharmacological treatments for PTSD (Puetz et al. 2015; Bossini et al. 2007; Brady et al. 2000; Bremner and Vermetten 2004; Tucker et al. 2001; Vermetten et al. 2003). There is also some evidence for the use of anticonvulsants, which act on the glutamatergic system (Bremner et al. 2005), tricylics (Puetz et al. 2015), and other agents both prophylactically and once PTSD is established (Steckler and Risbrough 2012). Given the cognitive deficits present in PTSD, neuropsychological function is a prime target for study in relation to pharmacotherapy for PTSD.

Research with paroxetine and sertraline suggests that long-term use of SSRI's ($\geq$9 months) results in subjective increases in cognition and work performance (Bremner and Vermetten 2004), and produces significant improvement in declarative memory ability (Bremner and Vermetten 2004; Vermetten et al. 2003). However, increased hippocampal volume has not been related to changes in memory performance, making it unclear whether changes in hippocampal volume are the mechanism for these beneficial effects (Vermetten et al. 2003). One study assessed the impact of anticonvulsant medication (phenytoin) on PTSD symptoms and neuropsychological ability (Bremner et al. 2005) and reported improvements in neuropsychological function that were limited to executive functioning (though were not statistically significant; (Bremner et al. 2005)). Interestingly, the level of improvement in executive functions related to the amount of change in hippocampal volume. While promising, further research is needed to confirm the reports from these small studies (with sample sizes of 9–28) and investigate whether other medications shown to be effective in the treatment of PTSD improve cognitive functioning.

3.4 Neuroimaging and PTSD Pharmacotherapy

Given that pharmacotherapy for PTSD specifically targets biological systems that regulate stress responses, it is reasonable to suspect that their beneficial effects may in part relate to functional and potentially even volumetric brain changes (Bossini et al. 2007; Bremner and Vermetten 2004; Vermetten et al. 2003; Bremner et al. 2005; Thomaes et al. 2014), the latter of which may be secondary to reorganization of cells (Lee et al. 2001), decreased neuronal death (perhaps due to changes in cortisol release) (Gage et al. 1980; Watanabe et al. 1992), and even neurogenesis (Brady et al. 2000; Tucker et al. 2001; Santarelli et al. 2003).

A subset of literature has investigated the impact of SSRI treatment for PTSD on brain function (Fani et al. 2011; Seedat et al. 2004). One study assessed the effects of 8 weeks of citalopram treatment using single-photon emission computed tomography (SPECT) with 11 patients (Seedat et al. 2004). Citalopram treatment resulted in reduced activation in the left medial temporal cortex, independent of PTSD symptom response, as well less deactivation of the medial PFC, the level of which related to amount of symptom change. A second study used positron emission tomography (PET) during script-driven imagery pre- and post-paroxetine treatment with 13 patients. The authors reported increased recruitment of the ACC (for both placebo and paroxetine treated patients) and increased OFC activation (specific to paroxetine) (Fani et al. 2011). The regions identified in these small studies are potentially important for supporting cognitive functions that go awry in PTSD, such as the involvement of the medial temporal cortex in memory (Squire and Zola-Morgan 1991), the OFC in decision-making (Bechara et al. 2000), the ACC in error monitoring and response inhibition (Ridderinkhof et al. 2004), and the medial PFC in the default mode network (the down-regulation of which may be important for supporting goal-directed cognitive tasks (Aupperle et al. 2016; Broyd et al. 2009)). However, there have been no studies directly assessing the impact of pharmacologic treatment on neural activation patterns during cognitive paradigms.

Similar to the psychotherapy literature, there have been repeated reports of increased grey matter volume, specifically within the hippocampus, following pharmacotherapy (Bossini et al. 2007; Bremner and Vermetten 2004; Vermetten et al. 2003; Bremner et al. 2005; Thomaes et al. 2014). Trials investigating long-term (≥9 months) use of SSRIs have reported increases in bilateral hippocampal volume in adults with PTSD (Bossini et al. 2007; Bremner and Vermetten 2004; Vermetten et al. 2003). Importantly, effects have been reported across two types of SSRIs including sertraline (Bossini et al. 2007) and paroxetine (Bremner and Vermetten 2004; Vermetten et al. 2003). Similar to SSRI findings, one study that investigated the impact of anticonvulsant medication (phenytoin) on PTSD symptoms and brain volume found that phenytoin administration was associated with an increase in right brain volume, including a non-significant increase within the hippocampus (Bremner et al. 2005). These findings are particularly intriguing given its role in memory function as well as the parallel reports from the animal literature that hippocampal neurogenesis may be one essential mechanism through which antidepressants exert their effects (Santarelli et al. 2003).

4 Neurocognitive-Informed Treatment Approaches for PTSD

In addition to understanding how standard PTSD treatments might lead to improved cognitive outcomes, growing interest has emerged in translating knowledge about aberrant neurocognitive processes and the risk cognition poses for poor treatment outcomes into novel intervention approaches. By focusing specifically on modifying cognition via therapeutic training programs that guide individuals through a set of exercises to facilitate practice and change within a specific cognitive process of interest, these paradigms allow for the evaluation of the causal relationships between cognitive processes and symptoms of PTSD. The information processing biases and executive functioning deficits often present in PTSD may be particularly appropriate targets for this type of intervention.

In addition to the cognitive deficits reviewed above, individuals with PTSD demonstrate preferential processing of negative, threatening information at the expense of more benign or positive cues, relative to healthy individuals, leading to negative biases in attention, interpretation, and memory (Buckley et al. 2000; Constans 2005; Pineles et al. 2009; Pineles et al. 2007). Theoretical accounts propose that these abnormalities are related to deficits in executive functioning (Aupperle et al. 2012a, b) and maintain psychopathology by leading to a heightened sense of threat and increased distress (Pineles et al. 2009), and thus correcting these automatic "mental habits" may lead to reduction in symptoms. Researchers have developed cognitive bias modification (CBM) programs to alter information processing biases by training cognitive processes away from negative information, using brief computer-administered programs (Amir et al. 2009). In anxious populations, these CBM procedures have altered cognitive processes and lead to a reduction in symptoms (Hakamata et al. 2010; Mathews and MacLeod 2002; Wilson et al. 2006), suggesting that experimental manipulation of these processes may be a promising treatment target for disorders such as PTSD.

To date, CBM procedures have been developed to retrain attentional biases in PTSD, but the efficacy of such training programs is limited and equivocal. Kuckertz and colleagues found that inpatient military personnel getting attention training showed less attentional bias and lower symptoms post-treatment than those receiving the control program (Kuckertz et al. 2014). However, two randomized controlled studies of stand-alone attention training did not find evidence for reductions in attention bias or symptoms in individuals receiving attention training including outpatients with chronic PTSD (Schoorl et al. 2013) and two samples of Veterans in the USA and Israel (Badura-Brack et al. 2015). Thus, at this time there is not a strong evidence base for attention training in PTSD using computer-based CBM as a stand-alone treatment. Importantly, it may be the case that attention training is best used as an adjunct intervention for PTSD, rather than an independent treatment, as there has been some work in the treatment of depression supporting this combined approach (Siegle et al. 2007). Future work is needed to delineate if and how this type of cognitive training approach will be helpful for altering cognition and ameliorating symptoms of this disorder.

CBM procedures aim to alter cognitive functions in the presence of threat. However, as noted above, neuropsychological abnormalities in PTSD are observed even in the absence of affectively valenced cues and stimuli. Recent theoretical accounts suggest that executive functioning, and specifically the ability to control interference, is critical for regulation of emotional information in working memory (Anderson and Levy 2009; Joormann et al. 2010) and may contribute to regulation of re-experiencing symptoms in PTSD (Bomyea and Lang 2015; Verwoerd et al. 2009). From a treatment perspective, utilizing training programs that improve the executive function deficits in PTSD (and thus strengthen prefrontal cortical circuits) may reduce PTSD symptoms (Etkin et al. 2013; McNally 2007).

Executive functioning training programs have demonstrated improvement in cognitive performance on trained and untrained cognitive tasks, as well as reductions in affective symptoms, in diverse clinical (Buschkuehl et al. 2008; Dickinson et al. 2010; Klingberg et al. 2005), and healthy populations (Klingberg et al. 2005; Jaeggi et al. 2008; Melby-Lervag and Hulme 2013; Salminen et al. 2012; Takeuchi et al. 2014). However, it must be noted that there have been some inconsistent findings regarding the ability for training to generalize to improvements in daily functioning. Functional neuroimaging studies support the potential for executive function training by demonstrating that training programs result in changes in neural activity (lateral prefrontal and parietal cortices, specifically) during task performance (Kane and Engle 2002; Olesen et al. 2004; Schweizer et al. 2013). Experimental analogue data suggests that manipulating executive functioning leads to changes in thought regulation ability and distress about trauma memories (Bomyea and Amir 2011; Callinan et al. 2015; Nassif and Wells 2014); however, only one study to date has adapted executive functioning training programs for use in PTSD populations. Bomyea and colleagues (Bomyea et al. 2015) administered a computerized executive functioning training program or a control program to women with PTSD, and found that those receiving training showed greater improvements in working memory capacity and a greater reduction in re-experiencing symptoms than those in the control program.

5 Conclusions

Learning and verbal memory, executive functioning, processing speed, attention, and working memory decrements are pervasive cognitive problems present in PTSD. Cognitive functioning prior to trauma can increase the risk for development of PTSD and can be further exacerbated by the experience of trauma. Preliminary data indicates that cognitive deficits present in individuals with PTSD can improve following psychotherapy and/or psychopharmacologic interventions, particularly in the domains of memory and executive functioning. Cognitive symptoms are, however, a risk factor for poorer response to standard interventions for PTSD. In concert with genetic and environmental elements, development and maintenance of PTSD is thought to be associated with changes in prefrontal (particularly OFC and ACC) and limbic (i.e., amygdala) structures. Cognitive behavioral therapies for PTSD have been reported to impact activity in the PFC and its regulation of amygdala activation (for example, when processing trauma-relevant stimuli), and in turn, the fear conditioning/extinction or cognitive reappraisal processes that are mediated by these brain circuits (Zantwoord et al. 2013). Because the PFC also guides working memory, attention, and executive functioning, it stands to reason that psychotherapy that effectively impacts these brain circuits also impacts objective neuropsychological performance.

The literature assessing neuropsychological changes associated with PTSD has primarily focused on declarative memory (Bremner and Vermetten 2004; Vermetten et al. 2003; Thomaes et al. 2014), which is thought to be partially reliant on hippocampal function (Tulving and Markowitsch 1998) and executive functions, mediated by the PFC. The hippocampus and the prefrontal cortex, particularly the ACC, are strongly implicated in cognitive dysfunction in PTSD. Though the research literature examining cognitive and neuroimaging outcomes following treatment is still nascent, psychotherapies and pharmacologic interventions for PTSD appear to positively impact memory and executive functioning, in particular. The current literature also suggests that treatment positively impacts hippocampal volume, but this is not a robust finding and changes in hippocampal volume do not always correspond to behavioral memory changes. Additional studies and replication are needed. Neuroimaging studies also strongly implicate PFC regions in development and maintenance of PTSD with evidence of improved functional connectivity between the insula and PFC regions from pre- to post-treatment in those whose PTSD had remitted at the conclusion of treatment. Other brain region involvement, such as the parietal cortex, has been less consistently found across imaging studies of PTSD or in studies examining the effects of psychotherapy. Taken together, the relationship between PTSD and hippocampal and PFC volume or activation, as well as changes in medial temporal and PFC volume or activation with PTSD treatment, lend strong support for the role of these brain regions in development and maintenance of the disorder. Future studies would benefit from employing functional connectivity analyses, as theories of PTSD-related dysfunction often center around problematic communication between areas or networks (for a review, see Garfinkel and Liberzon 2009) (e.g., top-down control of the PFC to inhibit insula or amygdala regions involved in processing emotionally relevant stimuli).

Directly intervening in cognitive processes holds promise as an alternative treatment approach for PTSD. There is evidence to support use of cognitive training approaches to improve not only cognitive functioning but also clinical symptoms, quality of life, and adaptive functioning across a variety of other conditions (e.g., schizophrenia, ADHD, anxiety, depression, substance abuse, aging, traumatic brain injury; Keshavan et al. 2014; Merzenich et al. 2014; Vinogradov et al. 2012). Importantly, studies of PTSD that used computerized training to enhance attentional/cognitive control found that it also reduced PTSD symptoms (Callinan et al. 2015; Bomyea et al. 2015). However, use of cognitive training programs as standalone treatments for PTSD requires more prospective clinical trials to ascertain the full depth and breadth of symptom and syndrome change.

Traditional psychotherapies for PTSD enhanced with cognitive training also appear efficacious in reducing both the clinical and cognitive symptoms of PTSD. Augmentation of CBT with CBM resulted in a greater reduction in PTSD and anxiety symptoms after treatment compared to a control condition (Kuckertz et al. 2014; Riemann et al. 2013) and integration of cognitive rehabilitation principles into CPT (Jak et al. 2015) improved both PTSD symptoms and cognitive functioning (Crocker et al. 2015). To build on this burgeoning literature, additional prospective trials that interweave cognitive training with traditional PTSD therapies are also warranted, particularly those that measure PTSD symptoms *and* neuropsychological functioning both pre- and post-intervention.

The relationship between PTSD and executive functioning is likely bidirectional, however. Deficits in executive functions are a risk factor that contribute to the development and maintenance of emotional and cognitive symptoms of PTSD (e.g., difficulty disengaging from trauma-related stimuli, problems inhibiting intrusive memories, emotion dysregulation, and biased attention for negative information), and in turn symptoms of PTSD exacerbate executive functioning impairments (Aupperle et al. 2012a, b; Verfaellie et al. 2012; Bomyea et al. 2012). Memory and executive functioning deficits in PTSD also can serve as a barrier to treatment engagement and effectiveness (Verfaellie et al. 2012). Cognitive training has been shown to increase mental health treatment completion rates in PTSD and other psychiatric conditions (Crocker et al. 2015; Grohman et al. 2006). Therefore, improving cognitive functioning prior to engaging in psychotherapy for PTSD could enhance treatment engagement and outcomes. Priming traditional interventions with cognitive training could address the negative impact of cognitive changes on treatment adherence. Randomized controlled trials examining if cognitive interventions, administered prior to standard psychotherapy, result in greater treatment gains than standard PTSD treatment are necessary.

Research utilizing cognitive training in PTSD could both increase knowledge about theoretical models proposing that cognitive functioning may be a causal factor in PTSD, and also add options for empirically supported treatments of PTSD that are cost effective and easily transportable. Although some studies suggest that attention and executive functioning training leads to changes in cognition and symptoms in PTSD, these studies are preliminary, and in the case of attention bias training, negative findings have also been reported. More work is required to better understand the neurocognitive mechanisms of action of CBMs. Future studies are needed to evaluate types and characteristics of training that may be effective (e.g., particular doses, stimuli types or other program parameters, and clinical settings) and for whom these programs might be best utilized (e.g., specific symptom severity ranges). Combining of executive functioning training with affective information, which has been shown to effectively improve emotion regulation ability (Schweizer et al. 2013), is also a future direction for this work as applied to PTSD.

Neuropsychological outcome data following psychotherapy for PTSD is limited and weaknesses of the extant literature include inclusion of both chronic and acute PTSD stemming from a multitude of causes (e.g., military, civilian, assault, accident, etc.), and samples spanning a broad age range leaving some studies vulnerable to age-related (as opposed to PTSD-specific) cognitive changes. The vast majority of studies examining cognitive functioning in PTSD in general or cognitive functioning following treatment for PTSD lack performance validity measures, which reduces confidence that the cognitive data are an accurate depiction of one's true performance capabilities. Future studies should routinely include performance and symptom validity measures.

Some of the mixed findings in the PTSD treatment and imaging literature, particularly in regard to which subregions of PFC are altered and the direction of activation (increased or decreased) subsequent to treatment, is likely due to heterogeneity across studies in terms of methodology used, specific types of treatments employed, varying sample sizes (with many small, likely leaving studies underpowered), differences in medication status, and the nature of the comparison group (e.g., healthy or combat controls versus a wait-list or other active treatment group). Many of the authors of the imaging studies suggested that psychotherapy normalizes structural and functional abnormalities in brain regions associated with PTSD, at least in those who benefitted from treatment and showed symptom improvement. However, a complicating issue in the functional imaging literature is the variation of tasks used across studies, ranging from tasks that involve evoking traumatic memories to elicit PTSD symptoms to more cognitive tasks that recruit regions involved in top-down control. Thus, some studies show that certain regions increase activation after psychotherapy, whereas others show decreased activation, depending on the nature of the task used (e.g., increased dorsal ACC activation in (Aupperle et al. 2013), decreased dorsal ACC activation in (Thomaes et al. 2012)). Thus, it is difficult to draw any firm conclusions, particularly regarding treatment mechanisms, with so few studies varying on several important factors. As additional studies are published to determine which findings replicate, an important consideration will be the nature of the psychological processes recruited by specific tasks and how those may map on to processes targeted by the treatments employed.

Results from pharmacotherapy research also suggest that deficits in neuropsychological functioning, brain function, and volume may be modifiable consequences of PTSD, and may serve as useful indicators of an individual's treatment response to particular classes of drugs. However, many questions remain regarding the effects of pharmacotherapy on neuropsychological and brain function, as well as brain volume. For example, it is unclear if effects of pharmacotherapy are transient or produce longer lasting effects and whether any effects would persist if medication use is discontinued. It is unknown if trauma type (e.g., natural disaster versus combat) may influence effects of pharmacotherapy, or if there is a critical time period for administration of pharmacotherapy for treatment effects to be evident. Lastly, it is unclear if pharmacotherapy impacts brain activity during cognitive tasks. Despite the limitations, there is promising evidence for the beneficial impact of pharmacotherapy on PTSD symptoms, as well as neurological changes that may result in improved neuropsychological function, as well as increased brain volume and brain function.

In summary, memory and executive functioning changes and corresponding dysfunction in medial temporal and prefrontal regions are strongly implicated in PTSD. The small but growing literature on cognitive outcomes of PTSD treatment provides evidence that memory and executive function, which are mediated by the PFC and hippocampus, improve with treatment. Greater understanding of the potential impact of PTSD treatment on neuropsychological functioning is essential to understanding neurocognition in PTSD. Ultimately, continued work to elucidate neuropsychological and neuroimaging outcomes following PTSD treatments has the potential to advance our understanding of the underlying neural mechanisms of cognitive dysfunction in PTSD, inform efforts to improve treatments, and better match individuals to specific treatments.

References

Amir N, Beard C, Burns M, Bomyea J (2009) Attention modification program in individuals with generalized anxiety disorder. J Abnorm Psychol 118(1):28–33. doi:10.1037/a0012589

Anderson MC, Levy BJ (2009) Suppressing unwanted memories. Curr Dir Psychol Sci 18 (4):189–194

Apfel BA, Ross J, Hlavin J, Meyerhoff DJ, Metzler TJ, Marmar CR, et al (2011) Hippocampal volume differences in Gulf War veterans with current versus lifetime posttraumatic stress disorder symptoms. Biol Psychiatry 69(6):541–548. doi:10.1016/j.biopsych.2010.09.044

Aupperle RL, Melrose AJ, Stein MB, Paulus MP (2012) Executive function and PTSD: disengaging from trauma. Neuropharmacology 62(2):686–694. doi:10.1016/j.neuropharm.2011.02.008

Aupperle RL, Allard CB, Grimes EM, Simmons AN, Flagan T, Behrooznia M, et al (2012) Dorsolateral prefrontal cortex activation during emotional anticipation and neuropsychological performance in posttraumatic stress disorder. Arch Gen Psychiatry 69(4):360–371. doi:10.1001/archgenpsychiatry.2011.1539

Aupperle RL, Allard CB, Simmons AN, Flagan T, Thorp SR, Norman SB, et al (2013) Neural responses during emotional processing before and after cognitive trauma therapy for battered women. Psychiatry Res 214(1):48–55. doi:10.1016/j.pscychresns.2013.05.001

Aupperle RL, Stillman A, Francisco A, Bruce J, Martin L, McDowd J, Simmons A (2014) Cognitive dysfunction in combat veterans is related to attenuated dorsal ACC activation during interference processing. Neuropsychopharmacology 39:S498–S499

Aupperle RL, Stillman AN, Simmons AN, Flagan T, Allard CB, Thorp SR, et al (2016) Intimate partner violence PTSD and neural correlates of inhibition. J Trauma Stress. doi:10.1002/jts.22068

Badura-Brack AS, Naim R, Ryan TJ, Levy O, Abend R, Khanna MM, et al (2015) Effect of attention training on attention bias variability and PTSD symptoms: randomized controlled trials in Israeli and U.S. combat veterans. Am J Psychiatry 172(12):1233–1241. doi:10.1176/appi.ajp.2015.14121578

Bailey CR, Cordell E, Sobin SM, Neumeister A (2013) Recent progress in understanding the pathophysiology of post-traumatic stress disorder: implications for targeted pharmacological treatment. CNS Drugs 27(3):221–232. doi:10.1007/s40263-013-0051-4

Bechara A, Damasio H, Damasio AR (2000) Emotion, decision making and the orbitofrontal cortex. Cereb Cortex 10(3):295–307

Bomyea J, Amir N (2011) The effect of an executive functioning training program on working memory capacity and intrusive thoughts. Cogn Ther Res 35(6):529–535. doi:10.1007/s10608-011-9369-8

Bomyea J, Lang AJ (2015) The role of executive functioning in PTSD and its treatment. Curr Psychiatr Rev 11(3):160–171

Bomyea J, Risbrough V, Lang AJ (2012) A consideration of select pre-trauma factors as key vulnerabilities in PTSD. Clin Psychol Rev 32(7):630–641. doi:10.1016/j.cpr.2012.06.008

Bomyea J, Stein MB, Lang AJ (2015) Interference control training for PTSD: a randomized controlled trial of a novel computer-based intervention. J Anxiety Disord 34:33–42. doi:10.1016/j.janxdis.2015.05.010

Bossini L, Tavanti M, Lombardelli A, Calossi S, Polizzotto NR, Galli R, et al (2007) Changes in hippocampal volume in patients with post-traumatic stress disorder after sertraline treatment. J Clin Psychopharmacol 27(2):233–235. doi:10.1097/01.jcp.0000264979.09862.a8

Boyd B, Rodgers CS, Aupperle RL, Jak AJ (2016) Case report on the effects of cognitive processing therapy on psychological, neuropsychological, and speech symptoms in comorbid PTSD and TBI. Cogn Behav Pract 23(2):173–183. doi:10.1016/j.cbpra.2015.10.001

Brady K, Pearlstein T, Asnis GM, Baker D, Rothbaum B, Sikes CR, Farfel GM (2000) Efficacy and safety of sertraline treatment of posttraumatic stress disorder: a randomized controlled trial. JAMA 283(14):1837–1844

Brandes D, Ben-Schachar G, Gilboa A, Bonne O, Freedman S, Shalev AY (2002) PTSD symptoms and cognitive performance in recent trauma survivors. Psychiatry Res 110:231–238. S0165178102001257 [pii]

Bremner JD, Vermetten E (2004) Neuroanatomical changes associated with pharmacotherapy in posttraumatic stress disorder. Ann N Y Acad Sci 1032:154–157. doi:10.1196/annals.1314.012

Bremner JD, Randall P, Scott TM, Bronen RA, Seibyl JP, Southwick SM, et al (1995) MRI-based measurement of hippocampal volume in patients with combat-related posttraumatic stress disorder. Am J Psychiatry 152(7):973–981. doi:10.1176/ajp.152.7.973

Bremner JD, Mletzko T, Welter S, Quinn S, Williams C, Brummer M, et al (2005) Effects of phenytoin on memory, cognition and brain structure in post-traumatic stress disorder: a pilot study. J Psychopharmacol 19(2):159–165

Brewin CR, Kleiner JS, Vasterling JJ, Field AP (2007) Memory for emotionally neutral information in posttraumatic stress disorder: a meta-analytic investigation. J Abnorm Psychol 116(3):448–463

Brewin CR (2011) The nature and significance of memory disturbance in posttraumatic stress disorder. Annu Rev Clin Psychol 7:203–227. doi:10.1146/annurev-clinpsy-032210-104544

Broyd SJ, Demanuele C, Debener S, Helps SK, James CJ, Sonuga-Barke EJ (2009) Default-mode brain dysfunction in mental disorders: a systematic review. Neurosci Biobehav Rev 33 (3):279–296. doi:10.1016/j.neubiorev.2008.09.002

Bryant RA, Felmingham KL, Kemp AH, Barton M, Peduto AS, Rennie C, et al (2005) Neural networks of information processing in posttraumatic stress disorder: a functional magnetic resonance imaging study. Biol Psychiatry 58(2):111–118. doi:10.1016/j.biopsych.2005.03.021

Buckley TC, Blanchard EB, Neill WT (2000) Information processing and PTSD: a review of the empirical literature. Clin Psychol Rev 20(8):1041–1065

Buschkuehl M, Jaeggi SM, Hutchison S, Pasqualina P, Dapp C, Muller M, et al (2008) Impact of working memory training on memory performance in old-old adults. Psychol Aging 23 (4):743–753

Callinan S, Johnson D, Wells A (2015) A randomised controlled study of the effects of the attention training technique on traumatic stress symptoms, emotional set shifting and flexibility. Cogn Ther Res 39:4–13

Constans JI (2005) Information-processing biases in PTSD. In: Vasterling JJ, Brewin C (eds) Neuropsychology of PTSD: biological, cognitive, and clinical perspectives. Guilford Press, New York, pp 105–129

Crocker LD, Heller W, Warren SL, O'Hare AJ, Infantolino ZP, Miller GA (2013) Relationships among cognition, emotion, and motivation: implications for intervention and neuroplasticity in psychopathology. Front Hum Neurosci 7:261. doi:10.3389/fnhum.2013.0026

Crocker LD, Jurick SM, Boyd B, Rodgers CS, Twamley EW, Schiehser DM, Jak AJ (2015) Treatment of veterans with comorbid PTSD and TBI using a hybrid approach. Paper presented at the annual meeting of the Anxiety and Depression Association of America conference, Miami, FL

Cusack K, Jonas DE, Forneris CA, Wines C, Sonis J, Middleton JC, et al (2016) Psychological treatments for adults with posttraumatic stress disorder: a systematic review and meta-analysis. Clin Psychol Rev 43:128–141

Diamond A, Ling DS (2016) Conclusions about interventions, programs, and approaches for improving executive functions that appear justified and those that, despite much hype, do not. Dev Cogn Neurosci 18:34–48

Dickinson D, Tenhula W, Morris S, Brown C, Peer J, Spencer K, et al (2010) A randomized, controlled trial of computer-assisted cognitive remediation for schizophrenia. Am J Psychiatry 167(2):170–180. doi:10.1176/appi.ajp.2009.09020264

Etkin A, Wager TD (2007) Functional neuroimaging of anxiety: a meta-analysis of emotional processing in PTSD, social anxiety disorder, and specific phobia. Am J Psychiatry 164:1476–1488

Etkin A, Gyurak A, O'Hara R (2013) A neurobiological approach to the cognitive deficits of psychiatric disorders. Dialogues Clin Neurosci 15(4):419–429

Falconer E, Bryant R, Felmingham KL, Kemp AH, Gordon E, Peduto A, et al (2008) The neural networks of inhibitory control in posttraumatic stress disorder. J Psychiatry Neurosci 33 (5):413–422

Fani N, Ashraf A, Afzal N, Jawed F, Kitayama N, Reed L, Bremner JD (2011) Increased neural response to trauma scripts in posttraumatic stress disorder following paroxetine treatment: a pilot study. Neurosci Lett 491(3):196–201. doi:10.1016/j.neulet.2011.01.037; S0304-3940(11) 00073-5 [pii]

Farrow TF, Hunter MD, Wilkinson ID, Gouneea C, Fawbert D, Smith R, et al (2005) Quantifiable change in functional brain response to empathic and forgivability judgments with resolution of posttraumatic stress disorder. Psychiatry Res 140(1):45–53. doi:10.1016/j.pscychresns.2005.05.012

Felmingham K, Kemp A, Williams L, Das P, Hughes G, Peduto A, Bryant R (2007) Changes in anterior cingulate and amygdala after cognitive behavior therapy of posttraumatic stress disorder. Psychol Sci 18(2):127–129. doi:10.1111/j.1467-9280.2007.01860.x

Foa EB, Keane TM, Friedman MJ, Cohen JA (2008) Effective treatments for PTSD: practice guidelines from the International Society for Traumatic Stress. Guilford Press, New York

Gage PW, Lonergan M, Torda TA (1980) Presynaptic and postsynaptic depressant effects of phenytoin sodium at the neuromuscular junction. Br J Pharmacol 69(1):119–121

Garfinkel SN, Liberzon I (2009) Neurobiology of PTSD: a review of neuroimaging findings. Psychiatr Ann 39:370–381

Geuze E, Vermetten E, Ruf M, de Kloet CS, Westenberg HG (2008) Neural correlates of associative learning and memory in veterans with posttraumatic stress disorder. J Psychiatr Res 42:659–669

Grohman K, Fals-Stewart W, Donnelly K (2006) Improving treatment response of cognitively impaired veterans with neuropsychological rehabilitation. Brain Cogn 60(2):203–204

Haaland KY, Sadek JR, Keller JE, Castillo DT (2016) Neurocognitive correlates of successful treatment of PTSD in female veterans. J Int Neuropsychol Soc 22(6):643–651. doi:10.1017/S1355617716000424

Hakamata Y, Lissek S, Bar-Haim Y, Britton JC, Fox NA, Leibenluft E, et al (2010) Attention bias modification treatment: a meta-analysis toward the establishment of novel treatment for anxiety. Biol Psychiatry 68(11):982–990. doi:10.1016/j.biopsych.2010.07.021

den Hartog HM, Derix MM, van Bemmel AL, Kremer B, Jolles J (2003) Cognitive functioning in young and middle-aged unmedicated out-patients with major depression: Testing the effort and cognitive speed hypotheses. Psychol Med 33:1443–1451

Hayes JP, Vanelzakker MB, Shin LM (2012) Emotion and cognition interactions in PTSD: a review of neurocognitive and neuroimaging studies. Front Integr Neurosci 6:1–14. doi:10.3389/fnint.2012.00089

Hayes JP, Hayes SM, Mikedis AM (2012b) Quantitative meta-analysis of neural activity in posttraumatic stress disorder. Biol Mood Anxiety Disord 2:9

Jaeggi SM, Buschkuehl M, Jonides J, Perrig WJ (2008) Improving fluid intelligence with training on working memory. Proc Natl Acad Sci U S A 105(19):6829–6833. doi:10.1073/pnas.0801268105

Jak AJ, Aupperle RL, Rodgers CS, Lang AJ, Schiehser DM, Norman SB, Twamley EW (2015) Evaluation of a hybrid treatment for veterans with comorbid traumatic brain injury and posttraumatic stress disorder: study protocol for a randomized controlled trial. Contemp Clin Trials 45:210–216

Jenkins MA, Langlais PJ, Delis DA, Cohen RA (2000) Attentional dysfunction associated with posttraumatic stress disorder among rape survivors. Clin Neuropsychol 14(1):7–12. doi:10.1076/1385-4046(200002)14:1;1-8;FT007

Joormann J, Yoon K, Siemer M (2010) Cognition and emotion regulation. In: Kring A, Sloan D (eds) Emotion regulation and psychopathology: a transdiagnostic approach to etiology and treatment. Guilford Press, New York, pp 174–203

Kane MJ, Engle RW (2002) The role of prefrontal cortex in working-memory capacity, executive attention, and general fluid intelligence: an individual-differences perspective. Psychon Bull Rev 9(4):637–671. doi:10.3758/Bf03196323

Kent M, Davis MC, Stark SL, Stewart LA (2011) A resilience-oriented treatment for posttraumatic stress disorder: results of a preliminary randomized clinical trial. J Trauma Stress 24(5):591–595

Keshavan MS, Vinogradov S, Rumsey J, Sherrill J, Wagner A (2014) Cognitive training in mental disorders: update and future directions. Am J Psychiatry 171(5):510–522. doi:10.1176/appi.ajp.2013.13081075

Kitayama N, Vaccarino V, Kutner M, Weiss P, Bremner JD (2005) Magnetic resonance imaging (MRI) measurement of hippocampal volume in posttraumatic stress disorder: a meta-analysis. J Affect Disord 88(1):79–86

Klingberg T, Fernell E, Olesen PJ, Johnson M, Gustafsson P, Dahlstrom K, et al (2005) Computerized training of working memory in children with ADHD – a randomized, controlled trial. J Am Acad Child Adolesc Psychiatry 44(2):177–186. doi:10.1097/00004583-200502000-00010

Koenen KC, Driver KL, Oscar-Berman M, Wolfe J, Folsom S, Huang MT, Schlesinger L (2001) Measures of prefrontal system dysfunction in posttraumatic stress disorder. Brain Cogn 45:64–78. doi:10.1006/brcg.2000.1256

Kuckertz JM, Amir N, Boffa JW, Warren CK, Rindt SE, Norman S, et al (2014) The effectiveness of an attention bias modification program as an adjunctive treatment for post-traumatic stress disorder. Behav Res Ther 63:25–35. doi:10.1016/j.brat.2014.09.002

Lagarde G, Doyon J, Brunet A (2010) Memory and executive dysfunctions associated with acute posttraumatic stress disorder. Psychiatry Res 177(1–2):144–149. doi:10.1016/j.psychres.2009.02.002

Lee HJ, Kim JW, Yim SV, Kim MJ, Kim SA, Kim YJ et al (2001) Fluoxetine enhances cell proliferation and prevents apoptosis in dentate gyrus of maternally separated rats. Mol Psychiatry 6(6):610, 725–728. doi:10.1038/sj.mp.4000947

Leskin LP, White PM (2007) Attentional networks reveal executive function deficits in posttraumatic stress disorder. Neuropsychology 21(3):275–284. doi:10.1037/0894-4105.21.3.275

Lindauer RJ, Olff M, van Meijel EP, Carlier IV, Gersons BP (2006) Cortisol, learning, memory, and attention in relation to smaller hippocampal volume in police officers with posttraumatic stress disorder. Biol Psychiatry 59(2):171–177. doi:10.1016/j.biopsych.2005.06.033

Lindauer RJ, Vlieger EJ, Jalink M, Olff M, Carlier IV, Majoie CB, Den Heeten GJ, Gersons BP (2005) Effects of psychotherapy on hippocampal volume in out-patients with post-traumatic stress disorder: a MRI investigation. Psychol Med 35(10):1421–1431

Levy-Gigi E, Szabó C, Kelemen O, Kéri S (2013) Association among clinical response, hippocampal volume, and FKBP5 gene expression in individuals with posttraumatic stress disorder receiving cognitive behavioral therapy. Biol Psychiatry 74(11):793–800

Marazziti D, Consoli G, Picchetti M, Carlini M, Faravelli L (2010) Cognitive impairment in major depression. Eur J Pharmacol 626:83–86

Mathews A, MacLeod C (2002) Induced processing biases have causal effects on anxiety. Cognit Emot 16(3):331–354. doi:10.1080/02699930143000518

McEwen BS, Sapolsky RM (1995) Stress and cognitive function. Curr Opin Neurobiol 5 (2):205–216

McNally RJ (2007) Mechanisms of exposure therapy: how neuroscience can improve psychological treatments for anxiety disorders. Clin Psychol Rev 27(6):750–759. doi:10.1016/j.cpr.2007.01.003

Melby-Lervag M, Hulme C (2013) Is working memory training effective? A meta-analytic review. Dev Psychol 49(2):270–291. doi:10.1037/a0028228

Merzenich MM, Van Vleet TM, Nahum M (2014) Brain plasticity-based therapeutics. Front Hum Neurosci 8:385. doi:10.3389/fnhum.2014.00385

Moores KA, Clark CR, McFarlane AC, Brown GC, Puce A, Taylor DJ (2008) Abnormal recruitment of working memory updating networks during maintenance of trauma-neutral information in post-traumatic stress disorder. Psychiatry Res 163(2):156–170. doi:10.1016/j.pscychresns.2007.08.011

Morey RA, Dolcos F, Petty CM, Cooper DA, Hayes JP, LaBar KS, McCarthy G (2009) The role of trauma-related distractors on neural systems for working memory and emotion processing in posttraumatic stress disorder. J Psychiatr Res 43(8):809–817. doi:10.1016/j.jpsychires.2008.10.014

Nassif Y, Wells A (2014) Attention training reduces intrusive thoughts cued by a narrative of stressful life events: a controlled study. J Clin Psychol 70(6):510–517. doi:10.1002/jclp.22047

Neylan TC, Lenoci M, Rothlind J, Metzler TJ, Schuff N, Du AT, et al (2004) Attention, learning, and memory in posttraumatic stress disorder. J Trauma Stress 17(1):41–46

Nijdam MJ, de Vries G, Gersons BPR, Olff M (2015) Response to psychotherapy for posttraumatic stress disorder: the role of pretreatment verbal memory performance. J Clin Psychiatry 76(8): e1023–e1028

Olesen PJ, Westerberg H, Klingberg T (2004) Increased prefrontal and parietal activity after training of working memory. Nat Neurosci 7(1):75–79. doi:10.1038/nn1165

Parslow RA, Jorm AF (2007) Pretrauma and postrauma neurocognitive functioning and PTSD symptoms in a community sample of young adults. Am J Psychiatry 164:509–515

Peres JF, Foerster B, Santana LG, Fereira MD, Nasello AG, Savoia M, et al (2011) Police officers under attack: resilience implications of an fMRI study. J Psychiatr Res 45(6):727–734. doi:10.1016/j.jpsychires.2010.11.004

Pineles SL, Shipherd JC, Welch LP, Yovel I (2007) The role of attentional biases in PTSD: is it interference or facilitation? Behav Res Ther 45(8):1903–1913. doi:10.1016/j.brat.2006.08.021

Pineles SL, Shipherd JC, Mostoufi SM, Abramovitz SM, Yovel I (2009) Attentional biases in PTSD: more evidence for interference. Behav Res Ther 47(12):1050–1057. doi:10.1016/j.brat.2009.08.001

Polak AR, Witteveen AB, Reitsma JB, Olff M (2012) The role of executive function in posttraumatic stress disorder: a systematic review. J Affect Disord 141(1):11–21. doi:10.1016/j.jad.2012.01.001

Puetz TW, Youngstedt SD, Herring MP (2015) Effects of pharmacotherapy on combat-related PTSD, anxiety, and depression: a systematic review and meta-regression analysis. PLoS One 10(5):e0126529

Ravindran LN, Stein MB (2009) Pharmacotherapy of PTSD: premises, principles, and priorities. Brain Res 1293:24–39. doi:10.1016/j.brainres.2009.03.037

Resick PA, Galovski TE, O'Brien Uhlmansiek M, Scher CD, Clum GA, et al (2008) A randomized clinical trial to dismantle components of cognitive processing therapy for posttraumatic stress disorder in female victims of interpersonal violence. J Consult Clin Psychol 76(2):243–258

Ridderinkhof KR, van den Wildenberg WP, Segalowitz SJ, Carter CS (2004) Neurocognitive mechanisms of cognitive control: the role of prefrontal cortex in action selection, response inhibition, performance monitoring, and reward-based learning. Brain Cogn 56(2):129–140. doi:10.1016/j.bandc.2004.09.016

Riemann BC, Kuckertz JM, Rozenman M, Weersing VR, Amir N (2013) Augmentation of youth cognitive behavioral and pharmacological interventions with attention modification: a preliminary investigation. Depress Anxiety 30(9):822–828. doi:10.1002/da.22127

van Rooij SJ, Geuze E, Kennis M, Rademaker AR, Vink M (2015a) Neural correlates of inhibition and contextual cue processing related to treatment response in PTSD. Neuropsychopharmacology 40(3):667–675. doi:10.1038/npp.2014.220

van Rooij SJ, Kennis M, Sjouwerman R, van den Heuvel MP, Kahn RS, Geuze E (2015b) Smaller hippocampal volume as a vulnerability factor for the persistence of post-traumatic stress disorder. Psychol Med 45:2737–2746

van Rooij SJ, Kennis M, Vink M, Geuze E (2016) Predicting treatment outcome in PTSD: a longitudinal functional MRI study on trauma-unrelated emotional processing. Neuropsychopharmacology 41(4):1156–1165. doi:10.1038/npp.2015.257

Roy MJ, Francis J, Friedlander J, Banks-Williams L, Lande RG, Taylor P, et al (2010) Improvement in cerebral function with treatment of posttraumatic stress disorder. Ann N Y Acad Sci 1208:142–149. doi:10.1111/j.1749-6632.2010.05689.x

Sailer U, Robinson S, Fischmeister FP, Konig D, Oppenauer C, Lueger-Schuster B, et al (2008) Altered reward processing in the nucleus accumbens and mesial prefrontal cortex of patients with posttraumatic stress disorder. Neuropsychologia 46(11):2836–2844. doi:10.1016/j.neuropsychologia.2008.05.022

Salminen T, Strobach T, Schubert T (2012) On the impacts of working memory training on executive functioning. Front Hum Neurosci 6:166. doi:10.3389/Fnhum.2012.00166

Samuelson KW (2011) Post-traumatic stress disorder and declarative memory functioning: a review. Dialogues Clin Neurosci 13(3):346–351

Samuelson KW, Neylan TC, Metzler TJ, Lenoci M, Rothlind J, Henn-Haase C, et al (2006) Neuropsychological functioning in posttraumatic stress disorder and alcohol abuse. Neuropsychology 20(6):716–726. doi:10.1037/0894-4105.20.6.716

Santarelli L, Saxe M, Gross C, Surget A, Battaglia F, Dulawa S, et al (2003) Requirement of hippocampal neurogenesis for the behavioral effects of antidepressants. Science 301 (5634):805–809. doi:10.1126/science.1083328

Schoorl M, Putman P, Van Der Does W (2013) Attentional bias modification in posttraumatic stress disorder: a randomized controlled trial. Psychother Psychosom 82(2):99–105. doi:10.1159/000341920

Schweizer S, Grahn J, Hampshire A, Mobbs D, Dalgleish T (2013) Training the emotional brain: improving affective control through emotional working memory training. J Neurosci 33 (12):5301–5311. doi:10.1523/jneurosci.2593-12.2013

Scott JC, Matt GE, Wrocklage KM, Crnich C, Jordan J, Southwick SM, et al (2015) A quantitative meta-analysis of neurocognitive functioning in posttraumatic stress disorder. Psychol Bull 141 (1):105–140. doi:10.1037/a0038039

Seedat S, Warwick J, van Heerden B, Hugo C, Zungu-Dirwayi N, Van Kradenburg J, Stein DJ (2004) Single photon emission computed tomography in posttraumatic stress disorder before and after treatment with a selective serotonin reuptake inhibitor. J Affect Disord 80(1):45–53. doi:10.1016/S0165-0327(03)00047-8

Shin LM, Whalen PJ, Pitman RK, Bush G, Macklin ML, Lasko NB, et al. (2001) An fMRI study of anterior cingulate function in posttraumatic stress disorder. Biol Psychiatry 50(12):932–942

Shin LM, Shin PS, Heckers S, Krangel TS, Macklin ML, Orr SP, Lasko N, Segal E, Makris N, Richert K, Levering J, Schacter DL, Alpert NM, Fischman AJ, Pitman RK, Rauch SL (2004) Hippocampal function in posttraumatic stress disorder. Hippocampus 14(3):292–300

Shucard JL, McCabe DC, Szymanski H (2008) An event-related potential study of attention deficits in posttraumatic stress disorder during auditory and visual Go/NoGo continuous performance tasks. Biol Psychol 79(2):223–233. doi:10.1016/j.biopsycho.2008.05.005

Siegle GJ, Ghinassi F, Thase ME (2007) Neurobehavioral therapies in the 21st century: summary of an emerging field and an extended example of cognitive control training for depression. Cogn Ther Res 31:235–262

Simmons AN, Matthews SC (2012) Neural circuitry of PTSD with or without mild traumatic brain injury: a meta-analysis. Neuropharmacology 62:598–606

Simmons AN, Norman SB, Spadoni AD, Strigo IA (2013) Neurosubstrates of remission following prolonged exposure therapy in veterans with posttraumatic stress disorder. Psychother Psychosom 82(6):382–389

Squire LR, Zola-Morgan S (1991) The medial temporal lobe memory system. Science 253 (5026):1380–1386

Steckler T, Risbrough V (2012) Pharmacological treatment of PTSD – established and new approaches. Neuropharmacology 62(2):617–627. doi:10.1016/j.neuropharm.2011.06.012

Stein MB, Koverola C, Hanna C, Torchia MG, McClarty B (1997) Hippocampal volume in women victimized by childhood sexual abuse. Psychol Med 27(4):951–959

Stillman, A. N., & Aupperle, R. L. (2015). Neuroanatomical correlates of PTSD: risk, resiliency, and sequelae.

Takeuchi H, Taki Y, Nouchi R, Hashizume H, Sekiguchi A, Kotozaki Y, et al (2014) Working memory training improves emotional states of healthy individuals. Front Syst Neurosci 8:200. doi:10.3389/fnsys.2014.00200

Thomaes K, Dorrepaal E, Draijer N, de Ruiter MB, Elzinga BM, van Balkom AJ, et al (2012) Treatment effects on insular and anterior cingulate cortex activation during classic and emotional Stroop interference in child abuse-related complex post-traumatic stress disorder. Psychol Med 42(11):2337–2349. doi:10.1017/s0033291712000499

Thomaes K, Dorrepaal E, Draijer N, Jansma EP, Veltman DJ, van Balkom AJ (2014) Can pharmacological and psychological treatment change brain structure and function in PTSD? A systematic review. J Psychiatr Res 50:1–15. doi:10.1016/j.jpsychires.2013.11.002

Tischler L, Brand SR, Stavitsky K, Labinsky E, Newmark R, Grossman R, et al (2006) The relationship between hippocampal volume and declarative memory in a population of combat veterans with and without PTSD. Ann N Y Acad Sci 1071:405–409. doi:10.1196/annals.1364.031

Tsourtos G, Thompson JC, Stough C (2002) Evidence of an early information processing speed deficit in unipolar major depression. Psychol Med 32:259–265

Tucker P, Zaninelli R, Yehuda R, Ruggiero L, Dillingham K, Pitts CD (2001) Paroxetine in the treatment of chronic posttraumatic stress disorder: results of a placebo-controlled, flexible-dosage trial. J Clin Psychiatry 62(11):860–868

Tulving E, Markowitsch HJ (1998) Episodic and declarative memory: role of the hippocampus. Hippocampus 8(3):198–204. doi:10.1002/(SICI)1098-1063(1998)8:3<198::AID-HIPO2>3.0.CO;2-G

Twamley EW, Allard CB, Thorp SR, Norman SB, Hami Cissell S, Hughes Berardi K, et al (2009) Cognitive impairment and functioning in PTSD related to intimate partner violence. J Int Neuropsychol Soc 15(6):879–887. doi:10.1017/S135561770999049X

Usui N, Haji T, Maruyama M, Katsuyama N, Uchida S, Hozawa A, et al (2009) Cortical areas related to performance of WAIS digit symbol test: a functional imaging study. Neurosci Lett 463(1):1–5. doi:10.1016/j.neulet.2009.07.048

Vasterling JJ, Brailey K, Constans JI, Sutker PB (1998) Attention and memory dysfunction in posttraumatic stress disorder. Neuropsychology 12(1):125–133

Verfaellie M, Amick MM, Vasterling JJ (2012) Effects of TBI-associated neurocognitive alterations on the course of PTSD. In: Vasterling JJ, Bryant RA, Keane TM (eds) PTSD and mild traumatic brain injury. Guilford Press, New York, pp 82–102

Vermetten E, Vythilingam M, Southwick SM, Charney DS, Bremner JD (2003) Long-term treatment with paroxetine increases verbal declarative memory and hippocampal volume in posttraumatic stress disorder. Biol Psychiatry 54(7):693–702. S0006322303006346 [pii]

Verwoerd J, Wessel I, de Jong PJ (2009) Individual differences in experiencing intrusive memories: the role of the ability to resist proactive interference. J Behav Ther Exp Psychiatry 40 (2):189–201. doi:10.1016/j.jbtep.2008.08.002

Vinogradov S, Fisher M, de Villers-Sidani E (2012) Cognitive training for impaired neural systems in neuropsychiatric illness. Neuropsychopharmacology 37:43–76

Vythilingam M, Luckenbaugh DA, Lam T, Morgan CA, Lipschitz D, Charney DS, et al (2005) Smaller head of the hippocampus in Gulf War-related posttraumatic stress disorder. Psychiatry Res 139(2):89–99. doi:10.1016/j.pscychresns.2005.04.003

Walter KH, Palmieri PA, Gunstad J (2010) More than symptom reduction: changes in executive functioning over the course of PTSD treatment. J Trauma Stress 23(2):292–295

Watanabe Y, Gould E, Cameron HA, Daniels DC, McEwen BS (1992) Phenytoin prevents stress- and corticosterone-induced atrophy of CA3 pyramidal neurons. Hippocampus 2(4):431–435. doi:10.1002/hipo.450020410

Wild J, Gur RC (2008) Verbal memory and treatment response in post-traumatic stress disorder. Br J Psychiatry 193:254–255

Wilson EJ, MacLeod C, Mathews A, Rutherford EM (2006) The causal role of interpretive bias in anxiety reactivity. J Abnorm Psychol 115(1):103–111. doi:10.1037/0021-843X.115.1.103

Woodward SH, Kaloupek DG, Grande LJ, Stegman WK, Kutter CJ, Leskin L, et al (2009) Hippocampal volume and declarative memory function in combat-related PTSD. J Int Neuropsychol Soc 15(6):830–839. doi:10.1017/S1355617709990476

Woon FL, Sood S, Hedges DW (2010) Hippocampal volume deficits associated with exposure to psychological trauma and posttraumatic stress disorder in adults: a meta-analysis. Prog Neuropsychopharmacol Biol Psychiatry 34(7):1181–1188. doi:10.1016/j.pnpbp.2010.06.016

Wu J, Ge Y, Shi Z, Duan X, Wang L, Sun X, Zhang K (2010) Response inhibition in adolescent earthquake survivors with and without posttraumatic stress disorder: a combined behavioral and ERP study. Neurosci Lett 486(3):117–121. doi:10.1016/j.neulet.2010.07.040

Yehuda R, LeDoux J (2007) Response variation following trauma: a translational neuroscience approach to understanding PTSD. Neuron 56:19–32

Zantwoord JB, Diehle J, Lindauer RJL (2013) Using neurobiological measures to predict and assess treatment outcome of psychotherapy in posttraumatic stress disorder: systematic review. Psychother Psychosom 82:142–151

Neurobiological Programming of Early Life Stress: Functional Development of Amygdala-Prefrontal Circuitry and Vulnerability for Stress-Related Psychopathology

Michelle R. VanTieghem and Nim Tottenham

Abstract Early adverse experiences are associated with heighted vulnerability for stress-related psychopathology across the lifespan. While extensive work has investigated the effects of early adversity on neurobiology in adulthood, developmental approaches can provide further insight on the neurobiological mechanisms that link early experiences and long-term mental health outcomes. In the current review, we discuss the role of emotion regulation circuitry implicated in stress-related psychopathology from a developmental and transdiagnostic perspective. We highlight converging evidence suggesting that multiple forms of early adverse experiences impact the functional development of amygdala-prefrontal circuitry. Next, we discuss how adversity-induced alterations in amygdala-prefrontal development are associated with symptoms of emotion dysregulation and psychopathology. Additionally, we discuss potential mechanisms through which protective factors may buffer the effects of early adversity on amygdala-prefrontal development to confer more adaptive long-term outcomes. Finally, we consider limitations of the existing literature and make suggestions for future longitudinal and translational research that can better elucidate the mechanisms linking early adversity, neurobiology, and emotional phenotypes. Together, these findings may provide further insight into the neuro-developmental mechanisms underlying the emergence of adversity-related emotional disorders and facilitate the development of targeted interventions that can ameliorate risk for psychopathology in youth exposed to early life stress.

Keywords Amygdala • Child/adolescent development • Early life stress • Prefrontal cortex • Psychopathology

M.R. VanTieghem (✉) and N. Tottenham
Department of Psychology, Columbia University, 406 Schermerhorn Hall, 1990 Amsterdam Ave, MC 5501, New York, NY 10027, USA
e-mail: mrv2115@columbia.edu

Curr Topics Behav Neurosci (2018) 38: 117–136
DOI 10.1007/7854_2016_42
Published Online: 24 April 2018

Contents

1 Introduction

Early life stress (ELS) is associated with higher incidence of mental health problems across the lifespan, accounting for 29% of health disorders worldwide [1–3]. Multiple forms of postnatal adversities confer vulnerability for stress-related psychopathology, including maltreatment, neglect, parental stress or psychopathology, trauma, family conflict, poverty-related stressors, and institutionalized care [3–6]. Although these adverse exposures often occur during infancy and/or childhood, emotional difficulties often continue to persist throughout development, with three quarters of stress-related mental health diagnoses made by the age of 24 [2, 7]. Given the robust epidemiological evidence linking ELS with long-lasting emotional difficulties, it is important to identify the neurobiological mechanisms through which early experiences "get under the skin" to increase risk for psychopathology.

Developmental mechanisms of adaptation play an important role in understanding the long-term links between ELS and mental health outcomes in adulthood. According to the Dynamic Systems Theory, development is experience-driven, emerging via interactions with the environment that unfold over time [8]. In the context of ELS, several developmental theories (Barker's hypothesis, Developmental Origins Theory, Adaptive Recalibration Model, Experiential Canalization) emphasize the role of adaptation in response to adversity, such that the organism develops in order to promote survival in the expected environment [9–12]. Similarly, the Stress-Acceleration Hypothesis posits that neurobiological changes in response to early adverse experiences are adaptive in the short-term, but may have long-term trade-offs in the functional integrity of neuro-affective circuitry and heighten vulnerability for maladaptive mental health outcomes later in life [13].

In line with this developmental perspective, the current review will discuss how early adverse experiences influence neuro-affective development to confer risk for stress-related emotion dysregulation. We will delineate how the amygdala-prefrontal circuit, implicated in threat-reactivity and emotion regulation, appears to be particularly sensitive to the effects of stress during early life. The current paper focuses on the functional development of amygdala-prefrontal circuitry, as stress-induced changes in structural development have been reviewed elsewhere [14]. Specifically, we will highlight converging evidence suggesting that multiple forms of ELS are characterized by similar functional phenotypes of neuro-affective circuitry across development: (1) heightened amygdala reactivity and (2) altered amygdala-prefrontal connectivity. Next, we will discuss how developmental changes in amygdala-prefrontal circuitry predict individual differences in symptoms of stress-related psychopathology. Finally, we will discuss potential protective factors that may buffer the effects of stress on neuro-affective development to confer more resilient long-term trajectories. Given that ELS increases risk across several, often comorbid psychiatric disorders [3, 15], this paper will focus on the neurobiology of emotion dysregulation from a transdiagnostic and dimensional perspective.

2 Target Neural Circuitry: Amygdala and Prefrontal Cortex

2.1 The Role of Amygdala-Prefrontal Circuitry in Emotion Regulation

Robust translational and clinical research has linked amygdala-prefrontal circuitry with symptoms of emotion dysregulation [16]. In adults, regulatory connections between amygdala and prefrontal cortex are critically implicated in learning and responding to emotional cues in the environment [17, 18]. The amygdala is involved in detecting salient information in the environment to initiate physiological responses to potential threat [17]. Top-down recruitment of medial prefrontal regions regulates amygdala reactivity to facilitate extinction learning [19, 20] whereas dorsolateral prefrontal regions implicated in more effortful processes, like cognitive reappraisal, modulate amygdala reactivity during emotion regulation [21]. Functional alterations of amygdala reactivity and amygdala-prefrontal connectivity have been identified in patients with internalizing and stress-related disorders, including anxiety, depression, and PTSD [22–24]. In the Research Domain Criteria (RDoC) recently outlined by the National Institutes of Mental Health [25], amygdala-prefrontal circuitry has been implicated in the psychological constructs of fear and sustained threat, highlighting its role in the neurobiological underpinnings of transdiagnostic dimensions of threat-reactivity and emotion regulation [26].

In humans, amygdala-prefrontal circuitry undergoes protracted development, with age-related changes observed across childhood, adolescence, and young adulthood. Several studies have observed heightened amygdala reactivity in response to emotionally salient cues in younger ages [27–31]. As amygdala reactivity declines with increasing age [27–32], the functional integrity of amygdala-mPFC circuitry continues to strengthen into young adulthood [33]. Importantly, age-related changes in amygdala reactivity and/or connectivity with the prefrontal cortex during cognitive reappraisal tasks correspond to the maturation of emotion regulation abilities across development [34–36]. Pediatric disorders of anxiety, depression, and PTSD are characterized by heightened amygdala reactivity and atypical amygdala-prefrontal connectivity during emotion processing tasks [37–42]. Moreover, altered patterns of age-related changes in amygdala-prefrontal connectivity have been shown in a cross-sectional sample of anxious youth and young adults [43] suggesting that deviations from the normative trajectory of amygdala-prefrontal development are associated with symptoms of emotional dysregulation in clinical samples.

2.2 Plasticity of Amygdala-PFC Circuitry in Early Life

Converging evidence across species suggests that amygdala-prefrontal circuitry is highly sensitive to environmental inputs, particularly during early life [44]. The amygdala is heavily innervated by glucocorticoid receptors [45], with the highest peak in corticotrophin releasing hormone (CRH) receptor density during the first few postnatal weeks [45]. Stress exposure during early life results in increased mRNA expression of CRH in the amygdala in rodents [46]. Importantly, the functional development of the amygdala is tightly linked to hypothalamic–pituitary–adrenal (HPA) axis function, such that increases in cortisol are associated with the developmental onset of amygdala reactivity and fear learning in rodents [47].

Several animal models of ELS (e.g., abusive maternal care, maternal separation, chronic restraint stress, and odor-shock conditioning) have shown that early adverse environments have enduring effects on amygdala structure and function [48–50]. Moreover, regulatory connections between amygdala and prefrontal cortex are highly susceptible to environmental influences during early life in rodent models. For example, chronic stress exposure during the juvenile stage causes dendritic atrophy in the prefrontal cortex (PFC; [48]) and alters the emergence of amygdala projections to the PFC, resulting in long-term imbalance of amygdala-prefrontal circuit function in adult rats [51]. In light of these findings, amygdala-prefrontal development may play an important role in the neurobiological etiology of emotion dysregulation in humans following ELS.

3 Effects of ELS on Amygdala-PFC Circuitry in Humans

When examining the effects of ELS on neurobiological development in humans, there are two important considerations that delineate the state of current research. First, aside from notable exceptions in which there is known timing and duration of adverse exposures (i.e., adoption from institutionalized care), many forms of ELS are chronic in nature, making it difficult to delineate the effects of stressors during specific time points across development (reviewed in [14]). Given cross-species evidence suggesting that amygdala development is most sensitive to environmental input early in life [44], the current review focuses on adverse experiences that occur during infancy and/or childhood. Second, recent theoretical frameworks have suggested that certain dimensions of adverse experiences (e.g., threat vs. deprivation) may have differential effects on neurobiological development [52]. Although early adversities are often complex exposures comprised of multiple dimensions of experience (e.g., abuse and neglect; [53]), many forms of ELS are considered threatening to children's physical or emotional well-being [52]. In the current review, we focus on research examining threat-related alterations in neuro-affective development following exposure to ELS. Specifically, we present converging evidence suggesting that amygdala-prefrontal circuitry, implicated in threat-reactivity and emotion regulation, is a common neurobiological target impacted by multiple forms of early adverse experiences.

3.1 Effects of ELS on Amygdala Reactivity

In adults, heightened amygdala reactivity to emotional cues has been identified across several domains of ELS reported retrospectively, including maltreatment [54, 55] emotional neglect [56, 57], and lower perceived social status [58]. Recent prospective longitudinal studies have corroborated these effects, showing that cumulative childhood stressors associated with low socioeconomic status have lasting effects on amygdala function in adulthood [59, 60]. For example, childhood poverty has been associated with increased amygdala reactivity to negative relative to positive emotional cues in adulthood [60]. In the same prospective cohort, cumulative risk exposure associated with childhood poverty was directly related to higher amygdala reactivity to neutral facial expressions, suggesting that stress-related increases in amygdala reactivity may not be specific to threat-related stimuli, also extends to neutral socio-emotional cues [59].

In accordance with studies in adult ELS samples, children and adolescents with a history of early adversity also show enhanced amygdala reactivity to emotional stimuli. Previously institutionalized (PI) youth with a history of institutional care exhibit heightened amygdala reactivity to threat-related facial expressions across childhood and adolescence [61–63]. Similarly, increased amygdala response to negative emotional stimuli has been identified in children and adolescents with prior exposure to maltreatment [64, 65], traumatic events [66], and family violence [67]. Moreover, greater levels of stressful life events have been associated with longitudinal increases in threat-related amygdala reactivity during adolescence, suggesting that heightened amygdala reactivity may represent a neural marker of previous stress exposure [68]. Importantly, McCrory et al. [64] found that children with earlier onset of maltreatment exposure showed higher levels of amygdala reactivity to pre-attentively presented emotional stimuli, suggesting a relationship between the timing of stress exposure onset and degree of amygdala reactivity. However, further research is needed to delineate whether stress-induced increases in amygdala reactivity are primarily driven by the developmental timing (i.e., age of onset) or the duration (i.e., chronic versus acute) of adverse experiences.

3.2 Effects of ELS on Amygdala-PFC Connectivity

In addition to heightened amygdala-reactivity, ELS has also been characterized by altered functional connectivity of the amygdala with prefrontal regions. Although the valence (i.e., positive or negative) and regional specificity (i.e., dorsolateral or medial regions of PFC) of amygdala-prefrontal connectivity findings are task-dependent and often vary across studies, ELS has been consistently associated with atypical connectivity patterns relative to non-stressed control groups. In a prospective study, young adults with a history of childhood maltreatment showed atypical connectivity between the amygdala and inferior frontal gyrus when processing threat-related emotional stimuli [69]. Childhood poverty has also been associated with alterations of amygdala-prefrontal connectivity in adulthood, such that lower family income during childhood is associated with reduced amygdala-ventrolateral PFC (vlPFC) connectivity during cognitive reappraisal [70]. Importantly, cumulative stress exposure mediated the effects of family income on vlPFC recruitment during reappraisal, suggesting that associations between childhood poverty and prefrontal dysregulation are driven by effects of chronic stress [70]. Together, these findings suggest that heightened emotional reactivity following ELS may emerge from impaired top-down prefrontal regulation of amygdala reactivity in response to emotional cues.

Given that ELS is associated with atypical amygdala-prefrontal function in adulthood, recent research has examined how these adversity-induced changes emerge across development. In a cross-sectional study from early childhood to late adolescence, PI youth showed an atypical trajectory of age-related changes in threat-related amygdala-mPFC connectivity relative to comparison youth, such that PI youth exhibited more mature (i.e., adult-like) connectivity at younger ages [61]. Youth with trauma exposure also show atypical amygdala-prefrontal function in response to emotional distractors, with weaker negative connectivity between the amygdala and pregenual ACC (pgACC) relative to comparison youth [66]. Moreover, the strength of amygdala-pgACC connectivity predicted performance on the emotional conflict task, suggesting that impaired regulation of emotional distractors in trauma-exposed youth may be related to altered circuit function [66]. Similarly, PTSD youth exhibit weaker amygdala-dACC connectivity and atypical age-related changes in amygdala-mPFC connectivity in response to threat-related stimuli [42]. Importantly, the youth diagnosed with PTSD in this sample were exposed to a wide range of early adverse experiences (e.g., trauma, abuse, neglect; [42]), suggesting evidence of equifinality with regard to neuro-affective phenotypes following exposure to different forms of ELS [71].

In addition to changes in task-elicited functional connectivity, ELS has also been associated with weaker resting-state amygdala-prefrontal connectivity across developmental stages, suggesting that early adversity has long-lasting impacts on the functional integrity of emotion regulation circuitry. In adults, self-reported history of childhood trauma is associated with weaker resting-state connectivity between amygdala and pregenual ACC (pgACC; [72]). Similarly, adolescents who experienced childhood maltreatment [73] and youth with history of trauma exposure [74] show weaker amygdala-subgenual anterior cingulate cortex (sgACC) connectivity at rest. In a younger cohort of children and young adolescents, higher levels of cumulative stress during childhood predicted weaker amygdala-ACC connectivity [75]. Importantly, ELS-induced changes in amygdala connectivity may be identifiable as early as infancy. At 6 months of age, family stress, as defined by high levels of interparental conflict, is associated with altered patterns of resting-state amygdala connectivity with posterior cingulate cortex, a regional hub of the default mode network [76]. Although further research is needed to delineate how early alterations in amygdala connectivity influence longitudinal neuro-affective development, these findings highlight the potential role of amygdala connectivity as a neurobiological marker for stress vulnerability as early as the first year of life [77].

4 Amygdala-PFC Circuitry and Individual Differences in Psychopathology Following ELS

In the previous section, we presented evidence suggesting that there is some degree of equifinality in neurobiological development following ELS [71], such that different types of early adverse experiences have converging effects on the development of emotion regulation circuitry, resulting in atypical amygdala-prefrontal circuit function. However, there is also evidence of multifinality, such that there is wide heterogeneity in long-term mental health outcomes following ELS [71]. For example, similar adverse experiences (e.g., institutional care) confer risk for multiple types of psychopathology across individuals [5, 15, 71]. In the context of developmental theory (Adaptive Calibration Model, Experiential Canalization, and Stress Acceleration Hypothesis), environmentally driven changes in neurobiology represent an ontogenetic response to adversity, and may confer adaptive or maladaptive behavioral outcomes in specific domains or contexts across development [10–13]. Given the heterogeneity in mental health outcomes associated with ELS, it is important to consider how individual trajectories of neuro-affective development predict risk or resilience following exposure to early adversity. The following discussion will review recent evidence linking adversity-induced changes in amygdala-prefrontal function with individual differences in psychopathology (i.e., anxiety, depression, PTSD).

4.1 Amygdala Reactivity and Psychopathology

Individual differences in amygdala reactivity predict dimensional measures of emotional functioning in both typically developing and stress-exposed youth. In typical children and adolescents, increased amygdala reactivity to sad facial expressions predicts level of concurrent internalizing symptoms [30] and depressive symptoms [78]. Youth with trauma exposure and post-traumatic stress symptoms have shown greater amygdala reactivity to emotional facial expressions relative to non-exposed youth [38] although there are mixed findings [42, 79]. A recent study examined the interaction of early trauma exposure and psychiatric status on amygdala reactivity to emotional stimuli during childhood [80]. Amygdala response varied as a function of both early trauma and concurrent levels of psychopathology, such that children with trauma exposure and current diagnosis of major depressive disorder exhibited the greatest levels of amygdala reactivity [80]. Moreover, recent evidence suggests that heightened amygdala reactivity predicts long-term increases in negative affect in both healthy and depressed preschool children [81]. Together, these studies suggest that amygdala reactivity may represent a neural marker for current and/or future levels of stress-related psychopathology during childhood and adolescence. However, further longitudinal studies are needed to delineate the specific effects of different types of stressors on amygdala reactivity phenotypes and long-term mental health outcomes.

4.2 Longitudinal Studies of Amygdala-PFC Connectivity and Psychopathology

Recent longitudinal findings also suggest that atypical amygdala-prefrontal connectivity may represent a neurobiological risk factor for the emergence of psychopathology following ELS. In adolescents with a history of childhood maltreatment, the strength of resting-state amygdala-sgACC connectivity mediated the relationship between maltreatment exposure and internalizing symptoms, such that weaker amygdala-sgACC connectivity conferred higher levels of anxiety and depressive symptoms [73]. In a recent study of cumulative childhood stress, Pagliaccio et al. [75] examined the relationship between resting-state amygdala-ACC connectivity and longitudinal assessments of internalizing psychopathology in children. Similar to Herringa et al. [73], weaker amygdala-ACC connectivity mediated the effect of stressful and traumatic life events on current symptoms of anxiety. Moreover, amygdala-prefrontal connectivity and concurrent symptom levels were both significant predictors of anxiety symptoms one year later, providing longitudinal evidence that stress-related changes in the functional integrity of amygdala-prefrontal circuitry confer vulnerability for future stress-related psychopathology [75].

Given that amygdala functional development is tightly linked to the HPA axis [82], cortisol reactivity may play an important role in the developmental cascade linking neuro-affective changes to long-term mental health outcomes following ELS. In a long-term prospective study, Burghy et al. [83] examined the effects of cumulative maternal stress on cortisol levels during childhood and resting-state amygdala-prefrontal connectivity in late adolescence. Greater levels of maternal stress during the first year of life were associated with heightened baseline cortisol levels during childhood, suggesting a dose-dependent response in the HPA axis response to ELS [83]. Although maternal stress did not directly predict amygdala-ventromedial PFC (vmPFC) connectivity, higher childhood baseline cortisol levels were associated with altered resting-state amygdala-vmPFC connectivity in adolescent females. Moreover, the strength of amygdala-vmPFC connectivity mediated the relationship between heightened cortisol and symptoms of depression and anxiety in adolescent females, albeit in different directions. Specifically, weaker amygdala-vmPFC connectivity predicted greater symptoms of anxiety, while stronger connectivity predicted greater symptoms of depression, suggesting that divergent trajectories of amygdala-prefrontal development following ELS confer risk for different forms of internalizing psychopathology. Overall, this study provides longitudinal evidence across multiple-levels of analysis that stress-related changes in HPA-axis regulation are associated with atypical amygdala-prefrontal connectivity and heightened vulnerability for internalizing psychopathology following ELS.

4.3 Cross-Sectional Studies of Amygdala-PFC Connectivity and Psychopathology

Cross-sectional studies have examined the effects of ELS on age-related changes in the developmental trajectory of amygdala-prefrontal circuit function. PI youth with a history of orphanage care showed atypical age-related changes in task-elicited amygdala-mPFC connectivity in response to fearful faces [61]. In typically developing youth, children showed more positive amygdala-mPFC connectivity, whereas adolescents showed negative amygdala-mPFC connectivity. However, PI children showed more mature (i.e., negative) connectivity at earlier ages relative to age-matched comparisons. In line with previous literature [83], cortisol levels mediated the relationship between ELS and amygdala-mPFC connectivity, supporting the role of the HPA axis in stress-related changes in neuro-affective development [61]. Importantly, amygdala-mPFC connectivity predicted current levels of psychopathology in the PI group, such that more mature connectivity conferred lower levels of anxiety. In the context of the Stress Acceleration Hypothesis [13], these findings suggest that earlier functional maturation of this circuitry may represent an adaptive response to previous stress exposure that reduces vulnerability for emotion dysregulation. However, given the cross-sectional nature of this study, further longitudinal research is needed to delineate whether these early stress-induced adaptations predict risk or resilience in the long-term.

Atypical amygdala-prefrontal functioning has also been identified in a cross-sectional study of PTSD youth with a history of early adversity [42]. Specifically, threat-related connectivity between the amygdala and dACC/dmPFC predicted severity of avoidant symptoms in PTSD youth. Moreover, they identified altered patterns of age-related connectivity phenotypes in the PTSD group, such that amygdala-vmPFC connectivity increased with age in typically developing youth, but decreased with age in PTSD youth [42]. Similar to PI children [61], children with PTSD exhibited a more mature pattern of amygdala-vmPFC connectivity, suggesting a developmental adaptation to compensate for heightened emotional reactivity following ELS. However, adolescents with PTSD showed less mature amygdala-vmPFC connectivity relative to age-matched controls. When considering the Stress Acceleration Hypothesis, these findings suggest that early maturation of this circuitry following ELS may be adaptive during childhood, but may result in reduced functional maturity of the circuit during adolescence. Although it is possible that exposure to traumatic events at earlier vs. later stages of development (i.e., childhood vs. adolescence) may differentially alter neuro-affective development, there were no reported effects of duration-since-exposure of adversity, nor the length of PTSD diagnosis on amygdala-vmPFC connectivity in this study [42]. Although the observed age-related changes in amygdala-vmPFC connectivity were not directly associated with PTSD symptoms, these findings highlight the importance of examining developmental trajectories when considering the effects of ELS on amygdala-prefrontal function and emotional disorders.

5 Protective Factors and Neuro-Affective Development Following ELS

Although ELS is associated with a higher incidence of stress-related psychopathology, many individuals exposed to early adversity do not develop clinical disorders [84]. Moreover, individuals with history of ELS may show difficulties in specific domains of socio-emotional functioning (e.g., anxiety), but show competence in other domains (e.g., social skills; [85]). A broad literature on resilience has identified factors at both the individual level (e.g., cognitive factors) and environmental level (e.g., family, community) that contribute to individual differences in mental health and well-being following ELS [85, 86]. Given the evidence of multifinality following ELS, it is important to identify how protective factors influence neurobiological development to reduce risk for stress-related psychopathology [87, 88]. For the purposes of the current review, we will focus on protective factors of the social environment that may ameliorate the effects of ELS on neuro-affective development via social buffering.

In behavioral studies, quality caregiving and family stability have been consistently shown to promote more resilient long-term outcomes following exposure to early adversity (reviewed in [89]). For example, in the Bucharest Early Intervention Project (BEIP), youth with stable foster-care placements following institutional care showed lower levels of internalizing symptoms during early adolescence relative to those who experienced disruptions in foster care [5]. Importantly, the two groups did not differ in the amount of time spent in institutional care or psychiatric history at age 4, suggesting that the observed difference in adolescent levels of psychopathology occurred as a function of caregiver stability, as opposed to earlier levels of trauma exposure or psychopathology [5]. Similarly, longitudinal studies of childhood maltreatment have shown that family level protective factors, such as caregiving stability [90], perceived parental care [91], and parental warmth [92] are associated with reduced risk for future psychopathology. Together, these findings suggest that positive and stable caregiving is associated with lower levels of emotional problems following multiple forms of early adverse experiences.

In light of strong evidence linking caregiver support and mental health outcomes, ample research has focused on identifying the neurobiological mechanisms underlying these social buffering effects [93, 94]. Evidence across species has shown that caregivers regulate emotional and neurobiological development (reviewed in [44]). In rodent pups, maternal presence has transient effects on cortisol release and amygdala function, such that maternal presence blocks stress reactivity and fear learning during the early stage of rat pup development [82]. Similar social buffering effects have been identified in humans; parent availability reduces cortisol response to social stress [95] and enhances emotion regulation abilities in children [96]. Moreover, parental stimuli can induce transient changes in functional connectivity of amygdala-mPFC circuitry, and these neurobiological changes predict the degree of parental buffering of children's emotion regulation abilities [96]. Together, these findings provide a plausible neurobiological mechanism through which caregivers can directly influence neuro-affective functioning during development.

Despite robust evidence of social buffering effects during typical neuro-affective development, no evidence to date has examined these effects on emotion regulation circuitry in youth with history of ELS. However, recent behavioral evidence suggests that interventions such as high-quality foster care may promote healthy emotional development in youth with a history of early institutional caregiving [97]. In the BEIP study, children with earlier placement into high-quality foster care showed greater attention bias to positive stimuli relative to children who experienced prolonged institutional rearing and typically developing children [97]. Importantly, positive attention bias in foster care youth predicted lower externalizing symptoms at age 8 and lower internalizing problems at age 12, suggesting that positivity-bias following early foster-care placement is associated with improved socio-emotional functioning in the long-term [97, 98]. However, a recent study of internationally adopted PI children and adolescents found that parental presence during a social stress task had no greater regulatory effect on cortisol reactivity relative to stranger presence, suggesting that social buffering mechanisms may exert differential effects on stress-related neurobiology depending on prior social experiences [99]. Moreover, animal models have shown that social buffering effects are diminished following atypical caregiving experiences (i.e., nursery rearing; reviewed in [94]). As such, further research is needed to investigate potential mechanisms through which protective factors such as positive parenting behaviors may be able to recalibrate the developmental trajectory of neuro-affective circuitry, and whether they exert effects over and above the effects of ELS to protect against future risk for stress-related psychopathology.

6 Limitations and Future Directions

While the current review focused on common phenotypes of neuro-affective circuitry associated with ELS, there are several directions of future research that will advance our understanding of how early adversity and protective factors influence neurobiological development and subsequent mental health outcomes. First, there is limited research examining the effects of timing and chronicity of stressors on neuro-affective functional development. Recent studies examining structural brain development have identified differential effects of adversity on amygdala volume depending on age of exposure [14, 100], and there is preliminary evidence linking the age of maltreatment exposure to degree of amygdala reactivity during childhood [64]. However, the complexity and chronicity of adverse experiences in the majority of human studies makes it challenging to differentiate whether stress-related effects on amygdala-prefrontal development occur as a function of the duration or timing of the stress exposure. Although international adoption studies can provide insight into the effects of ELS (e.g., institutional care) that occurs during a discrete developmental window, there may be limitations in its generalizability. These limitations highlight the important role of preclinical studies that use animal models of ELS. While there will always be the ethical limitations in studying stress exposure in humans, animal studies can experimentally manipulate age of onset, chronicity, and severity of ELS to allow for greater conclusions of causality. Moreover, translational research can provide more precise examination of the underlying neurobiological mechanisms associated with early adverse experiences that cannot be accessed through human neuroimaging studies.

Second, recent theoretical frameworks have emphasized importance of examining specific dimensions of early adverse experiences, such as threat and neglect, and how they influence different aspects of neurobiological development [52]. Although the current review focused specifically on threat-related alterations in amygdala-prefrontal circuitry, other dimensions of early experience may target different neural circuits (e.g., cortico-striatal circuitry) and neuro-cognitive domains (e.g., reward learning, executive functions; [52, 101]). Further longitudinal research is needed to compare how certain dimensions of adverse experiences differentially alter neurobiological circuitry to confer risk for specific domains of psychopathology.

In addition to protective factors of the social environment, genetic factors play an important role in moderating risk for emotional psychopathology following ELS [102, 103]. For example, genetic polymorphisms in neuroplasticity genes (e.g., BDNF) have been associated with ELS-related changes in neurobiological development and emotion regulation [104]. More recent work has shown that cumulative risk profiles across several HPA-related genetic alleles moderate the association between amygdala-prefrontal connectivity and anxiety symptoms in children exposed to stressful life events [75]. Importantly, genetic factors are often correlated with variability in the early environment in human studies, representing a significant challenge for researchers to differentiate the effects of genetics (e.g., parent psychopathology) from the effects of ELS (e.g., family conflict). This can include studies of adoption and foster-care cohorts, as children who display more emotional difficulties at a young age may experience greater disruptions in family placements [105]. Despite these potential confounds, not all individuals with genetic predispositions (e.g., family history of psychopathology) will develop an emotional disorder, and emerging research suggests that environmentally induced epigenetic modifications in gene expression also predict vulnerability for psychopathology [106]. For example, low socioeconomic status has been associated with longitudinal increases in promotor methylation of the serotonin transporter gene during adolescence [106]. Importantly, these epigenetic changes were associated with enhanced threat-related amygdala reactivity, which in turn predicted longitudinal increases in depressive symptoms in adolescents with a family history of depression [106]. These findings emphasize the critical role of early experiences on the developmental trajectories of neuro-affective circuitry and risk for stress-related psychopathology.

7 Conclusion

In summary, emerging research has begun to identify the developmental pathways through which early adverse experiences alter emotion regulation circuitry to increase risk for stress-related psychopathology. However, little is known regarding the differential effects of adversity on amygdala-prefrontal function during different developmental stages (i.e., infancy, childhood, adolescence) and different dimensions of exposure (i.e., maltreatment vs. neglect). Further research delineating the effects of timing and type of adversities, as well as their interplay with genetic and epigenetic factors, is needed to advance our understanding of the neurodevelopmental mechanisms implicated in vulnerability for psychopathology following ELS. This research will be facilitated by the incorporation of translational studies that directly compare human studies with animal models of ELS to provide further insight into the mechanisms underlying the link between early experiences and neuro-affective development. By applying a dimensional and developmental framework to future research, we can also begin to elucidate how and when protective factors can buffer the effects of ELS on neurobiological development to mitigate long-term risk for psychopathology. Ultimately, such research will be informative for developing policies and targeted interventions to improve mental health outcomes for individuals who have experienced early adversity.

Acknowledgements This work was supported by the NIMH under grant R01MH091864 (N. Tottenham, PI), an NSF Conference Grant conference grant BCS-1439258 (N. Tottenham, co-I), and the Dana Foundation (N. Tottenham, PI), and the NSF Graduate Research Fellowship Program under Grant No. DGE-16-44869 (M. R. VanTieghem). The content is solely the responsibility of the authors and does not necessarily represent the official views of the National Institute of Mental Health, the National Institutes of Health, the National Science Foundation, or the Dana Foundation.

References

1. Green JG, Mclaughlin KA, Berglund PA, Gruber MJ, Sampson NA, Zaslavsky AM, Kessler RC (2010) Childhood adversities and adult psychiatric disorders in the national comorbidity survey replication I: associations with first onset of DSM-IV disorders. Arch Gen Psychiatry 67(2):113–123, http://doi.org/10.1001/archgenpsychiatry.2009.187
2. Kessler RC, Berglund P, Demler O, Jin R, Merikangas KR, Walters EE (2005) Lifetime prevalence and age-of-onset distributions of DSM-IV disorders in the National Comorbidity Survey Replication. Arch Gen Psychiatry 62(6):593–602, http://doi.org/10.1001/archpsyc.62.6.593
3. Kessler RC, McLaughlin KA, Green JG, Gruber MJ, Sampson NA, Zaslavsky AM et al (2010) Childhood adversities and adult psychopathology in the WHO World Mental Health Surveys. Br J Psychiatry 197(5):378–385, http://doi.org/10.1192/bjp.bp.110.080499
4. Essex MJ, Shirtcliff EA, Burk LR, Ruttle PL, Klein MH, Slattery MJ et al (2011) Influence of early life stress on later hypothalamic-pituitary-adrenal axis functioning and its covariation with mental health symptoms: a study of the allostatic process from childhood into adolescence. Dev Psychopathol 23(4):1039–1058, http://doi.org/10.1017/S0954579411000484
5. Humphreys KL, Gleason MM, Drury SS, Miron D, Nelson CA, Fox NA, Zeanah CH (2015) Effects of institutional rearing and foster care on psychopathology at age 12 years in Romania: follow-up of an open, randomised controlled trial. Lancet Psychiatry 2 (7):625–634, http://doi.org/10.1016/S2215-0366(15)00095-4
6. Lansford JE, Dodge KA, Pettit GS, Bates JE, Crozier J, Kaplow J (2014) A 12-year prospective study of the long-term effects of early child physical maltreatment on psychological. Arch Pediatr Adolesc Med 156:824–830
7. Merikangas KR, He J, Burstein M, Swanson SA, Avenevoli S, Cui L et al (2010) Lifetime prevalence of mental disorders in U.S. adolescents: results from the National Comorbidity Survey Replication–Adolescent Supplement (NCS-A). J Am Acad Child Adolesc Psychiatry 49(10):980–989, http://doi.org/10.1016/j.jaac.2010.05.017

8. Smith LB, Thelen E (2003) Development as a dynamic system. Trends Cogn Sci 7 (8):343–348, http://doi.org/10.1016/S1364-6613(03)00156-6
9. Barker DJP (2007) The origins of the developmental origins theory. J Intern Med 261 (5):412–417, http://doi.org/10.1111/j.1365-2796.2007.01809.x
10. Blair C, Raver CC (2012) Child development in the context of adversity: experiential canalization of brain and behavior. Am Psychol 67(4):309–318, http://doi.org/10.1037/a0027493
11. Del Giudice M, Ellis BJ, Shirtcliff EA (2011) The adaptive calibration model of stress responsivity. Neurosci Biobehav Rev 35(7):1562–1592
12. Wadhwa P, Buss C, Entringer S, Swanson M (2010) Developmental origins of health and disease: brief history of the approach and current focus on epigenetic mechanisms. Semin Reprod Med 27(5):358–368, http://doi.org/10.1055/s-0029-1237424.Developmental
13. Callaghan BL, Tottenham N (2016) The stress acceleration hypothesis: effects of early-life adversity on emotion circuits and behavior. Curr Opin Behav Sci 7:76–81, http://doi.org/10.1016/j.cobeha.2015.11.018
14. Tottenham N, Sheridan MA (2009) A review of adversity, the amygdala and the hippocampus: a consideration of developmental timing. Front Hum Neurosci 3:68, http://doi.org/10.3389/neuro.09.068.2009
15. De Bellis MD, Broussard ER, Herring DJ, Wexler S, Moritz G, Benitez JG (2001) Psychiatric co-morbidity in caregivers and children involved in maltreatment: a pilot research study with policy implications. Child Abuse Negl 25(7):923–944, http://doi.org/10.1016/S0145-2134(01)00247-2
16. Hariri AR, Holmes A (2015) Finding translation in stress research. Nat Neurosci 18 (10):1347–1352, http://doi.org/10.1038/nn.4111
17. Davis M, Whalen PJ (2001) The amygdala: vigilance and emotion. Mol Psychiatry 6 (1):13–34. Retrieved from http://www.ncbi.nlm.nih.gov/pubmed/11244481
18. Kim JH, Hamlin AS, Richardson R (2009) Fear extinction across development: the involvement of the medial prefrontal cortex as assessed by temporary inactivation and immunohistochemistry. J Neurosci 29(35):10802–10808, http://doi.org/10.1523/JNEUROSCI.0596-09.2009
19. Milad MR, Rauch SL, Pitman RK, Quirk GJ (2006) Fear extinction in rats: implications for human brain imaging and anxiety disorders. Biol Psychol 73:61–71, http://doi.org/10.1016/j.biopsycho.2006.01.008
20. Phelps EA, LeDoux JE (2005) Contributions of the amygdala to emotion processing: from animal models to human behavior. Neuron 48(2):175–187, http://doi.org/10.1016/j.neuron.2005.09.025
21. Buhle JT, Silvers JA, Wager TD, Lopez R, Onyemekwu C, Kober H et al (2014) Cognitive reappraisal of emotion: a meta-analysis of human neuroimaging studies. Cereb Cortex 24:2981–2990, http://doi.org/10.1093/cercor/bht154
22. Etkin A, Klemenhagen KC, Dudman JT, Rogan MT, Hen R, Kandel ER, Hirsch J (2004) Individual differences in trait anxiety predict the response of the basolateral amygdala to unconsciously processed fearful faces. Neuron 44(6):1043–1055, http://doi.org/10.1016/j.neuron.2004.12.006
23. Koenigs M, Grafman J (2009) Posttraumatic stress disorder: the role of medial prefrontal cortex and amygdala. Neuroscientist 15(5):540–548
24. Murray EA, Wise SP, Drevets WC (2011) Localization of dysfunction in major depressive disorder: prefrontal cortex and amygdala. Biol Psychiatry 69(12):e43–e54, http://doi.org/10.1016/j.biopsych.2010.09.041
25. Morris SE, Cuthbert BN (2012) State of the art. Dialogues Clin Neurosci 14:29–37
26. Dillon DG, Rosso IM, Pechtel P, Killgore WDS, Rauch SL, Pizzagalli DA (2014) Peril and pleasure: an rdoc-inspired examination of threat responses and reward processing in anxiety and depression. Depress Anxiety 31(3):233–249, http://doi.org/10.1002/da.22202.Peril
27. Gee DG, Humphreys KL, Flannery J, Goff B, Telzer EH, Shapiro M et al (2013) A developmental shift from positive to negative connectivity in human amygdala-prefrontal circuitry. J Neurosci 33(10):4584–4593, http://doi.org/10.1523/JNEUROSCI.3446-12.2013

28. Guyer AE, Monk CS, McClure-Tone EB, Nelson EE, Roberson-Nay R, Adler AD et al (2008) A developmental examination of amygdala response to facial expressions. J Cogn Neurosci 20(9):1565–1582, http://doi.org/10.1162/jocn.2008.20114
29. Hwang S, White SF, Nolan ZT, Sinclair S, Blair RJR (2014) Neurodevelopmental changes in the responsiveness of systems involved in top down attention and emotional responding. Neuropsychologia 62:1–9, http://doi.org/10.1016/j.neuropsychologia.2014.08.003
30. Swartz JR, Carrasco M, Wiggins JL, Thomason ME, Monk CS (2014) Age-related changes in the structure and function of prefrontal cortex-amygdala circuitry in children and adolescents: a multi-modal imaging approach. Neuroimage 86:212–220, http://doi.org/10.1016/j.neuroimage.2013.08.018
31. Vink M, Derks JM, Hoogendam JM, Hillegers M, Kahn RS (2014) Functional differences in emotion processing during adolescence and early adulthood. Neuroimage 91:70–76, http://doi.org/10.1016/j.neuroimage.2014.01.035
32. Decety J, Michalska KJ, Kinzler KD (2012) The contribution of emotion and cognition to moral sensitivity: a neurodevelopmental study. Cereb Cortex 22(1):209–220, http://doi.org/10.1093/cercor/bhr111
33. Gabard-Durnam LJ, Flannery J, Goff B, Gee DG, Humphreys KL, Telzer E et al (2014) The development of human amygdala functional connectivity at rest from 4 to 23 years: a cross-sectional study. Neuroimage 95:193–207, http://doi.org/10.1016/j.neuroimage.2014.03.038
34. Dougherty LR, Blankenship SL, Spechler PA, Padmala S, Pessoa L (2015) An fMRI pilot study of cognitive reappraisal in children: divergent effects on brain and behavior. J Psychopathol Behav Assess 37(4):634–644, http://doi.org/10.1007/s10862-015-9492-z
35. McRae K, Gross JJ, Weber J, Robertson ER, Sokol-Hessner P, Ray RD et al (2012) The development of emotion regulation: an fMRI study of cognitive reappraisal in children, adolescents and young adults. Soc Cogn Affect Neurosci 7(1):11–22, http://doi.org/10.1093/scan/nsr093
36. Silvers JA, Shu J, Hubbard AD, Weber J, Ochsner KN (2015) Concurrent and lasting effects of emotion regulation on amygdala response in adolescence and young adulthood. Dev Sci 18 (5):771–784, http://doi.org/10.1111/desc.12260
37. Gaffrey MS, Luby JL, Belden AC, Hirshberg JS, Volsch J, Barch DM (2011) Association between depression severity and amygdala reactivity during sad face viewing in depressed preschoolers: an fMRI study. J Affect Disord 129(1-3):364–370, http://doi.org/10.1016/j.jad.2010.08.031
38. Garrett AS, Carrion V, Kletter H, Karchemskiy A, Weems CF, Reiss A (2012) Brain activation to facial expressions in youth with PTSD symptoms. Depress Anxiety 29 (5):449–459, http://doi.org/10.1002/da.21892
39. Pagliaccio D, Luby J, Gaffrey M, Belden A, Botteron K, Gotlib IH, Barch DM (2012) Anomalous functional brain activation following negative mood induction in children with pre-school onset major depression. Dev Cogn Neurosci 2(2):256–267, http://doi.org/10.1016/j.dcn.2011.11.008
40. Pine DS, Guyer AE, Leibenluft E (2008) Functional magnetic resonance imaging and pediatric anxiety. J Am Acad 47(11):1217–1221, http://doi.org/10.1097/CHI.0b013e318185dad0.FUNCTIONAL
41. Roy AK, Fudge JL, Kelly C, Perry JSA, Daniele T, Carlisi C et al (2013) Intrinsic functional connectivity of amygdala-based networks in adolescent generalized anxiety disorder. J Am Acad Child Adolesc Psychiatry 52(3):290–299.e2. http://doi.org/10.1016/j.jaac.2012.12.010
42. Wolf RC, Herringa RJ (2016) Prefrontal-amygdala dysregulation to threat in pediatric post-traumatic stress disorder. Neuropsychopharmacology 41(3):822–831, http://doi.org/10.1038/npp.2015.209
43. Kujawa A, Wu M, Klump H, Pine DS, Swain JE, Fitzgerald KD et al (2016) Altered development of amygdala-anterior cingulate cortex connectivity in anxious youth and young adults. Biol Psychiatry Cogn Neurosci Neuroimaging 1(4):345–352, http://doi.org/10.1017/CBO9781107415324.004

44. Callaghan BL, Sullivan RM, Howell B, Tottenham N (2014) The international society for developmental psychobiology Sackler symposium: early adversity and the maturation of emotion circuits-A cross-species analysis. Dev Psychobiol 56(8):1635–1650, http://doi.org/10.1002/dev.21260
45. Avishai-Eliner S, Yi SJ, Baram TZ (1996) Developmental profile of messenger RNA for the corticotropin-releasing hormone receptor in the rat limbic system. Brain Res Dev Brain Res 91(2):159–163, http://doi.org/0165380695001581
46. Hatalski CG, Guirguis C, Baram TZ (2012) Corticotropin releasing factor mRNA expression in the hypothalamic paraventricular nucleus and the central nucleus of the amygdala is modulated by repeated acute stress in the immature rat. J Neuroendocrinol 10(9):663–669
47. Moriceau S, Wilson DA, Levine S, Sullivan RM (2006) Dual circuitry for odor-shock conditioning during infancy: corticosterone switches between fear and attraction via amygdala. J Neurosci 26(25):6737–6748, http://doi.org/10.1523/JNEUROSCI.0499-06.2006
48. Eiland L, Ramroop J, Hill MN, Manley J, McEwen BS (2012) Chronic juvenile stress produces corticolimbic dendritic architectural remodeling and modulates emotional behavior in male and female rats. Psychoneuroendocrinology 37(1):39–47, http://doi.org/10.1016/j.psyneuen.2011.04.015
49. Malter Cohen M, Jing D, Yang RR, Tottenham N, Lee FS, Casey BJ (2013) Early-life stress has persistent effects on amygdala function and development in mice and humans. Proc Natl Acad Sci U S A 110(45):18274–18278, http://doi.org/10.1073/pnas.1310163110
50. Raineki C, Cortés MR, Belnoue L, Sullivan RM (2012) Effects of early-life abuse differ across development: infant social behavior deficits are followed by adolescent depressive-like behaviors mediated by the amygdala. J Neurosci 32(22):7758–7765, http://doi.org/10.1523/JNEUROSCI.5843-11.2012
51. Ishikawa J, Nishimura R, Ishikawa A (2015) Early-life stress induces anxiety-like behaviors and activity imbalances in the medial prefrontal cortex and amygdala in adult rats. Eur J Neurosci 41(4):442–453, http://doi.org/10.1111/ejn.12825
52. McLaughlin KA, Sheridan MA, Lambert HHK (2014) Childhood adversity and neural development: deprivation and threat as distinct dimensions of early experience. Neurosci Biobehav Rev 47:578–591, http://doi.org/10.1016/j.neubiorev.2014.10.012
53. Arata CM, Langhinrichsen-Rohling J, Bowers D, O'Brien N (2007) Differential correlates of multi-type maltreatment among urban youth. Child Abuse Negl 31(4):393–415, http://doi.org/10.1016/j.chiabu.2006.09.006
54. Dannlowski U, Kugel H, Huber F, Stuhrmann A, Redlich R, Grotegerd D et al (2013) Childhood maltreatment is associated with an automatic negative emotion processing bias in the amygdala. Hum Brain Mapp 34(11):2899–2909, http://doi.org/10.1002/hbm.22112
55. van Harmelen A-L, van Tol M-J, Demenescu LR, van der Wee NJA, Veltman DJ, Aleman A et al (2013) Enhanced amygdala reactivity to emotional faces in adults reporting childhood emotional maltreatment. Soc Cogn Affect Neurosci 8(4):362–369, http://doi.org/10.1093/scan/nss007
56. Bogdan R, Williamson DE, Hariri AR (2012) Mineralocorticoid receptor Iso/Val (rs5522) genotype moderates the association between previous childhood emotional neglect and amygdala reactivity. Am J Psychiatry 169:515–522, http://doi.org/10.1176/appi.ajp.2011.11060855
57. White MG, Bogdan R, Fisher PM, Muñoz KE, Williamson DE, Hariri AR (2012) FKBP5 and emotional neglect interact to predict individual differences in amygdala reactivity. Genes Brain Behav 11(7):869–878, http://doi.org/10.1111/j.1601-183X.2012.00837.x
58. Gianaros PJ, Horenstein JA, Hariri AR, Sheu LK, Manuck SB, Matthews KA, Cohen S (2008) Potential neural embedding of parental social standing. Soc Cogn Affect Neurosci 3(2):91–96, http://doi.org/10.1093/scan/nsn003
59. Evans GW, Swain JE, King AP, Wang X, Javanbakht A, Ho SS et al (2016) Childhood cumulative risk exposure and adult amygdala volume and function. J Neurosci Res 94(6):535–543, http://doi.org/10.1002/jnr.23681

60. Javanbakht A, King AP, Evans GW, Swain JE, Angstadt M, Phan KL, Liberzon I (2015) Childhood poverty predicts adult amygdala and frontal activity and connectivity in response to emotional faces. Front Behav Neurosci 9(154), http://doi.org/10.3389/fnbeh.2015.00154
61. Gee DG, Gabard-Durnam LJ, Flannery J, Goff B, Humphreys KL, Telzer EH et al (2013) Early developmental emergence of human amygdala-prefrontal connectivity after maternal deprivation. Proc Natl Acad Sci U S A 110(39):15638–15643, http://doi.org/10.1073/pnas.1307893110
62. Maheu FS, Dozier M, Guyer AE, Mandell D, Peloso E, Poeth K, Jenness J (2010) Preliminary study of medial temporal lobe function in youths with a history of caregiver deprivation and emotional neglect. Cogn Affect Behav Neurosci 10(1):34–49, http://doi.org/10.3758/CABN.10.1.34.A
63. Tottenham N, Hare TA, Millner A, Gilhooly T, Zevin JD, Casey BJ (2011) Elevated amygdala response to faces following early deprivation. Dev Sci 14(2):190–204, http://doi.org/10.1111/j.1467-7687.2010.00971.x
64. McCrory EJ, De Brito SA, Kelly PA, Bird G, Sebastian CL, Mechelli A et al (2013) Amygdala activation in maltreated children during pre-attentive emotional processing. Br J Psychiatry 202(4):269–276, http://doi.org/10.1192/bjp.bp.112.116624
65. McLaughlin KA, Peverill M, Gold AL, Alves S, Sheridan MA (2015) Child maltreatment and neural systems underlying emotion regulation. J Am Acad Child Adolesc Psychiatry 54 (9):753–762, http://doi.org/10.1016/j.jaac.2015.06.010
66. Marusak HA, Martin KR, Etkin A, Thomason ME (2014) Childhood trauma exposure disrupts the automatic regulation of emotional processing. Neuropsychopharmacology 40 (5):1250–1258, http://doi.org/10.1038/npp.2014.311
67. McCrory EJ, De Brito SA, Sebastian CL, Mechelli A, Bird G, Kelly PA, Viding E (2011) Heightened neural reactivity to threat in child victims of family violence. Curr Biol 21(23): R947–R948, http://doi.org/10.1016/j.cub.2011.10.015
68. Swartz JR, Williamson DE, Hariri AR (2015) Developmental change in amygdala reactivity during adolescence: effects of family history for depression and stressful life events. Am J Psychiatry 172(3):276–283, http://doi.org/10.1176/appi.ajp.2014.14020195
69. Jedd K, Hunt RH, Cicchetti D, Hunt E, Cowell RA, Rogosch FA et al (2015) Long-term consequences of childhood maltreatment: altered amygdala functional connectivity. Dev Psychopathol 27:1577–1589, http://doi.org/10.1017/S0954579415000954
70. Kim P, Evans GW, Angstadt M, Ho SS, Sripada CS, Swain JE (2013) Effects of childhood poverty and chronic stress on emotion regulatory brain function in adulthood. Proc Natl Acad Sci U S A 110(46):18442–18447, http://doi.org/10.1073/pnas.1308240110/-/DCSupplemental. www.pnas.org/cgi/doi/10.1073/pnas.1308240110
71. Cicchetti D, Rogosch FA (1996) Equifinality and multifinality in developmental psychopathology. Dev Psychopathol 8:597–600. Retrieved from file://localhost/Users/lornaquandt/Documents/Papers/2007/Unknown/2007-3.pdf\npapers://4f210845-804c-495f-b761-84503ca2694d/Paper/p757
72. Fan Y, Herrera-Melendez AL, Pestke K, Feeser M, Aust S, Otte C et al (2014) Early life stress modulates amygdala-prefrontal functional connectivity: implications for oxytocin effects. Hum Brain Mapp 35(10):5328–5339, http://doi.org/10.1002/hbm.22553
73. Herringa RJ, Birn RM, Ruttle PL, Burghy CA, Stodola DE, Davidson RJ, Essex MJ (2013) Childhood maltreatment is associated with altered fear circuitry and increased internalizing symptoms by late adolescence. Proc Natl Acad Sci U S A 110(47):19119–19124, http://doi.org/10.1073/pnas.1310766110
74. Thomason ME, Marusak HA, Tocco MA, Vila AM, McGarragle O, Rosenberg DR (2015) Altered amygdala connectivity in urban youth exposed to trauma. Soc Cogn Affect Neurosci 10(11):1460–1468, http://doi.org/10.1093/scan/nsv030
75. Pagliaccio D, Luby JL, Bogdan R, Agrawal A, Gaffrey MS, Belden AC et al (2015) Amygdala functional connectivity, HPA axis genetic variation, and life stress in children and relations to anxiety and emotion regulation. J Abnorm Psychol 124(4):817–833

76. Graham AM, Pfeifer JH, Fisher PA, Carpenter S, Fair DA (2015) Early life stress is associated with default system integrity and emotionality during infancy. J Child Psychol Psychiatry 56(11):1212–1222, http://doi.org/10.1111/jcpp.12409
77. Graham AM, Pfeifer JH, Fisher PA, Lin W, Gao W, Fair DA (2014) The potential of infant fMRI research and the study of early life stress as a promising exemplar. Dev Cogn Neurosci 12:12–39, http://doi.org/10.1016/j.dcn.2014.09.005
78. Pagliaccio D, Luby JL, Luking KR, Belden AC, Barch DM (2014) Brain–behavior relationships in the experience and regulation of negative emotion in healthy children: Implications for risk for childhood depression. Dev Psychopathol 26:1289–1303
79. Crozier JC, Wang L, Huettel SA, De Bellis MD (2014) Neural correlates of cognitive and affective processing in maltreated youth with posttraumatic stress symptoms: does gender matter? Dev Psychopathol 26(2):491–513, http://doi.org/10.1017/S095457941400008X
80. Suzuki H, Luby JL, Botteron KN, Dietrich R, McAvoy MP, Barch DM (2014) Early life stress and trauma and enhanced limbic activation to emotionally valenced faces in depressed and healthy children. J Am Acad Child Adolesc Psychiatry 53(7):800–813.e10. http://doi.org/10.1016/j.jaac.2014.04.013
81. Gaffrey MS, Barch DM, Luby JL (2016) Amygdala reactivity to sad faces in preschool children: an early neural marker of persistent negative affect. Dev Cogn Neurosci 17:94–100, http://doi.org/10.1016/j.dcn.2015.12.015
82. Moriceau S, Sullivan RM (2006) Maternal presence serves as a switch between learning fear and attraction in infancy. Nat Neurosci 9(8):1004–1006, http://doi.org/10.1038/nn1733
83. Burghy CA, Stodola DE, Ruttle PL, Molloy EK, Armstrong JM, Oler JA et al (2012) Developmental pathways to amygdala-prefrontal function and internalizing symptoms in adolescence. Nat Neurosci 15(12):1736–1741, http://doi.org/10.1038/nn.3257
84. McGloin JM, Widom CS (2001) Resilience among abused and neglected children grown up. Dev Psychopathol 13(4):1021–1038, http://doi.org/10.1017/S095457940100414X
85. Masten AS (2004) Regulatory processes, risk, and resilience in adolescent development. Ann N Y Acad Sci 1021:310–319, http://doi.org/10.1196/annals.1308.036
86. Jaffee SR, Caspi A, Moffitt TE, Polo-Tomás M, Taylor A (2007) Individual, family, and neighborhood factors distinguish resilient from non-resilient maltreated children: a cumulative stressors model. Child Abuse Negl 31(3):231–253, http://doi.org/10.1016/j.chiabu.2006.03.011
87. Cicchetti D, Blender JA (2006) A multiple-levels-of-analysis perspective on resilience: implications for the developing brain, neural plasticity, and preventive interventions. Ann N Y Acad Sci 1094:248–258, http://doi.org/10.1196/annals.1376.029
88. McLaughlin KA (2016) Future directions in childhood adversity and youth psychopathology. J Clin Child Adolesc Psychol 45(3):361–382, http://doi.org/10.1080/15374416.2015.1110823
89. Afifi TO, MacMillan HL (2011) Resilience following child maltreatment: a review of protective factors. Can J Psychiatry 56(5):266–272
90. DuMont KA, Widom CS, Czaja SJ (2007) Predictors of resilience in abused and neglected children grown-up: the role of individual and neighborhood characteristics. Child Abuse Negl 31(3):255–274, http://doi.org/10.1016/j.chiabu.2005.11.015
91. Collishaw S, Pickles A, Messer J, Rutter M, Shearer C, Maughan B (2007) Resilience to adult psychopathology following childhood maltreatment: evidence from a community sample. Child Abuse Negl 31(3):211–229, http://doi.org/10.1016/j.chiabu.2007.02.004
92. Miller-Graff LE, Cater ÅK, Howell KH, Graham-Bermann SA (2016) Parent-child warmth as a potential mediator of childhood exposure to intimate partner violence and positive adulthood functioning. Anxiety Stress Coping 29(3):259–273, http://doi.org/10.1080/10615806.2015.1028030
93. Hennessy MB, Kaiser S, Sachser N (2009) Social buffering of the stress response: diversity, mechanisms, and functions. Front Neuroendocrinol 30(4):470–482, http://doi.org/10.1016/j.yfrne.2009.06.001

94. Kikusui T, Winslow JT, Mori Y (2006) Social buffering: relief from stress and anxiety. Philos Trans R Soc B Biol Sci 361(1476):2215–2228, http://doi.org/10.1098/rstb.2006.1941
95. Hostinar CE, Johnson AE, Gunnar MR (2014) Parent support is less effective in buffering cortisol stress reactivity for adolescents compared to children. Dev Sci 2:281–297, http://doi.org/10.1111/desc.12195
96. Gee DG, Gabard-Durnam L, Telzer EH, Humphreys KL, Goff B, Shapiro M et al (2014) Maternal buffering of human amygdala-prefrontal circuitry during childhood but not during adolescence. Psychol Sci 25(11):2067–2078, http://doi.org/10.1177/0956797614550878
97. Troller-Renfree S, Mcdermott JM, Nelson CA, Zeanah CH, Fox NA (2015) The effects of early foster care intervention on attention biases in previously institutionalized children in Romania. Dev Sci 18(5):713–722, http://doi.org/10.1111/desc.12261
98. Troller-renfree S, Mclaughlin KA, Sheridan MA, Nelson CA, Zeanah CH, Fox NA (2016) The benefits of a positive attention bias amongst children with a history of psychosocial deprivation. Biol Psychol. http://doi.org/10.1016/j.biopsycho.2016.04.008
99. Hostinar CE, Johnson AE, Gunnar MR (2015) Early social deprivation and the social buffering of cortisol stress responses in late childhood: an experimental study. Dev Psychol 51(11):1597–1608
100. Pechtel P, Lyons-Ruth K, Anderson CM, Teicher MH (2014) Sensitive periods of amygdala development: the role of maltreatment in preadolescence. Neuroimage 97:236–244, http://doi.org/10.1016/j.neuroimage.2014.04.025
101. Goff B, Tottenham N (2015) Early-life adversity and adolescent depression: mechanisms involving the ventral striatum. CNS Spectr 20(4):337–345, http://doi.org/10.1017/S1092852914000674
102. Heim C, Binder EB (2012) Current research trends in early life stress and depression: review of human studies on sensitive periods, gene-environment interactions, and epigenetics. Exp Neurol 233(1):102–111, http://doi.org/10.1016/j.expneurol.2011.10.032
103. Uher R, McGuffin P (2008) The moderation by the serotonin transporter gene of environmental adversity in the aetiology of mental illness: review and methodological analysis. Mol Psychiatry 13(2):131–146, http://doi.org/10.1038/sj.mp.4002067
104. Casey BJ, Glatt CE, Tottenham N, Soliman F, Bath K, Amso D et al (2009) Brain-derived neurotrophic factor as a model system for examining gene by environment interactions across development. Neuroscience 164(1):108–120, http://doi.org/10.1016/j.neuroscience.2009.03.081
105. Scarr S, Mccartney K (1983) How people make their own environments: a theory of genotype → environment effects author (s): Sandra Scarr and Kathleen McCartney Published by: Wiley on behalf of the Society for Research in Child Development Stable URL: http://www.jstor.org/stable/. Child Dev 54(2):424–435
106. Swartz JR, Hariri AR, Williamson DE (2016) An epigenetic mechanism links socioeconomic status to changes in depression-related brain function in high-risk adolescents. Mol Psychiatry. http://doi.org/10.1038/mp.2016.82

Subanesthetic Dose Ketamine in Posttraumatic Stress Disorder: A Role for Reconsolidation During Trauma-Focused Psychotherapy?

Cato Veen, Gabriel Jacobs, Ingrid Philippens, and Eric Vermetten

Abstract Despite efforts to develop more effective therapies, PTSD remains a difficult disorder to treat. Insight into the dynamic nature of memory formation and its required molecular machinery can provide an opportunity to target pathological memories for emotionally arousing events. As memories become labile upon retrieval, novel information can update the strength and course of these consolidated memories. Targeting the process of reconsolidation may offer a relevant approach to attenuate fearful and traumatic memories. Specific molecular mechanisms that are required for reconsolidation of arousing information include an intact functioning of the glutamatergic signaling pathways and, more specifically, the integrity of NMDA receptors. Ketamine, a noncompetitive NMDA-receptor antagonist, is receiving increasing interest for a variety of psychiatric indications. This compound can also be an interesting candidate for targeting emotional memories. We explore whether single intravenous infusion of a subanesthetic dose of ketamine can be considered as a viable augmentation strategy for trauma-focused psychotherapy in patients with PTSD. As a consequence, a systematic approach is needed to assess the pharmacodynamic effects of ketamine in relation to both psychotherapy and its pharmacokinetics prior to

C. Veen (✉)
Department of Psychiatry, Leiden University Medical Center, Leiden, The Netherlands
e-mail: c.veen@lumc.nl

G. Jacobs
Department of Psychiatry, Leiden University Medical Center, Leiden, The Netherlands

Centre for Human Drug Research, Leiden, The Netherlands

I. Philippens
Biomedical Primate Research Centre, Rijswijk, The Netherlands

E. Vermetten
Department of Psychiatry, Leiden University Medical Center, Leiden, The Netherlands

Arq Psychotrauma Research Group, Diemen, The Netherlands

Curr Topics Behav Neurosci (2018) 38: 137–162
DOI 10.1007/7854_2017_34
Published Online: 31 March 2018

its application in patient populations. By using a "question-based drug development plan," we can explore such aspects for novel drugs, and we formulated five additional topics that need to be addressed concerning the psychotherapeutic approach and phase orientation of pharmacological assisted psychotherapy.

Keywords Consolidation • Extinction learning • Ketamine • Memory • Pharmacological assisted psychotherapy • Posttraumatic stress disorder • Reconsolidation

Contents

1 Introduction

Posttraumatic stress disorder (PTSD) can be considered as a disorder in which emotional memory plays a key role (Yehuda and LeDoux 2007). Despite efforts to develop more effective therapies, it remains a difficult disorder to treat (Jonas et al. 2013). Targeting the strength of highly affective memories has shown to be a novel and promising approach to attenuate fearful and traumatic memories. Over the last decade, this has reflected in an abundance of research on memory resulting in the discovery that memories become labile upon retrieval. In this phase novel information can be integrated and can alter the nature and strength of consolidated fear memories (Quirk and Mueller 2008). This approach has opened opportunities for exploration of a new strategy with compounds that could lead to significant clinical improvements in treating PTSD.

In this review, we provide a rationale for targeting pathological memory formation with a pharmacological approach embedded in a psychological treatment. We will specifically highlight the role of *N*-methyl-D-aspartate (NMDA)-receptor functioning. We will argue that ketamine, a NMDA-receptor antagonist, potentially holds promise in treatment of PTSD when administered as a single intravenous (IV) infusion of a subanesthetic dose, especially as augmentation of trauma-focused psychotherapy by attenuating reconsolidation of traumatic memories. Additionally, we will provide a framework to systematically address all relevant topics for drug-assisted psychotherapies, and we will apply this for ketamine-assisted psychotherapy in PTSD.

2 Memory Formation and Glutamatergic Signaling Through the NMDA Receptor

A well-studied translational model to understand formation of trauma-related memories is the Pavlovian fear-conditioning paradigm illustrated in Fig. 1 (Maren et al. 2013; Pavlov 1927; Pedreira and Maldonado 2003; Suzuki et al. 2004).

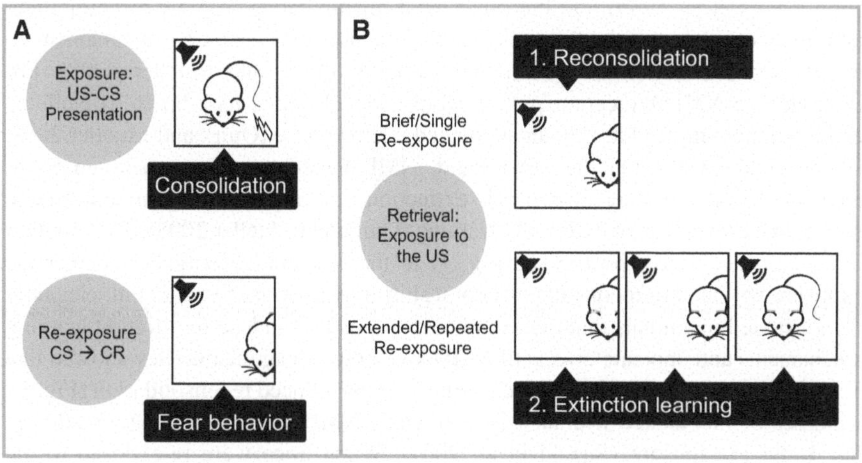

Fig. 1 A translational model to investigate trauma-related memories; a Pavlovian fear-conditioning paradigm. (**a**) An animal is exposed to a cue (tone), the conditioned stimulus (CS), followed by an aversive stimulus (shock), the unconditioned stimulus (US), several times. This US is an aversive stimulus and subsequently elicits fear behavior in the animals, the conditioned response (CR). This behavior represents a memory of the connection between the CS and the US, consolidation of the fear memory. During presentation of the CS without the US (re-exposure), the animal now exhibits fear behavior. A new (fear) memory is represented by a "memory trace" or "memory engram." (**b**) Depending on the frequency and duration of re-exposure, it can induce retrieval and reconsolidation of the primary memory trace or result in extinction learning. **1.** During brief re-exposure to the CS without US (re-exposure), the fear memory is retrieved and returned to its unstable state and subsequently restabilized (reconsolidation). When the memory is unstable, it is possible to integrate new information into the same memory trace. **2.** Extinction learning occurs if the duration or frequency of re-exposure is sufficient to induce extinction learning. Extinction learning does not affect the "old memory trace," but rather it creates a "new memory trace." Extinction learning results in decrease of fear behavior exposed by the animals by antagonizing the primary formed memory. Both reconsolidation and extinction learning can be involved in adaption of fear memory and behavior (CR) in animals after re-exposure

Consolidation, reconsolidation, and extinction learning have both common and distinct steps in their biological pathways (Miller and Sweatt 2006; Nader 2015; Suzuki et al. 2004). As consolidated memories become labile upon retrieval, novel information can update the strength and course of these consolidated memories; this process is reconsolidation (Miller and Sweatt 2006; Nader 2015). Extinction learning does not affect the "old memory trace" like in reconsolidation, but rather it creates a "new memory trace," and it is thought that the new memory antagonizes the previous formed fear memory (Berman and Dudai 2001; Quirk and Mueller 2008). The glutamatergic signaling pathway is important in synaptic plasticity and has a crucial role during several stages of learning such as consolidation, retrieval, reconsolidation, and extinction learning (Miller and Sweatt 2006; Quirk and Mueller 2008). More specifically, NMDA-receptor functioning has shown to be essential during all these phases of learning (Miller and Sweatt 2006; Quirk and Mueller 2008). Disruptions in NMDA signaling can lead to deficits, e.g., chronic stress exposure in rodents can disrupt NMDA-dependent hippocampal long-term potentiation (Shors et al. 1989). On the other hand, manipulating NMDA-receptor functioning after exposure to traumatic stress is a potential candidate mechanism to modify reconsolidation and extinction learning (Quirk and Mueller 2008). For example, the NMDA-receptor partial agonist D-cycloserine has shown to facilitate extinction learning and lead to an attenuated fear response (Quirk and Mueller 2008; Weber et al. 2007). On the other hand, NMDA-receptor antagonists can block reconsolidation but may also block extinction learning resulting in undesirable effects like prevention of extinction learning (Quirk and Mueller 2008). The duration of re-exposure (CS exposure see Fig. 1) to the fear cue determines whether the exposure will lead to reconsolidation or a shift to extinction. Suzuki et al. elegantly demonstrated in animals that the extent of re-exposure can lead to different learning mechanisms and opposite effects of NMDA-receptor antagonism. They showed that a 3-min re-exposure to a conditioned stimulus (CS) induced reconsolidation (Fig. 1), which could be blocked with an injection with a NMDA-receptor antagonist during the 3-min re-exposure (Suzuki et al. 2004). When mice were re-exposed to the CS for much longer, 30 min, the learning shifted to extinction. After a 30-min re-exposure, the animals would normally show reduced fear behavior, but injection of a NMDA-receptor antagonist prevented reduction of fear behavior and resulted in a blockade of extinction learning (Duclot et al. 2016; Suzuki et al. 2004). The potential dynamic of the NMDA receptor is complex since there are several binding sites for antagonistic compounds. Competitive antagonism at different bindings sites, like the glutamate or the glycine binding site, has been shown to block reconsolidation (Das et al. 2013; Nader 2015). Thus, pharmacological compounds that (temporarily) promote or inhibit the function of the NMDA receptor appear to be a promising strategy to attenuate fear memory either by facilitating extinction or blocking reconsolidation, respectively.

3 Treatment Opportunities of Combining Psychotherapy and Pharmacotherapy in PTSD

Evidence-based psychotherapies for PTSD are commonly divided as either trauma-focused or non-trauma-focused. A key element in trauma-focused psychotherapy is the exposure of the traumatic memory through speech, writing, or visualization. This is considered a common and essential intervention because it is responsible for the processing of traumatic memories (Schnyder et al. 2015). Targeting functional processes like extinction learning and desensitization are capable to reduce the fear response related to the traumatic memories (Garakani et al. 2006; Rauch et al. 2012). While both trauma-focused and non-trauma-focused psychotherapies are considered state-of-the-art treatments (Lancaster et al. 2016), almost all PTSD treatment guidelines recommend trauma-focused psychotherapies, and clinicians generally accept these as the most effective treatment strategy (Bisson et al. 2007; Rauch et al. 2012).

Several central nervous system (CNS)-penetrating agents, such as antidepressants and antipsychotics, are used as pharmacological treatments for patients with PTSD. Overall, the aforementioned drugs are hypothesized to exert their therapeutic effects by primarily modulating central monoaminergic neurocircuits (Casey 1997; Owens et al. 1997). Some of these drugs have additional effects on the autonomic nervous system (ANS) by either suppressing sympathetic activation or boosting parasympathetic activity (Casey 1997; Davidson 2015). These drugs influence clinical symptom clusters in PTSD such as anxiety, depressed mood and cognitive processes by modulating central monoaminergic circuits, and hyperarousal and sleep disruption via central and/or peripheral ANS effects in patients with PTSD (Davidson 2015). Yet these mechanisms are often insufficient for symptom relief in PTSD since these drugs alone or in different combinations have been shown to have only modest therapeutic effects when compared to psychotherapeutic interventions (Jonas et al. 2013; Lee et al. 2016). We argue that this might be partly due to the fact that these drugs do not target the key mechanism of the disorder, pathological memory formation (Yehuda and LeDoux 2007). Therefore, new strategies for pharmacotherapy that focus on pathological memory formation with pharmacological compounds, manipulating the excitatory glutamate system, seem to be very promising (Garakani et al. 2006; Rauch et al. 2012).

In everyday practice, it is common that patients are treated concurrently with psychotherapy and pharmacotherapy. However, it is not yet established on which of the treatments the efficacy is based (Hetrick et al. 2010). Moreover, therapeutic combinations involving both, therapy and current drugs (antidepressants and antipsychotics), have not been designed to target the etiopathophysiology of the disorder. Hence, new studies are needed to determine how psycho- and pharmacotherapy may have a synergistic effect. The possibility to manipulate consolidated traumatic memories (McGaugh 1966; Vermetten et al. 2014) may provide an opportunity to manipulate pathological memories even long after the incubation of PTSD. The addition of pharmacological compounds could target underlying memory mechanisms during psychotherapy and provide an opportunity to augment therapeutic efficiency (see Fig. 3c, d) (Vermetten and Krugers 2016). Currently, patients undergo repeated exposure sessions to establish an improvement that is relying partly on extinction learning (Craske 2015). Pharmacological compounds could very well facilitate extinction learning that results in a faster reduction of the fear response (Quirk and Mueller 2008). Another intervention strategy in this respect is to block post-retrieval reconsolidation of a traumatic memory (Nader et al. 2013). Theoretically, blocking reconsolidation could lead to a faster reduction of symptoms in comparison with enhanced extinction learning (Nader 2015; Nader et al. 2013). It may be so that fear reduction could possibly be established just after one session, while several sessions are needed to achieve extinction learning.

The concept of enhancing or augmenting psychotherapy with pharmacotherapeutics is known as pharmacological or medication-assisted psychotherapy (Vermetten et al. 2014). Preclinical research has discovered several compounds with different receptor profiles (e.g., propranolol, cortisol, D-cycloserine, ketamine, oxytocin, and MDMA) that have been identified as potential candidates to enhance the effect of psychotherapy through several different mechanisms (Vermetten and Krugers 2016). This review focuses in particular on the possibility of ketamine to block reconsolidation of fear memories through NMDA-receptor antagonism. Furthermore, a background on ketamine as a CNS drug for augmenting psychotherapies in PTSD will be discussed.

4 Ketamine as a Central Nervous System Drug

4.1 Clinical Pharmacology of Ketamine

Ketamine is a pharmacological compound that was developed in the 1960s as an anesthetic agent (Domino et al. 1965). It easily crosses the blood-brain barrier (BBB) and affects a great variety of receptors. Ketamine often refers to a racemic mixture of two enantiomers, S(+)-ketamine and R(−)-ketamine, but is also available as S(+)-ketamine alone which is much more potent (Mion and Villevieille 2013). Demethylation of ketamine occurs within 2–3 min after IV infusion, and the metabolites (norketamine and hydroxynorketamine among others) have pharmacological properties as well. Relevant information about the chemical structure, pharmacokinetics, and pharmacodynamics is given in Box 1. Ketamine has a variety of dose-dependent pharmacodynamic effects on the human brain that are mediated through different receptors (Mion and Villevieille 2013; Oye et al. 1992). Its noncompetitive antagonism on the NMDA receptor is the most studied effect causing changes in opening time and opening frequency of the channel pore (Li and Vlisides 2016; Orser et al. 1997). When administered in subanesthetic dosages ($\leq$0.5 mg/kg), ketamine has an impact on mood, perception, and cognition (Bowdle et al. 1998; Krystal et al. 1994). Mood is first described to be flattened with some drowsiness, while with higher doses anxiety can occur (Li and Vlisides 2016). Ketamine can alter visual (heightened, dulled, and distorted), auditory, and tactile perception of the world and the self

(Bowdle et al. 1998; Pomarol-Clotet et al. 2006). Cognition can be affected by ketamine in a variety of ways, including vigilance and disrupted executive functioning (Krystal et al. 1994). Additionally, ketamine can induce referential thinking and delusions such as paranoia and subjective changes in thinking (Bowdle et al. 1998; Pomarol-Clotet et al. 2006). It was also shown that ketamine can induce amnesia (Oye et al. 1992) and disruption of recall (Krystal et al. 1994). Due to these effects, ketamine is used as a recreational drug especially among young people (Bearn and O'Brien 2015). Besides the central effects, ketamine has several systemic effects like increased heart rate and blood pressure through activation of the sympathetic ANS. All these effects are transitory and are dependent of time to peak concentration and elimination time (Kirby 2015; Li and Vlisides 2016).

Box 1: Pharmacological Structure, Pharmacokinetics, and Neuropharmacology of Ketamine

Chemical Structure

Ketamine is phencyclidine derivate (CI-581, 2-chlorophenyl-2-methylamino-cyclohexanone, $C_{13}H_{16}ClNO$) (Domino et al. 1965). Ketamine primarily existed of a racemic mixture of two enantiomers S(+)-ketamine and R(−)-ketamine. Where S(+)-ketamine has a CH3 group, R(−)-ketamine had H alone. Nowadays also S(+)-ketamine alone has become available. S(+)-ketamine is at least two times more potent compared to racemic ketamine (Mion and Villevieille 2013).

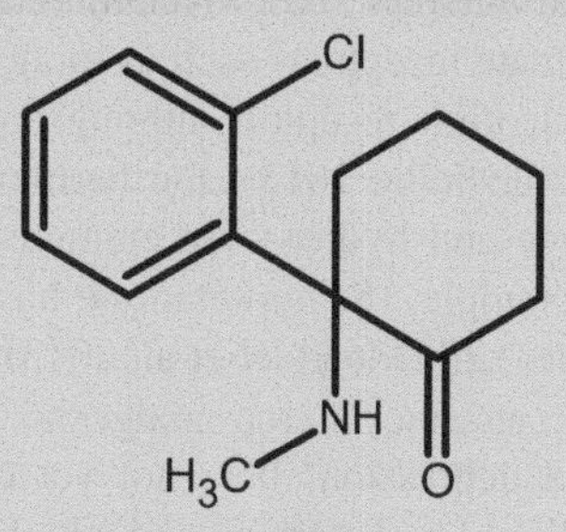

Chemical structure of S(+)ketamine

Pharmacokinetics

Ketamine is a lipid-soluble molecule with a large volume of distribution which rapidly crosses the blood-brain barrier (BBB). There is low binding of 10–40% to plasma proteins (Mion and Villevieille 2013). The bioavailability

(continued)

Box 1 (continued)

of ketamine is intramuscular 93%, oral 17–29%, and intranasal 8–45% (Li and Vlisides 2016). Via demethylation, approximately 80% of ketamine is metabolized into norketamine and 15% into 6-hydroxynorketamine (Woolf and Adams 1987). Subsequently, norketamine is metabolized by liver microsomal cytochrome P450. CYP3A4 is the major enzyme involved in ketamine's *N*-demethylation (Hijazi and Boulieu 2002). Demethylation of ketamine occurs within 2–3 min after injection of an intravenous (IV) bolus. Clearance of ketamine is high and it has an elimination halftime of 2–3 h (Domino et al. 1984). Norketamine is detectable after 2–3 min with a peak concentration after 30 min. The elimination of norketamine is slower than ketamine, and continuous IV administration of ketamine will result in an accumulation of norketamine (Mion and Villevieille 2013).

Pharmacodynamics

The primary central nervous system effects are thought to be mediated by binding on glutamate receptors (Mion and Villevieille 2013). The most known mechanism of action is the noncompetitive antagonism of the NMDA receptor through changes in opening time and opening frequency (Orser et al. 1997). The binding affinity for the NMDA receptor is high; Ki values are S(+)-ketamine 0.69 and for R(−)-ketamine 2.57 (Moaddel et al. 2013). Binding affinity for dopamine d2 receptor (Ki value =0.5 μM) and serotonin 5-HT2 receptor (Ki value = 15 μM) has been found within the pharmacological relevant range (Kapur and Seeman 2002). Ketamine also has the potential to bind to others (non-glutamate receptors) such as monoaminergic, cholinergic, opioid, muscarinic, and nicotinic receptors although with a much lower binding affinity (Mion and Villevieille 2013). Furthermore, metabolites of ketamine, such as norketamine and hydroxynorketamine, have pharmacological properties as well. For example, (S)-norketamine has a high affinity for the NMDA receptor, Ki value 2.25 (Moaddel et al. 2013).

Ketamine is used for anesthesia, for analgesia, and more recently for psychiatric disorders like depression. It is not yet completely clear which part of the effect is established by ketamine and which part by its metabolites. Norketamine is thought to be involved in nociception but has a three times smaller effect compared to ketamine (Li and Vlisides 2016). Recently hydroxynorketamine has been showed to be responsible for the antidepressant effect through binding on the AMPA receptors (Zanos et al. 2016).

Dosage depends strongly on the indication. Anesthesia is induced with dosages that range from 1 to 4.5 mg/kg IV, and a coma can be maintained with repeated dosages of 0.5–1 mg/kg IV. Because a coma is not induced by ketamine, 0.5 mg/kg IV is considered a subanesthetic dose (Li and Vlisides 2016).

4.2 *Ketamine in Search of an Indication in Psychiatry*

Because of its psychotropic effects (Krystal et al. 1994) and its ability to easily cross the BBB (Peltoniemi et al. 2016), research of ketamine has expanded to the field of psychiatry. Due to its effect on cognition and perception, ketamine can mimic psychosis-like symptoms and be used as a model of schizophrenia in healthy individuals without a prior or current history of psychotic disorder (Domino and Luby 2012). Increased interest in the use of ketamine as a potential treatment for psychiatric disorders emerged after the discovery of NMDA-receptor antagonism reversing depressive-like behaviors in animals (Papp and Moryl 1994; Przegalinski et al. 1997; Trullas and Skolnick 1990). Only two decades ago, Berman et al. published the first paper about the rapid effects of ketamine on patients with depressive episodes (Berman et al. 2000). They showed that a single IV infusion of 0.5 mg/kg ketamine resulted in greater reduction of depressive symptoms compared to placebo (Berman et al. 2000). This was replicated by Zarate et al. with a second randomized controlled trial who reported the antidepressant effect lasted for 1 week (Zarate et al. 2006). Since then, research on ketamine as a treatment strategy for depression has greatly expanded (Caddy et al. 2014). To extend the durability of the antidepressant effect, the protocol has been adjusted with repeated IV infusion of ketamine instead of a single IV infusion. A study using this protocol reported a prolonged effect of more than 2 weeks (18 days) (Murrough et al. 2013b). Although many studies (RCTs with passive and active placebo, open-label studies, case reports) published that ketamine exerts a rapid effect on symptom reduction (Caddy et al. 2014), the most recent meta-analysis concluded that this evidence is quite promising but limited due to risk of bias and small study samples (Caddy et al. 2015). In addition to the interest in ketamine as treatment for depression, studies have started to investigate the potential of ketamine as an acute treatment for suicidal ideation. The first reports are positive and these studies conclude a beneficial effect (Mallick and McCullumsmith 2016; Murrough et al. 2015; Price et al. 2009). Although a lot of research has already been performed, more research is needed to strengthen the evidence on ketamine as therapeutic drug in mood disorders and suicidal ideations. This research needs to address the question about optimal dosing and route of administration in order to establish a prolonged and sustained effect. Additionally, it should aim to unravel the underlying biological mechanism through which ketamine exerts its effect on mood symptoms. New insights have shown that the antidepressant effect of ketamine is partially mediated through its metabolite hydroxynorketamine, which appears to be NMDA receptor independent and accomplished through the α-amino-3-hydroxy-5-methyl-4-isoxazolepropionic acid (AMPA) receptor (Zanos et al. 2016). Via this type of research, a more targeted treatment approach with fewer side effects could be developed.

4.3 *Single IV Infusion of a Subanesthetic Dose Ketamine as Treatment in PTSD*

In 2014 Feder et al. published the first randomized controlled trial using ketamine as treatment in patients with PTSD. In this proof of concept study, they compared a single IV infusion of racemic ketamine of 0.5 mg/kg for 40 min with IV midazolam as active comparator in 41 patients (Feder et al. 2014). The protocol used for this symptom-focused study is similar to the first studies that investigated the effect of ketamine on depression (R. M. Berman et al. 2000; Murrough et al. 2013a). Feder et al. reported a rapid and significantly increased improvement of PTSD symptoms in patients who received ketamine compared to midazolam. Reduction of subjective stress was measured 24 h after infusion with a self-report questionnaire (Impact of Event Scale – Revised). This questionnaire used subscales on symptoms of intrusion, avoidance, and hyperarousal. The reduction was significant for all three subscales. As patients with PTSD often suffer from comorbid depression, the study also discussed the effect of ketamine on depressive symptoms. They showed that ketamine infusion resulted in a significantly larger reduction in comorbid depressive symptoms when compared to midazolam. It is important to note that baseline depressive symptoms and the effects of treatment on depressive symptoms did not predict the effect of ketamine on PTSD symptoms. Therefore, Feder et al. argued that the improvement in symptoms of PTSD was not due to the effect on comorbid depressive symptoms (Feder et al. 2014). Supportive evidence exists from three case reports that demonstrated that ketamine decreases symptoms of PTSD in patients for a period up to 2 weeks (D'Andrea and Andrew Sewell 2013; Donoghue et al. 2015; Womble 2013). Another group highlighted the effect of ketamine on PTSD symptoms with a subanalysis of comorbid PTSD in patients with depression (Zeng et al. 2013).

An important contribution of the Feder study is that they were the first to show that ketamine is safe and well tolerated in the treatment of PTSD. They do report transient dissociative symptoms (Feder et al. 2014). This is specifically important because ketamine can induce (in a dose-dependent manner) dissociative symptoms and paranoid thoughts, which are symptoms reported by many patients with PTSD. However, it is not reported to what extent patients already experienced dissociative or cognitive symptoms at baseline. Furthermore, it should be noted that the Feder study is performed with racemic ketamine (mixture of both enantiomers S(+)-ketamine and R(−)-ketamine) and that side effects might be more prominent if only S(+)-ketamine would have been used (Peltoniemi et al. 2016). To strengthen these first promising results, a replication study is needed. The Feder study raises questions concerning the underlying mechanisms. Are these effects mediated solely through NMDA-receptor antagonism and how does it affect pathological memory formation? Additional studies need to answer these questions.

4.4 *The Effects of Ketamine on Memory, Development of Memory, and Stress Disorders*

The impact of ketamine on the different stages of memory formation in humans has not been systematically addressed. There is some circumstantial evidence on the relationship between ketamine and its effect on memory and stress in humans from retrospective observation studies in battlefield anesthetics (McGhee et al. 2008, 2014). They initially reported that perioperative use of ketamine would protect against the development of PTSD. However, in a larger study, this was not replicated, and they concluded that perioperative ketamine did not protect against acute and delayed stress symptoms (McGhee et al. 2008, 2014). On the other hand, another group reported that perioperative ketamine would even lead to a higher percentage of PTSD symptoms (Schonenberg et al. 2005). These studies could potentially give information whether ketamine can promote or inhibit memory consolidation. However, interpretation is difficult because the exact time window between the traumatic event (and thus the start of consolidation of the memory) and the administration of ketamine is not mentioned. Even though another study showed elevated symptoms of acute stress (dissociation, reexperiencing, hyperarousal, avoidance) 3 days after administration of perioperative ketamine, it is difficult to interpret this effect as an effect on traumatic memory due to the dissociative side effects (Schonenberg et al. 2008). Due to this conflicting evidence, the relation of perioperative ketamine administration after trauma and PTSD symptomology linked to pathological memories remains unclear. However, these studies did identify that something profound is going on with ketamine treatment after traumatic events, which is emphasized by a qualitative study that interviewed clinicians who used ketamine as an anesthetic (Wilson and Pokorny 2012). To obtain better knowledge about the effect of ketamine on human memory formation, dose-range studies during specific stages need to be performed.

5 Novel Use of Ketamine to Target Traumatic Memories in PTSD

5.1 *Ketamine Targeting Pathological Memory Formation*

The different stages of memory formation provide several treatment targets for PTSD and can be distinguished as the event-based "golden hours" and exposure-based "golden hours" (Fig. 2) (Vermetten and Krugers 2016). The event-based "golden hours" provide an opportunity to investigate the potential of ketamine as primary or secondary prevention of PTSD in high-risk groups. Thus, can ketamine

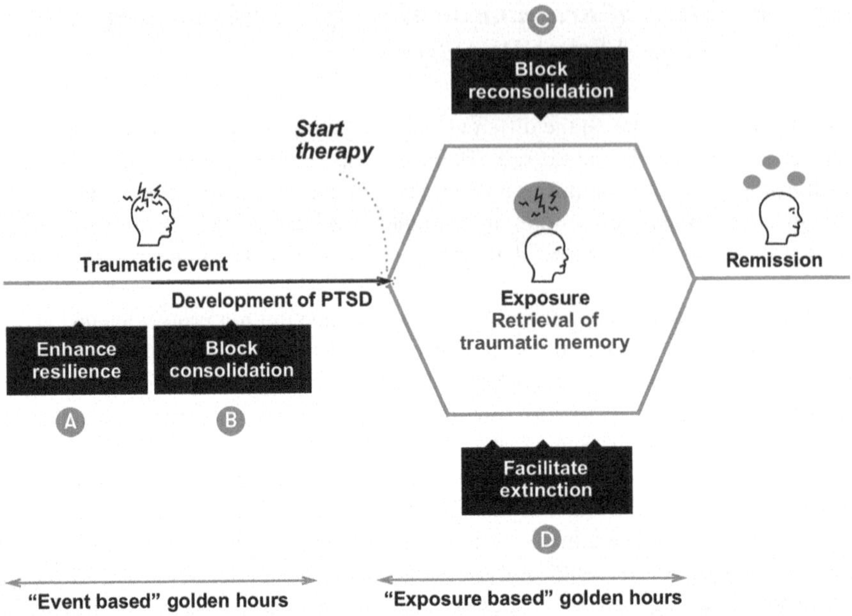

Fig. 2 This figure illustrates the development of traumatic memories over time. Four time points are identified for potential pharmacological interference with traumatic memory (formation). A. Prior to the exposure and formation of the traumatic memory, the compound could enhance resilience. B. Directly after exposure of a traumatic event, consolidation of the traumatic memory can be blocked. C. Blocking of reconsolidation directly after exposure within the time window of 0–6 h. D. Facilitation of extinction learning of traumatic memories

enhance resilience prior to a traumatic event or alter consolidation directly after a traumatic event?

A proof-of-mechanism study in mice showed the potential of ketamine to *enhance resilience* (Fig. 2A) to a stress exposure (Brachman et al. 2016; McGowan et al. 2017). They reported that injection of a subanesthetic dosage of ketamine prior to exposure of the CS + US in a fear-conditioning paradigm (Fig. 1) (in addition to the obvious effects on US perception) reduced fear behaviors in mice (McGowan et al. 2017). It was shown that this protective effect was only present within a specific time window. Injection of ketamine 1 week before CS + US exposure also enhanced resilience. However, injection of ketamine 1 h or 1 month before CS + US exposure did not result in this protective effect (McGowan et al. 2017). Furthermore, it is reported that the effect lasted beyond the half-life period of ketamine (Brachman et al. 2016; McGowan et al. 2017). Therefore the authors suggested that ketamine treatment might be implemented as a vaccine-like strategy in high-risk groups (Brachman et al. 2016; McGowan et al. 2017). The exact time window of this putative protective effect in humans would need to be more carefully investigated. For a successful translation of this approach to enhance resilience with ketamine, many other aspects, such as the dosage and dissociative side effects, need to be carefully considered. Administration of ketamine directly after the traumatic event, within *the window of consolidation* (Fig. 2B), is another opportunity. However preclinical studies show inconsistent result of the effect that ketamine administration has on consolidation of fear memories (Groeber Travis et al. 2015; Ito et al. 2015; Juven-Wetzler et al. 2014; Suzuki et al. 2004). Due to this conflicting evidence, it is currently too early to implement ketamine as an effective prevention strategy for adverse memory formation after a traumatic event. Possible preclinical studies in healthy humans could help to determine the effect of ketamine on consolidation of fear-related and emotional memory.

Another novel and promising approach is to implement ketamine in trauma-focused psychotherapies for PTSD, during exposure-based "golden hours" (Fig. 2). During reconsolidation and extinction learning, NMDA-receptor functioning is essential. We argued in Sect. 3 that *blocking reconsolidation* (Fig. 2C) might have a more rapid effect compared to *facilitation of extinction learning* (Fig. 2D). There is convincing evidence that NMDA-receptor antagonism blocks reconsolidation in preclinical studies (Das et al. 2013; Duclot et al. 2016; Suzuki et al. 2004). Via antagonistic binding to the NMDA receptor, ketamine has the potential to adapt NMDA-receptor functioning (Orser et al. 1997). Therefore, we hypothesize that targeting reconsolidation of pathological memory formation with ketamine during the exposure-based "golden hours" is a potential treatment strategy for PTSD.

To our knowledge, only a few preclinical studies have investigated the effect of ketamine on reconsolidation in a fear-conditioning paradigm in rodents (Duclot et al. 2016; Honsberger et al. 2015). Honsberger et al. found that an injection of ketamine directly after re-exposure did not lead to reduction of fear behavior in an auditory fear memory paradigm (Honsberger et al. 2015). However, Duclot et al. showed that ketamine administration directly after re-exposure induced disruption of contextual fear memory and led to a reduction of fear behavior. This study also showed that the reduction of fear is dependent on the dosage of ketamine. A lower dose would lead to fear reduction, whereas a higher dose of ketamine would not. This study showed that ketamine is effective in blocking reconsolidation of fear memory in rodents in an inverted U-shaped dose-response manner (Duclot et al. 2016). Girgenti et al. also reported about their experiments with ketamine in a fear extinction paradigm. Results showed that ketamine administration before re-exposure of 30 s reduced fear behavior in animals (Girgenti et al. 2017). However, this study remains inconclusive whether this effect is due to blocking reconsolidation or facilitation of extinction. Most evidence reports that extinction learning requires NMDA-receptor functioning (Quirk and Mueller 2008), and Suzuki et al. reported that re-exposure paradigm of 3 min induces reconsolidation (Suzuki et al. 2004). Therefore, we argue that it seems more feasible that the effect that Girgenti et al. showed is the result of NMDA-receptor-dependent blockade of reconsolidation. Preliminary results in nonhuman primates also showed a reduction of fear behavior after ketamine was administered during the reconsolidation phase (Philippens et al. 2017). These studies trigger the demand to investigate the effect of ketamine on reconsolidation of fearful memories in humans and, more specifically, to answer the question if ketamine is able to block reconsolidation of traumatic memories in patients with PTSD in relation to re-exposure during psychotherapy.

Ketamine is a promising candidate in the strategy of pharmacological assisted psychotherapy. Ketamine may indeed be able to reinforce the effect of psychotherapies. Patients might experience lower stress levels during therapy, and dropout rates could potentially be reduced. As far as we know, no studies have been performed with ketamine as additional treatment during trauma-focused psychotherapy. At the time of writing, two clinical trials are registered for the investigation of ketamine in combination with psychotherapy (The Cooper Health System n.d.).

In the next section, the steps that need to be taken to use ketamine as a method to specifically target pathological memory via reconsolidation are discussed.

5.2 How to Investigate Ketamine-Assisted Psychotherapy for PTSD That Targets Reconsolidation

Evidence that ketamine can modulate traumatic memory is predominantly circumstantial and originates mainly from preclinical animal data (Sect. 5.1) and retrospective observational studies in humans (Sect. 4.4). The implementation of ketamine as a viable pharmacological strategy to enhance psychotherapies in PTSD therefore requires further evaluation. To illustrate this point, the NMDA-receptor agonist D-cycloserine has recently been proposed as a promising candidate for pharmacologically augmented fear extinction during trauma-focused psychotherapy for PTSD (Rothbaum et al. 2014). Although numerous studies with D-cycloserine have been conducted, they lack methodological standardization such as dose selection, timing of administration relative to the psychotherapeutic invention, and primary outcome parameters; for review, see Mataix-Cols et al. (2017). Together, these issues contribute to wide range of compelling findings among research groups (for instance, post hoc classifications of good and bad therapy session response) and raise questions regarding its reliability, as a consequence, limited applicability in research and clinical practice. Thus the D-cycloserine case illustrates that designing a study to investigate pharmacologically augmented fear extinction during psychotherapy is not straightforward, since the compound's pharmacology, the psychotherapeutic intervention, and the functional CNS processes that underlie memory formation that are targeted all need to be considered.

In this context, ketamine can be conceived of as a novel pharmacological compound for the treatment of PTSD in general and the modulation of pathological traumatic memories in particular. As a consequence, a systematic approach is needed to assess its pharmacodynamic effects in relation to both psychotherapy and its pharmacokinetics prior to its application in patient populations. The "question-based drug development" approach has been proposed as a model to explore such aspects for novel drugs (Cohen et al. 2015). This approach consists of six questions that systematically address the most relevant aspects during drug development of a novel pharmacological treatment (Cohen et al. 2015). In addition, we formulated five topics that need to be addressed concerning the psychotherapeutic approach and phase orientation of pharmacological assisted psychotherapy (see Box 2). We believe that addressing these questions and topics represents a crucial step in designing a study for drug-assisted psychotherapy, which systematically addresses below.

Box 2: Five Essential Topics for Designing Pharmacological Assisted Psychotherapy, in Addition to the "Question-Based Drug Development" Approach

Concerning the psychotherapeutic intervention

- Describe the therapeutic process that is targeted with the medication (*memory processing).
- Outline the psychotherapeutic approach for the combined approach (*trauma-focused psychotherapy).
- Define the essential therapy conditions in order to affect the target process (*prediction error, proper amount of emotional involvement).

Concerning phase orientation

- Define the timing of drug administration in relation to the psychotherapeutic intervention (*after retrieval of the traumatic memory within the reconsolidation window).
- Define the proper frequency of the combined intervention (*hypothesized one time, since repeated infusion might prevent extinction).

*Short answers for ketamine-assisted psychotherapy to block reconsolidation which is extensively explained in Sect. 5.4.

5.3 "Question-Based Drug Development" Approach

5.3.1 Does Ketamine Reach the Site of Action?

When administered intravenously in humans, ketamine has a high bioavailability, readily crosses the BBB, and dose-dependently induces transient psychomimetic effects in healthy humans (Krystal et al. 1994) and patients with mood disorders (Berman et al. 2000; Feder et al. 2014) around the maximal ketamine plasma concentration (Cmax) (Kleinloog et al. 2015). In addition, ketamine displays relatively high in vitro binding affinity for the NMDA receptor (see Box 1), with S(+)-ketamine having a higher affinity than R(−)-ketamine or the racemic mixture. Together, these findings provide confidence that ketamine reaches the NMDA receptors located within the CNS. It should be noted though that ketamine's active metabolites norketamine and hydroxynorketamine have also been demonstrated to have CNS effects. In fact, they have been implicated to contribute to both the antidepressant and the nociceptive effects of ketamine (Box 1) (Mion and Villevieille 2013; Zanos et al. 2016). The exact extent to which the pharmacodynamic effects of ketamine in humans can be attributed to ketamine itself or to these metabolites still remains to be established and is subject to ongoing investigation.

5.3.2 Does Ketamine Exert an On-Target Pharmacological Effect?

For ketamine to effectively block reconsolidation of traumatic memories, it relies on its properties as central NMDA-receptor antagonist and possibly on some of its active metabolites, such as norketamine (Moaddel et al. 2013) (Box 1). Several potentially viable pharmacological biomarkers for NMDA-receptor antagonism have been suggested in literature. For example, gamma-band electroencephalography (EEG) has been proposed as a putative translational biomarker for antagonism of the NMDA receptor (Sanacora et al. 2014). In addition, in healthy volunteers ketamine demonstrates concentration-dependent decreases in saccadic peak velocity and smooth pursuit eye movements and increases of the total Positive and Negative Syndrome Scale (PANSS) (Kleinloog et al. 2015). Taken together, several candidate biomarkers for NMDA-receptor antagonism exist which may be applied to quantify ketamine's NMDA-receptor antagonistic properties in health and disease.

5.3.3 Does Ketamine Display Off-Target Pharmacological Effects?

Ketamine and its metabolites can bind to several non-glutamate receptors which may result in off-target pharmacological effects. Binding affinity for other receptors than the NMDA-receptor is generally much lower and can therefore be considered pharmacologically irrelevant (Hirota et al. 2002; Moaddel et al. 2013), with the exception of the dopamine 2 receptor (D2) and serotonin 5-HT2 receptor (Kapur and Seeman 2002). Ketamine is associated with concentration-dependent increases in peripheral serum prolactin (Kleinloog et al. 2015). Since prolactin release is modulated by dopamine in central tuberoinfindibular projections, these findings suggest that ketamine has potent D2 antagonistic properties. Ketamine is associated with the occurrence of transient increases in both systolic and diastolic blood pressure. The precise mechanism however is poorly understood and assumed to result from centrally mediated ANS activation.

5.3.4 Does Ketamine Display On-Target Pathophysiological Effects?

Central to this question is whether ketamine modulates memory reconsolidation by antagonizing the NMDA receptor and by doing that can enhance the effect of psychotherapy in PTSD. Preclinical research has shown that ketamine can block reconsolidation in rodents (Duclot et al. 2016) and nonhuman primates (Philippens et al. 2017). Investigators observed a reduction of fear behavior in a contextual fear paradigm and concluded that reconsolidation of the fear memory was blocked. Whether ketamine can block reconsolidation of fear memory in humans is yet to be studied. No reliable biomarker that quantifies the disruption of memory reconsolidation by an intervention, whether psychotherapeutic or pharmacological, yet exists and will need to be developed. The on-target pathophysiological response in patients with PTSD to ketamine would be the reduction of fear related to the traumatic memory, which is most commonly measured with questionnaires. An elegant example of a stepped approach is the study by the group of Kindt investigating the effect that propranolol (noradrenergic β-blocker) has on reconsolidation (Kindt et al. 2009; Soeter and Kindt 2015). First a fear-conditioning model induced emotional (fear) memories in healthy human subjects, by applying painful shocks related to specific visual cues. They showed that after re-exposure to these specific cues, reconsolidation of the (fear) memories can be blocked by propranolol administration (Kindt et al. 2009; Schwabe et al. 2012). They measured fear reduction with the startle response, which is one of the most robust biological methods to measure fear in healthy humans (Kindt et al. 2009). For the next step, they successfully translated these results for patients with spider phobia. They showed if patients are treated with propranolol concurrent with exposure to real spiders, patients displayed a reduction of fear behavior toward real spiders. Thus this study elegantly showed that propranolol blocked reconsolidation of fear memories in patients with spider phobia (Soeter and Kindt 2015). An important next step now is to investigate if propranolol can block reconsolidation of traumatic memories in PTSD (Brunet et al. 2008; Kindt and van Emmerik 2016). Thus, a fear memory paradigm in healthy humans provides a more objective marker to determine if ketamine can indeed block reconsolidation of human memories and will help to determine the on-target pharmacological effect in patients with PTSD.

5.3.5 Does Ketamine Have Any Off-Target Pathophysiological Effects?

Ketamine has a high in vitro affinity for the central D2 receptor (see Box 1) and induces transient increases in systolic and diastolic blood pressure. D2-mediated sustained hyperprolactinemia with antipsychotic drugs is associated with the occurrence of untoward effects such as galactorrhea and sexual dysfunction (Bostwick et al. 2009). However, since the incidental administration of ketamine during psychotherapy in PTSD is not expected to result in sustained hyperprolactinemia, unwanted clinical symptoms are not anticipated. Also, transient blood pressure elevations are not expected to cause safety issues, provided that PTSD patients with hypertension are carefully selected and blood pressure is monitored during trials with ketamine. Both clinicians and patients should be aware that ketamine is an addictive drug that has a strong potential to be abused (Kirby 2015). This is specifically important since numbers of substance abuse are high in patients with PTSD (Kessler et al. 1995).

5.3.6 What Is Ketamine's "Therapeutic Window" for Pharmacologically Assisted Psychotherapy Targeting Reconsolidation?

The most frequently applied dosing regimen for both depression and PTSD currently is 0.5 mg/kg ketamine over 40 min (Feder et al. 2014; Zarate et al. 2006). Notably, these dosages are based on consensus derived from clinical experience with ketamine in depression research and dose-ranging studies with ketamine as anesthetic agent (Berman et al. 2000; Domino et al. 1984; Zarate et al. 2006). In addition a single study in depression demonstrated similar antidepressant effects for both the 0.2 and 0.5 mg/kg dose (Singh et al. 2016). Together, these findings raise the issue of dose selection for pharmacologically assisted psychotherapy that targets reconsolidation in PTSD. Moreover, an inadequate dose may be associated with no effect, while supratherapeutic doses may be associated with burdensome side effects or loss of effect due to a specific stimulatory effects and/or dissociative symptoms that might obscure desirable effects on memory. It is not yet known what the required dose is to block reconsolidation of a memory and more specifically of traumatic memory in PTSD.

The issues raised by this question-based approach might challenge researchers to think thoroughly about the design of studies with ketamine in PTSD. We emphasize the benefits of a stepped approach to first investigate the potential of ketamine to block reconsolidation in healthy humans with a fear memory paradigm. This provides an opportunity to determine if ketamine can indeed block reconsolidation of memories in humans in a controlled setting. During this experiment, it might be considered to add a dose-ranging element in order to determine the plasma concentration that is required to block reconsolidation of memories in healthy humans. Simultaneously the occurrence of side effects can be monitored for these plasma concentrations. A subsequent step is to examine if reconsolidation of traumatic memories can be blocked after re-exposure during psychotherapy in patients with PTSD. Notably, the sensitivity to ketamine and the robustness of the traumatic memories in patients with PTSD can differ greatly from trauma controls. An important step is to determine the sensitivity of patients with PTSD to ketamine, compared with the sensitivity of healthy controls. EEG or saccadic peak velocity and smooth pursuit eye movements combined with the PANSS might be considered as biomarkers for the on-target effects of NMDA-receptor antagonism (Kleinloog et al. 2015; Mion and Villevieille 2013; Sanacora et al. 2014). These biomarkers can be helpful to determine the differences in sensitivity by measuring NMDA-receptor antagonism of ketamine in patients with PTSD and compare the results with healthy controls. When done for various doses and plasma concentrations, this provides information on how to translate results from a fear-conditioning paradigm in healthy humans to a study design for patients with PTSD. The results from a healthy subject design can also be useful for other specific anxiety disorders, which are somewhat easier to investigate due to less generalization of fear and more straightforward retrieval of memories. Thus, a stepped approach to unravel the optimal dosing for blockade of reconsolidation can be helpful before implementing ketamine as additional treatment during psychotherapy in patients with PTSD. Remaining topics that need to be addressed for pharmacological assisted psychotherapy concern the psychotherapeutic approach and phase orientation of both interventions, and these will be discussed in the next paragraph.

5.4 *Ketamine Embedded in a Psychotherapeutic Process*

To investigate if ketamine has an effect on reconsolidation of traumatic memories, it is important to expand the scope from the "question-based drug development" approach and consider also clinical aspects of pharmacological assisted psychotherapy. Researchers need to carefully think about how to best optimize the psychotherapeutic process and how medication and therapy may fit for optimal effect. A proper framework to carefully design studies that investigate medication-assisted psychotherapy is missing. We formulated five topics that need to be addressed in any study design of medication-assisted psychotherapy, also for ketamine-assisted psychotherapy. These issues need to be framed.

5.4.1 Describe the Therapeutic Process That Is Targeted with the Medication

For ketamine this is memory processing. With ketamine the goal is to change the memory engram of the consolidated traumatic memory by interfering with reconsolidation. It also may have an additional effect on affective processing, since the memories may be revisited in a slightly less fearful state, allowing more ease to revisit them. Perhaps we do not know enough yet to make a clear-cut distinction between manipulations of reconsolidation as opposed to extinction learning in current psychotherapies.

5.4.2 Outline the Psychotherapeutic Approach for the Combined Approach

For ketamine a trauma-focused psychotherapy is preferred. All specific techniques that are essential for re-exposure to the traumatic memory can be used, CPT, EMDR, or other exposure-based interventions. There are several protocols that describe these techniques (Schnyder et al. 2015).

5.4.3 Define the Essential Therapy Conditions in Order to Affect the Target Process

During re-exposure the aim is to retrieve and destabilize the traumatic memory. Only then it is possible to update the memory engram or interfere with subsequent reconsolidation. Memory "de"stabilization can occur merely under the circumstances that there is something new to learn during retrieval (Sevenster et al. 2012). Based on prior experience and patterns of response, the brain expects (or predicts) what will happen with a certain stimulus or situation. A memory enters the unstable (labile) phase in the case that the outcome of "what is about to happen" is not fully predictable (Krawczyk et al. 2017). For that reason the time and the magnitude of the exposure should be planned in a proper way in order to be able to block reconsolidation of the memory. Yet, the intensity/quantity of emotional involvement of patients during exposure needs to be addressed in order for this strategy to be well evaluated; over-involvement could lead to dissociation and anxious flashbacks, while under-involvement may lead to avoidance. An experienced therapist is able to establish balanced and well-tolerated spot-on exposure and prevent avoidance, which would be preferred over a more standardized design, such as listening to a tape of the traumatic event when cognitive avoidance may be more likely to occur. To identify the right "hot spots" for retrieval and destabilization of traumatic memories in PTSD is quite a task, and to establish the right dose of exposure in such a study could perhaps prove to be one of the most challenging aspects. Since our therapeutic strategy targets reconsolidation of traumatic memories, a critical parameter is that a significant level of distress in patients should originate from these intrusive traumatic memories and flashbacks. Notably, this strategy may not necessarily need to affect all symptoms of PTSD, like feelings of distrust and guilt that are often reported in patients.

5.4.4 Define the Timing of Drug Administration in Relation to the Psychotherapeutic Intervention

In this strategy it concerns the relation between timing of ketamine administration and opportunity of proper reconsolidation of the memory. There is some evidence that NMDA-receptor antagonism by ketamine can interfere with (or even prevent) retrieval of the memory (Milton et al. 2013), and therefore ketamine should not be administered before retrieval, thus not before exposure. In addition, the time-specific nature of the reconsolidation window needs to be taken into account, which is generally considered 0–6 h after retrieval (Nader et al. 2000). These aspects together with the pharmacological knowledge of ketamine (Tmax, elimination time, effect of metabolites, etc.) determine the best timing of administration.

5.4.5 Define the Proper Frequency of the Combined Intervention

The frequency of administration relies on the clinical goals. Overall, this strategy aims to make psychotherapies for PTSD more effective, e.g., an increased symptom reduction or obtaining a more robust and sustainable effect. It can be argued that one exposure session and a single infusion of ketamine should be investigated first. In this way, the effect of ketamine on reconsolidation of memory after exposure can be identified. However, reintegration sessions, independent of ketamine-assisted re-exposure session, are also important to integrate new learning into the patient's life. On another note, assuming that NMDA-receptor antagonism blocks extinction learning and if ketamine is integrated in multiple sessions, long-term studies can also induce undesirable effects. At this moment, there is no empirical evidence for the frequency in a psychotherapeutic approach.

6 Conclusion

To improve the treatment of PTSD, research should focus on targeting one of the core pathophysiological mechanisms, disrupted memory formation. Targeting NMDA receptors by using ketamine can be effective to change the retention of emotional and traumatic memories. There are promising first results for ketamine infusion without memory retrieval in PTSD. We argue that ketamine infusion after memory retrieval to block reconsolidation is another promising strategy for the treatment of PTSD. We propose the "question-based drug development plan" and the additional five topics as a framework for developing medication-assisted psychotherapy. By using this framework for ketamine-assisted psychotherapy to block reconsolidation in patients with PTSD, we showed that ketamine has potential but as well the gaps in the knowledge that need to be investigated. Additional preclinical research is needed to expand the understanding of the contribution of the NMDA receptor in the modulation, acquisition, and storage of memories for traumatic events before and after consolidation. This information will help to clarify whether targeting NMDA receptors with ketamine can indeed successfully contribute to the treatment of patients with PTSD.

Acknowledgments We are grateful to Dr. A. Feder and J. Boonstra for their insightful comments. We are grateful to Elien Phernambucq for her support in designing the illustrations.

References

Bearn J, O'Brien M (2015) "Addicted to euphoria": the history, clinical presentation, and management of party drug misuse. Int Rev Neurobiol 120:205–233. https://doi.org/10.1016/bs.irn.2015.02.005

Berman D, Dudai Y (2001) Memory extinction, learning anew, and learning the new: dissociations in the molecular machinery of learning in cortex. Science 291(5512):2417–2419. https://doi.org/10.1126/science.1058165

Berman RM, Cappiello A, Anand A, Oren DA, Heninger GR, Charney DS, Krystal JH (2000) Antidepressant effects of ketamine in depressed patients. Biol Psychiatry 47(4):351–354

Bisson JI, Ehlers A, Matthews R, Pilling S, Richards D, Turner S (2007) Psychological treatments for chronic post-traumatic stress disorder. Systematic review and meta-analysis. Br J Psychiatry 190:97–104. https://doi.org/10.1192/bjp.bp.106.021402

Bostwick JR, Guthrie SK, Ellingrod VL (2009) Antipsychotic-induced hyperprolactinemia. Pharmacotherapy 29(1):64–73. https://doi.org/10.1592/phco.29.1.64
Bowdle TA, Radant AD, Cowley DS, Kharasch ED, Strassman RJ, Roy-Byrne PP (1998) Psychedelic effects of ketamine in healthy volunteers: relationship to steady-state plasma concentrations. Anesthesiology 88(1):82–88
Brachman RA, McGowan JC, Perusini JN, Lim SC, Pham TH, Faye C et al (2016) Ketamine as a prophylactic against stress-induced depressive-like behavior. Biol Psychiatry 79(9):776–786. https://doi.org/10.1016/j.biopsych.2015.04.022
Brunet A et al (2008) Effect of post-retrieval propranolol on psychophysiologic responding during subsequent script-driven traumatic imagery in post-traumatic stress disorder. J Psychiatr Res 42 (6):503–506
Caddy C, Giaroli G, White TP, Shergill SS, Tracy DK (2014) Ketamine as the prototype glutamatergic antidepressant: pharmacodynamic actions, and a systematic review and meta-analysis of efficacy. Ther Adv Psychopharmacol 4(2):75–99. https://doi.org/10.1177/2045125313507739
Caddy C, Amit BH, McCloud TL, Rendell JM, Furukawa TA, McShane R et al (2015) Ketamine and other glutamate receptor modulators for depression in adults. Cochrane Database Syst Rev 9:CD011612. https://doi.org/10.1002/14651858.CD011612.pub2
Casey DE (1997) The relationship of pharmacology to side effects. J Clin Psychiatry 58 (Suppl 10):55–62
Cohen AF, Burggraaf J, van Gerven JMA, Moerland M, Groeneveld GJ (2015) The use of biomarkers in human pharmacology (phase I) studies. Annu Rev Pharmacol Toxicol 55:55–74. https://doi.org/10.1146/annurev-pharmtox-011613-135918
The Cooper Health System. TIMBER psychotherapy and ketamine single infusion in chronic PTSD. https://ClinicalTrials.gov/show/NCT02766192
Craske M (2015) Optimizing exposure therapy for anxiety disorders: an inhibitory learning and inhibitory regulation approach. Verhaltenstherapie 25(2):134–143
D'Andrea D, Andrew Sewell R (2013) Transient resolution of treatment-resistant posttraumatic stress disorder following ketamine infusion. Biol Psychiatry 74(9):e13–e14. https://doi.org/10.1016/j.biopsych.2013.04.019
Das RK, Freeman TP, Kamboj SK (2013) The effects of N-methyl D-aspartate and B-adrenergic receptor antagonists on the reconsolidation of reward memory: a meta-analysis. Neurosci Biobehav Rev 37(3):240–255. https://doi.org/10.1016/j.neubiorev.2012.11.018
Davidson J (2015) Vintage treatments for PTSD: a reconsideration of tricyclic drugs. J Psychopharmacol 29(3):264–269. https://doi.org/10.1177/0269881114565143
Domino EF, Luby ED (2012) Phencyclidine/schizophrenia: one view toward the past, the other to the future. Schizophr Bull 38(5):914–919. https://doi.org/10.1093/schbul/sbs011
Domino EF, Chodoff P, Corssen G (1965) Pharmacologic effects of Ci-581, a new dissociative anesthetic, in man. Clin Pharmacol Ther 6:279–291
Domino EF, Domino SE, Smith RE, Domino LE, Goulet JR, Domino KE, Zsigmond EK (1984) Ketamine kinetics in unmedicated and diazepam-premedicated subjects. Clin Pharmacol Ther 36(5):645–653
Donoghue AC, Roback MG, Cullen KR (2015) Remission from behavioral dysregulation in a child with PTSD after receiving procedural ketamine. Pediatrics 136(3):e694–e696. https://doi.org/10.1542/peds.2014-4152
Duclot F, Perez-Taboada I, Wright KN, Kabbaj M (2016) Prediction of individual differences in fear response by novelty seeking, and disruption of contextual fear memory reconsolidation by ketamine. Neuropharmacology 109:293–305. https://doi.org/10.1016/j.neuropharm.2016.06.022
Feder A, Parides MK, Murrough JW, Perez AM, Morgan JE, Saxena S et al (2014) Efficacy of intravenous ketamine for treatment of chronic posttraumatic stress disorder: a randomized clinical trial. JAMA Psychiat 71(6):681–688. https://doi.org/10.1001/jamapsychiatry.2014.62
Garakani A, Mathew SJ, Charney DS (2006) Neurobiology of anxiety disorders and implications for treatment. Mt Sinai J Med 73(7):941–949

Girgenti MJ, Ghosal S, LoPresto D, Taylor JR, Duman RS (2017) Ketamine accelerates fear extinction via mTORC1 signaling. Neurobiol Dis 100:1–8. https://doi.org/10.1016/j.nbd.2016.12.026

Groeber Travis CM, Altman DE, Genovese RF (2015) Ketamine administration diminishes operant responding but does not impair conditioned fear. Pharmacol Biochem Behav 139 (Part A):84–91. https://doi.org/10.1016/j.pbb.2015.10.013

Harps-Rotem I. Intensive 7-day treatment for PTSD combining ketamine with exposure therapy. https://ClinicalTrials.gov/show/NCT02727998

Hetrick SE, Purcell R, Garner B, Parslow R (2010) Combined pharmacotherapy and psychological therapies for post traumatic stress disorder (PTSD). Cochrane Database Syst Rev 7:CD007316. https://doi.org/10.1002/14651858.CD007316.pub2

Hijazi Y, Boulieu R (2002) Contribution of CYP3A4, CYP2B6, and CYP2C9 isoforms to N-demethylation of ketamine in human liver microsomes. Drug Metab Dispos 30(7):853–858

Hirota K, Hashimoto Y, Lambert DG (2002) Interaction of intravenous anesthetics with recombinant human M1-M3 muscarinic receptors expressed in chinese hamster ovary cells. Anesth Analg 95(6):1607–1610

Honsberger MJ, Taylor JR, Corlett PR (2015) Memories reactivated under ketamine are subsequently stronger: a potential pre-clinical behavioral model of psychosis. Schizophr Res 164 (1–3):227–233. https://doi.org/10.1016/j.schres.2015.02.009

Ito W, Erisir A, Morozov A (2015) Observation of distressed conspecific as a model of emotional trauma generates silent synapses in the prefrontal-amygdala pathway and enhances fear learning, but ketamine abolishes those effects. Neuropsychopharmacology 40(11):2536–2545. https://doi.org/10.1038/npp.2015.100

Jonas DE, Cusack K, Forneris CA, Wilkins TM, Sonis J, Middleton JC et al (2013) Psychological and pharmacological treatments for adults with posttraumatic stress disorder (PTSD). Agency for Healthcare Research and Quality, Rockville

Juven-Wetzler A, Cohen H, Kaplan Z, Kohen A, Porat O, Zohar J (2014) Immediate ketamine treatment does not prevent posttraumatic stress responses in an animal model for PTSD. Eur Neuropsychopharmacol 24(3):469–479. https://doi.org/10.1016/j.euroneuro.2013.08.007

Kapur S, Seeman P (2002) NMDA receptor antagonists ketamine and PCP have direct effects on the dopamine D(2) and serotonin 5-HT(2)receptors-implications for models of schizophrenia. Mol Psychiatry 7(8):837–844. https://doi.org/10.1038/sj.mp.4001093

Kessler RC, Sonnega A, Bromet E, Hughes M, Nelson CB (1995) Posttraumatic stress disorder in the National Comorbidity Survey. Arch Gen Psychiatry 52(12):1048–1060

Kindt M, van Emmerik A (2016) New avenues for treating emotional memory disorders: towards a reconsolidation intervention for posttraumatic stress disorder. Ther Adv Psychopharmacol 6(4):283–295

Kindt M et al (2009) Beyond extinction: erasing human fear responses and preventing the return of fear. Nat Neurosci 12(3):256–258

Kirby T (2015) Ketamine for depression: the highs and lows. Lancet Psychiatry 2(9):783–784. https://doi.org/10.1016/S2215-0366(15)00392-2

Kleinloog D, Uit den Boogaard A, Dahan A, Mooren R, Klaassen E, Stevens J et al (2015) Optimizing the glutamatergic challenge model for psychosis, using S+-ketamine to induce psychomimetic symptoms in healthy volunteers. J Psychopharmacol 29(4):401–413. https://doi.org/10.1177/0269881115570082

Krawczyk MC, Fernandez RS, Pedreira ME, Boccia MM (2017) Toward a better understanding on the role of prediction error on memory processes: from bench to clinic. Neurobiol Learn Mem 142(Pt A):13–20. https://doi.org/10.1016/j.nlm.2016.12.011

Krystal JH, Karper LP, Seibyl JP, Freeman GK, Delaney R, Bremner JD et al (1994) Subanesthetic effects of the noncompetitive NMDA antagonist, ketamine, in humans. Psychotomimetic, perceptual, cognitive, and neuroendocrine responses. Arch Gen Psychiatry 51(3):199–214

Lancaster C, Teeters J, Gros D, Back S (2016) Posttraumatic stress disorder: overview of evidence-based assessment and treatment. J Clin Med 5(11):105

Lee DJ, Schnitzlein CW, Wolf JP, Vythilingam M, Rasmusson AM, Hoge CW (2016) Psychotherapy versus pharmacotherapy for posttraumatic stress disorder: systemic review and meta-analyses to determine first-line treatments. Depress Anxiety 33(9):792–806

Li L, Vlisides PE (2016) Ketamine: 50 years of modulating the mind. Front Hum Neurosci 10:612. https://doi.org/10.3389/fnhum.2016.00612

Mallick F, McCullumsmith CB (2016) Ketamine for treatment of suicidal ideation and reduction of risk for suicidal behavior. Curr Psychiatry Rep 18(6):61. https://doi.org/10.1007/s11920-016-0680-7

Maren S, Phan KL, Liberzon I (2013) The contextual brain: implications for fear conditioning, extinction and psychopathology. Nat Rev Neurosci 14(6):417–428. https://doi.org/10.1038/nrn3492

Mataix-Cols D, Fernandez de la Cruz L, Monzani B, Rosenfield D, Andersson E, Perez-Vigil A et al (2017) D-Cycloserine augmentation of exposure-based cognitive behavior therapy for anxiety, obsessive-compulsive, and posttraumatic stress disorders: a systematic review and meta-analysis. JAMA Psychiat. https://doi.org/10.1001/jamapsychiatry.2016.3955

McGaugh JL (1966) Time-dependent processes in memory storage. Science 153(3742):1351–1358

McGhee LL, Maani CV, Garza TH, Gaylord KM, Black IH (2008) The correlation between ketamine and posttraumatic stress disorder in burned service members. J Trauma 64(2 Suppl): S195–S198.; Discussion S197–198. https://doi.org/10.1097/TA.0b013e318160ba1d

McGhee LL, Maani CV, Garza TH, Slater TM, Petz LN, Fowler M (2014) The intraoperative administration of ketamine to burned U.S. service members does not increase the incidence of post-traumatic stress disorder. Mil Med 179(8 Suppl):41–46. https://doi.org/10.7205/MILMED-D-13-00481

McGowan JC, LaGamma CT, Lim SC, Tsitsiklis M, Neria Y, Brachman RA, Denny CA (2017) Prophylactic ketamine attenuates learned fear. Neuropsychopharmacology. https://doi.org/10.1038/npp.2017.19

Miller CA, Sweatt JD (2006) Amnesia or retrieval deficit? Implications of a molecular approach to the question of reconsolidation. Learn Mem 13(5):498–505. https://doi.org/10.1101/lm.304606

Milton AL, Merlo E, Ratano P, Gregory BL, Dumbreck JK, Everitt BJ (2013) Double dissociation of the requirement for GluN2B- and GluN2A-containing NMDA receptors in the destabilization and restabilization of a reconsolidating memory. J Neurosci 33(3):1109–1115. https://doi.org/10.1523/JNEUROSCI.3273-12.2013

Mion G, Villevieille T (2013) Ketamine pharmacology: an update (pharmacodynamics and molecular aspects, recent findings). CNS Neurosci Ther 19(6):370–380. https://doi.org/10.1111/cns.12099

Moaddel R, Abdrakhmanova G, Kozak J, Jozwiak K, Toll L, Jimenez L et al (2013) Sub-anesthetic concentrations of (R,S)-ketamine metabolites inhibit acetylcholine-evoked currents in alpha7 nicotinic acetylcholine receptors. Eur J Pharmacol 698(1–3):228–234. https://doi.org/10.1016/j.ejphar.2012.11.023

Murrough JW, Iosifescu DV, Chang LC, Al Jurdi RK, Green CE, Perez AM et al (2013a) Antidepressant efficacy of ketamine in treatment-resistant major depression: a two-site randomized controlled trial. Am J Psychiatry 170(10):1134–1142. https://doi.org/10.1176/appi.ajp.2013.13030392

Murrough JW, Perez AM, Pillemer S, Stern J, Parides MK, aan het Rot M et al (2013b) Rapid and longer-term antidepressant effects of repeated ketamine infusions in treatment-resistant major depression. Biol Psychiatry 74(4):250–256. https://doi.org/10.1016/j.biopsych.2012.06.022

Murrough JW, Soleimani L, DeWilde KE, Collins KA, Lapidus KA, Iacoviello BM et al (2015) Ketamine for rapid reduction of suicidal ideation: a randomized controlled trial. Psychol Med 45 (16):3571–3580. https://doi.org/10.1017/S0033291715001506

Nader K (2015) Reconsolidation and the dynamic nature of memory. Cold Spring Harb Perspect Biol 7(10):a021782. https://doi.org/10.1101/cshperspect.a021782

Nader K, Schafe GE, LeDoux JE (2000) Fear memories require protein synthesis in the amygdala for reconsolidation after retrieval. Nature 406:722–726

Nader K, Hardt O, Lanius R (2013) Memory as a new therapeutic target. Dialogues Clin Neurosci 15(4):475–486

Orser BA, Pennefather PS, MacDonald JF (1997) Multiple mechanisms of ketamine blockade of N-methyl-D-aspartate receptors. Anesthesiology 86(4):903–917

Owens MJ, Morgan WN, Plott SJ, Nemeroff CB (1997) Neurotransmitter receptor and transporter binding profile of antidepressants and their metabolites. J Pharmacol Exp Ther 283(3):1305–1322

Oye I, Paulsen O, Maurset A (1992) Effects of ketamine on sensory perception: evidence for a role of N-methyl-D-aspartate receptors. J Pharmacol Exp Ther 260(3):1209–1213

Papp M, Moryl E (1994) Antidepressant activity of non-competitive and competitive NMDA receptor antagonists in a chronic mild stress model of depression. Eur J Pharmacol 263(1–2):1–7

Pavlov IP (1927) Conditioned reflexes: an investigation of the physiological activity of the cerebral cortex. Oxford University Press, Oxford

Pedreira ME, Maldonado H (2003) Protein synthesis subserves reconsolidation or extinction depending on reminder duration. Neuron 38(6):863–869

Peltoniemi MA, Hagelberg NM, Olkkola KT, Saari TI (2016) Ketamine: a review of clinical pharmacokinetics and pharmacodynamics in anesthesia and pain therapy. Clin Pharmacokinet 55(9):1059–1077. https://doi.org/10.1007/s40262-016-0383-6

Philippens I, Draaisma L, Krugers H, Vermetten E (2017) Ketamine treatment during reconsolidation of traumatic stress exposure: molecular biological validation of reduced avoidant behavior in marmoset monkeys. Poster presented at 56th meeting of the American College of Neuropsychopharmacology

Pomarol-Clotet E, Honey GD, Murray GK, Corlett PR, Absalom AR, Lee M et al (2006) Psychological effects of ketamine in healthy volunteers. Phenomenological study. Br J Psychiatry 189:173–179. https://doi.org/10.1192/bjp.bp.105.015263

Price RB, Nock MK, Charney DS, Mathew SJ (2009) Effects of intravenous ketamine on explicit and implicit measures of suicidality in treatment-resistant depression. Biol Psychiatry 66 (5):522–526. https://doi.org/10.1016/j.biopsych.2009.04.029

Przegalinski E, Tatarczynska E, Deren-Wesolek A, Chojnacka-Wojcik E (1997) Antidepressant-like effects of a partial agonist at strychnine-insensitive glycine receptors and a competitive NMDA receptor antagonist. Neuropharmacology 36(1):31–37

Quirk GJ, Mueller D (2008) Neural mechanisms of extinction learning and retrieval. Neuropsychopharmacology 33(1):56–72. https://doi.org/10.1038/sj.npp.1301555

Rauch SA, Eftekhari A, Ruzek JI (2012) Review of exposure therapy: a gold standard for PTSD treatment. J Rehabil Res Dev 49(5):679–687

Rothbaum BO, Price M, Jovanovic T, Norrholm SD, Gerardi M, Dunlop B et al (2014) A randomized, double-blind evaluation of D-cycloserine or alprazolam combined with virtual reality exposure therapy for posttraumatic stress disorder in Iraq and Afghanistan War veterans. Am J Psychiatry 171(6):640–648. https://doi.org/10.1176/appi.ajp.2014.13121625

Sanacora G, Smith MA, Pathak S, Su HL, Boeijinga PH, McCarthy DJ, Quirk MC (2014) Lanicemine: a low-trapping NMDA channel blocker produces sustained antidepressant efficacy with minimal psychotomimetic adverse effects. Mol Psychiatry 19(9):978–985. https://doi.org/10.1038/mp.2013.130

Schnyder U, Ehlers A, Elbert T, Foa EB, Gersons BP, Resick PA et al (2015) Psychotherapies for PTSD: what do they have in common? Eur J Psychotraumatol 6:28186. https://doi.org/10.3402/ejpt.v6.28186

Schonenberg M, Reichwald U, Domes G, Badke A, Hautzinger M (2005) Effects of peritraumatic ketamine medication on early and sustained posttraumatic stress symptoms in moderately injured accident victims. Psychopharmacology 182(3):420–425. https://doi.org/10.1007/s00213-005-0094-4

Schonenberg M, Reichwald U, Domes G, Badke A, Hautzinger M (2008) Ketamine aggravates symptoms of acute stress disorder in a naturalistic sample of accident victims. J Psychopharmacol 22(5):493–497. https://doi.org/10.1177/0269881107082481

Schwabe L, Nader K, Wolf OT, Beaudry T, Pruessner JC (2012) Neural signature of reconsolidation impairments by propranolol in humans. Biol Psychiatry 71(4):380–338

Sevenster D, Beckers T, Kindt M (2012) Retrieval per se is not sufficient to trigger reconsolidation of human fear memory. Neurobiol Learn Mem 97(3):338–345. https://doi.org/10.1016/j.nlm.2012.01.009

Shors TJ, Seib TB, Levine S, Thompson RF (1989) Inescapable versus escapable shock modulates long-term potentiation in the rat hippocampus. Science 244(4901):224–226

Singh JB, Fedgchin M, Daly E, Xi L, Melman C, De Bruecker G et al (2016) Intravenous esketamine in adult treatment-resistant depression: a double-blind, double-randomization, placebo-controlled study. Biol Psychiatry 80(6):424–431. https://doi.org/10.1016/j.biopsych.2015.10.018

Soeter M, Kindt M (2015) An abrupt transformation of phobic behavior after a post-retrieval amnesic agent. Biol Psychiatry 78(12):880–886

Suzuki A, Josselyn SA, Frankland PW, Masushige S, Silva AJ, Kida S (2004) Memory reconsolidation and extinction have distinct temporal and biochemical signatures. J Neurosci 24(20):4787–4795. https://doi.org/10.1523/JNEUROSCI.5491-03.2004

Trullas R, Skolnick P (1990) Functional antagonists at the NMDA receptor complex exhibit antidepressant actions. Eur J Pharmacol 185(1):1–10

Vermetten E, Krugers H (2016) A roadmap to golden hour intervention for posttraumatic stress disorder. Int J Neuropsychopharmacol 19:141–141

Vermetten E, Zhohar J, Krugers HJ (2014) Pharmacotherapy in the aftermath of trauma; opportunities in the 'golden hours'. Curr Psychiatry Rep 16(7):455. https://doi.org/10.1007/s11920-014-0455-y

Weber M, Hart J, Richardson R (2007) Effects of D-cycloserine on extinction of learned fear to an olfactory cue. Neurobiol Learn Mem 87(4):476–482. https://doi.org/10.1016/j.nlm.2006.12.010

Wilson JT, Pokorny ME (2012) Experiences of military CRNAs with service personnel who are emerging from general anesthesia. AANA J 80(4):260–265

Womble AL (2013) Effects of ketamine on major depressive disorder in a patient with posttraumatic stress disorder. AANA J 81(2):118–119

Woolf TF, Adams JD (1987) Biotransformation of ketamine, (Z)-6-hydroxyketamine, and (E)-6-hydroxyketamine by rat, rabbit, and human liver microsomal preparations. Xenobiotica 17 (7):839–847. https://doi.org/10.3109/00498258709043993

Yehuda R, LeDoux J (2007) Response variation following trauma: a translational neuroscience approach to understanding PTSD. Neuron 56(1):19–32. https://doi.org/10.1016/j.neuron.2007.09.006

Zanos P, Moaddel R, Morris PJ, Georgiou P, Fischell J, Elmer GI et al (2016) NMDAR inhibition-independent antidepressant actions of ketamine metabolites. Nature 533(7604):481–486. https://doi.org/10.1038/nature17998

Zarate CA Jr, Singh JB, Carlson PJ, Brutsche NE, Ameli R, Luckenbaugh DA et al (2006) A randomized trial of an N-methyl-D-aspartate antagonist in treatment-resistant major depression. Arch Gen Psychiatry 63(8):856–864. https://doi.org/10.1001/archpsyc.63.8.856

Zeng MC, Niciu MJ, Luckenbaugh DA, Ionescu DF, Mathews DC, Richards EM et al (2013) Acute stress symptoms do not worsen in posttraumatic stress disorder and abuse with a single subanesthetic dose of ketamine. Biol Psychiatry 73(12):e37–e38. https://doi.org/10.1016/j.biopsych.2012.10.017

Emerging Approaches to Neurocircuits in PTSD and TBI: Imaging the Interplay of Neural and Emotional Trauma

Andrea D. Spadoni, Mingxiong Huang, and Alan N. Simmons

Abstract Posttraumatic stress disorder (PTSD) and traumatic brain injury (TBI) commonly co-occur in general and military populations and have a number of overlapping symptoms. While research suggests that TBI is risk factor for PTSD and that PTSD may mediate TBI-related outcomes, the mechanisms of these relationships are not well understood. Neuroimaging may help elucidate patterns of neurocircuitry both specific and common to PTSD and TBI and thus help define the nature of their interaction, refine diagnostic classification, and may potentially yield opportunities for targeted treatments. In this review, we provide a summary of some of the most common and the most innovative neuroimaging approaches used to characterize the neural circuits associated with PTSD, TBI, and their comorbidity. We summarize the state of the science for each disorder and describe the few studies that have explicitly attempted to characterize the neural substrates of their shared and dissociable influence. While some promising targets in the medial frontal lobes exist, there is not currently a comprehensive understanding of the neurocircuitry mediating the interaction of PTSD and TBI. Future studies should exploit innovative neuroimaging approaches and longitudinal designs to specifically target the neural mechanisms driving PTSD-TBI-related outcomes.

A. D. Spadoni (✉) and A. N. Simmons
Research Service, VA San Diego Healthcare System, San Diego, CA, USA

Department of Psychiatry, University of California, San Diego, La Jolla, CA, USA
e-mail: aspadoni@ucsd.edu; alan@ucsd.edu

M. Huang
Radiology and Research Services, VA San Diego Healthcare System, San Diego, CA, USA

Department of Radiology, University of California, San Diego, La Jolla, CA, USA
e-mail: mxhuang@ucsd.edu

Curr Topics Behav Neurosci (2018) 38: 163–192
DOI 10.1007/7854_2017_35
Published Online: 8 January 2018

Keywords Diffusion tensor imaging (DTI) • fMRI • Functional neuroimaging • Magnetic resonance imaging • Magnetoencephalography (MEG) • Neuroimaging • Positron-emission tomography (PET) • Posttraumatic stress disorder • PTSD • Traumatic brain injury (TBI)

Contents

1 Introduction

Posttraumatic stress disorder (PTSD) and traumatic brain injury (TBI) are highly overlapping disorders (Stein and McAllister 2009). Based on a sample of 1965 OEF/OIF/OND veterans, an estimated near two thirds of those with PTSD and nearly one third had comorbid TBI (Tanielian and Jaycox 2008). Given their prevalence, TBI and PTSD have been termed the "signature wounds" of veterans returning from recent conflicts (Tanielian and Jaycox 2008). Moreover, TBI is thought to be an important risk factor for the development of PTSD. A recent study on 4,645 American veterans deployed to Afghanistan showed that those who experienced a TBI had 1.81 increased odds of having PTSD within 3 months and 1.48 at 9 months (Stein et al. 2015; Yurgil et al. 2014). Research also indicates that PTSD may mediate the impact of TBI-related symptoms. A survey of 2,525 post-deployment Army soldiers indicated that while TBI was associated with more physical health problems, PTSD and depressive symptoms mediated this relationship (Hoge et al. 2008). However, the mechanisms for these relationships are as yet not fully understood, and few imaging studies have focused on this comorbidity (Table 1). Thus, there is a high demand for more advanced and powerful methodology that may shed light on the relationship between TBI and PTSD for the purposes of (1) clarifying the psychiatric nosology specific to each disorder and their overlap and (2) providing opportunities for improved psychological and pharmacologic treatments. In this manuscript, we will review the methodologies and the findings from these modalities and discuss how these could potentially be applied to further our understanding of these commonly coupled disorders.

Table 1 Percentage of all published works per imaging method that examine each disorder and their comorbidity

Method	TBI/method (%)	PTSD/method (%)	PTSD + TBI/method (%)
MRI	0.23	0.17	0.01
DTI	2.61	0.60	0.15
fMRI	0.22	0.18	0.01
MEG	0.30	0.45	0.04
PET	0.15	0.11	0.01

These data suggest that only very small percentage of work addresses the crossover of PTSD and TBI. Data accessed as of 9/25/2017 at 5 p.m. EST

1.1 Neurocircuits in PTSD and TBI

There is a large and heterogeneous set of findings noted in a broad array of imaging studies with divergence both in structural and functional modalities describing PTSD and TBI. In addition, many functional imaging findings do not find results that are unique to these specific disorders. Indeed, meta-analyses consistently show substantial overlap across *numerous* psychiatric disorders. For example, studies in both anxiety and depression report increased activation in the amygdala, anterior insula, and anterior cingulate. There have been numerous meta-analyses specific to PTSD (O'Doherty et al. 2015; Patel et al. 2012; Simmons and Matthews 2012) and TBI (Eierud et al. 2014). Together these findings suggest that much of the brain imaging work may be tied to similar key functions and networks that regulate brain behavior, potentially with the insula, amygdala, and cingulate as important hubs (Menon 2011). While clinically and symptomatically PTSD and TBI show substantial overlap, the constructs point to a potential dissociable etiology. However, there is no currently accepted way to apportion such etiologies beyond clinical judgment. In an effort to understand the scope and the nature of this overlap, much of the early work on PTSD and TBI has focused on different aspects, approaches, and methodologies.

In this article, we will focus primarily on providing some familiarity in the range of imaging methodologies that are utilized for the inspection and dissociation of PTSD and TBI diagnoses, with special emphasis on novel methodologies or approaches that may gain greater prominence in the coming years. A brief introduction to numerous methods is given in the section Measurement of Neural Circuits and is followed by a Summary of Current Findings. Each section is divided into major imaging methodologies including magnetic resonance imaging (MRI), functional magnetic resonance imaging (fMRI), diffusion tensor imaging (DTI), positron-emission tomography (PET), magnetoencephalography (MEG), and others. In the summary section, we suggest future directions or analyses that may yield areas for growth in future work, with the goal of clarifying existing diagnostic criteria for TBI as separable from PTSD and ultimately elucidating mechanistic processes that may lead to efficacious treatments.

2 Measurement of Neural Circuits

2.1 MRI

Magnetic resonance imaging (MRI) was initially utilized for medical use in 1977. Since then, the application of this technology, and its ability to provide noninvasive imaging of internal structures, has seen particular growth and utilization for understanding the neural correlates of mental health diagnoses. By placing the brain within a standing magnetic field, the absorption and release of radiofrequency waves can be predicted and measured to determine structure and location of specific substances (Haacke et al. 1999). This highly prevalent technology is broadly available in hospitals and clinics and has allowed for numerous studies focused on finding structural differences in PTSD and TBI. The most often weighting of structural images is T1 weighted (short radiofrequency echo times, bright white matter, and dark cerebral spinal fluids (CSF)), although T2 weighted (longer radiofrequency echo times, dark white matter, and bright CSF) can be helpful for TBI or when taken in combination with T1.

As applied to psychiatric research, two major analysis paths have been employed to quantify brain structure either looking at the quantity or thickness of the gray matter. Voxel-based morphometry (VBM) uses signal intensities in MRI to estimate the probable density of gray matter in a region of the brain and the probable size of known cortical areas (Ashburner and Friston 2000). Conversely, these gray matter maps can be seen as two meshes, one on the inner surface and one on the external surface; from the distance between these two meshes, a cortical thickness can be derived (Dale et al. 1999; Fischl et al. 1999).

An additional recent advancement in methodology is macromolecular proton fraction (MPF) mapping, which provides information on the relative quantity of immobile macromolecular protons involved in the magnetization exchange with mobile water protons, and is characterized by marked distinctions between white matter and gray matter (Naumova et al. 2017). Recent work in humans has used MPF mapping for quantitative assessment of microscopic demyelination in both white and gray matter brain tissues (Yarnykh et al. 2015). The application of MPF in gray matter, in particular, may provide an additional sensitive clinical index of TBI and or PTSD burden.

2.2 fMRI

Functional MRI utilizes the methods of MRI, but by interpretation of distortion of the signal due to blood flow, it allows for inference of hemodynamic change (Ogawa et al. 1990). While the core of this technology has stayed relatively static in the last 20 years, there have been a few advancements in the collection capacity and substantial advancement in the statistical decomposition of the signal.

A notable technological advancement has been in the utilization of multiband sequences in the collection of MRI data, an advance that benefits primarily fMRI and DTI approaches. Principally, the number of coils used to emit and collect the radio frequencies for the measurement of change in blood flow has increased. A new approach has been to split these coils to run independently to allow for simultaneous data acquisition. This, in addition to other streamlining advances, such as multi-echo sequences and increased strength of the static field, has allowed for a profound increase in the speed and resolution of the acquired scans.

Functional MRI data have traditionally been collected in three-dimensional matrices of the brain (i.e., a set of voxels) repeatedly, over a duration of interest. This provides a four-dimensional matrix (i.e., three spatial and time), where the individual voxels show fluctuations over time. These fluctuations in signal are associated with either an external stimulus (using a correlation approach) (Friston and Penny 2003; Josephs et al. 1997; Penny and Friston 2003) or an internal stimulus (using a connectivity approach) (Friston et al. 1996). The seed-based connectivity approach takes the activation in a region of interest and observes other areas of the brain that show related activation. Another approach that can be taken is to decompose this signal into core components using independent component analysis (Beckmann et al. 2005). These regions can then be linked to known temporal or spatial patterns for further interpretation.

Recently, additional methods of note have emerged, e.g., graph theory has been applied to seed-based connectivity to provide a data driven, whole brain analytic approach that describes how brain networks interact together instead of simply within or between segregated regions (Bullmore and Sporns 2009). Such interactions may provide important information in distinguishing how these pathologies interact and diverge. This has led to the notion that areas acting in conjunction have local hubs and nodes.

Another interesting methodological advancement is multi-voxel pattern analysis (Norman et al. 2006) and multivariate Bayesian decoding (Friston et al. 2008) which provide a way to integrate multiple voxels in predicting or classifying brain state.

2.3 DTI

Diffusion-weighted imaging is a variant of MRI that characterizes water movements, or Brownian motion, within tissues. First introduced in the mid-1980s and early 1990s (Le Bihan et al. 1986), it has undergone significant refinement over the past three decades to accommodate improved description of white matter neural microstructure. Diffusion tensor imaging (DTI) was the first application of this technology, providing indirect information regarding the structural orientation of white matter fiber tracts, and is the most common method used to characterize the "integrity" or directional coherence of white matter (i.e., fractional anisotropy). Voxel-wise calculations of fractional anisotropy (FA) and mean diffusivity (MD) are the primary measures of white matter integrity and indicate the degree of restriction allowing for greater water diffusion along the length of an axon versus perpendicular to it. While these are the primary measures that have been used for assessment of TBI, other diffusivity measures, such as axial diffusivity (AD) and radial (RD) diffusivity, may be helpful in measurement of TBI (Mohammadian et al. 2017). Healthy tissue is generally associated with high FA and low MD values (Beaulieu 2002). Other qualitative metrics such as regional tract count (i.e., number of efferent and afferent tracts) and visualization of white matter tracts can also be obtained. Two common DTI analysis pathways include tract-based spatial statistics (TBSS) which allows for the creation of structural FA maps (Smith et al. 2006) and tractography which allows for the visualization of diffusion directional preference applied to map the white matter fiber pathways between specific regions (Basser et al. 2000). Recently, tractography has been utilized to perform graph theory analysis (Bullmore and Sporns 2009). These traditional measures of DTI are limited to conveying a single fiber orientation in each voxel. In regions of complex fiber geometries, this may result in ambiguous orientation estimates and subsequent failure of tractography. High angular resolution diffusion imaging (HARDI) (and similar e.g., "Q-ball imaging") was developed to better handle situations where multiple crossing white matter fibers/tensors mimic free water. HARDI and the orientation distribution function (ODF) (Tuch 2004) do not constrain the shape of diffusion (i.e., not necessarily elliptical) and thus allow for a measure of generalized fractional anisotropy (GFA) that summarizes integrity across multiple directions within an MRI voxel (Assemlal et al. 2007). This approach may be superior for the assessment of crossing fibers or areas in which axons converge or diverge.

2.4 PET

The first positron-emission tomography (PET) device used for large-scale cerebral scanning was developed in the 1950s for the detection of brain tumors (Portnow et al. 2013). Modern-era PET provides mapping of functional utilization of specific compounds by putting a radioactive tracer on a compound and measuring the release of positrons as the compound is metabolized in the body (Worsley et al. 1992). While traditionally sugar molecules are used as a tracer, such that energy utilization can be quantified, a plethora of markers have been developed to look at the transmission of serotonin (Drevets et al. 1999), dopamine (Elsinga et al. 2006), and other transmitters that have been relevant to psychiatric conditions.

2.5 MEG

Magnetoencephalography (MEG) is a noninvasive functional imaging technique that directly measures the magnetic signal due to neuronal activation in gray matter with high temporal resolution (<1 ms) and spatial localization accuracy (2–3 mm at cortical level) (Leahy et al. 1998). Although first introduced over 40 year ago (Cohen 1968), innovations over the past decade have dramatically improved its implementation. MEG measures neuronal activity from the gray matter with a population about 100,000 neurons (Hamalainen et al. 1993). Recent development of high-resolution MEG source imaging techniques such as the VESTAL (vector-based spatiotemporal analysis of L1 minimum norm) algorithm (Huang et al. 2006, 2014a, 2016a) allows the application of MEG to many psychiatric and neurological disorders such as PTSD and TBI.

2.6 Summary

Each imaging modality has a unique set of strengths and weaknesses important to acknowledge in the attempt to describe the interplay of PTSD and TBI. Inherent limitations of spatial and temporal resolution (Fig. 1) as well as cost, storage, and design (Table 2) support a multimodal approach with the goal of adequately describing the overlap of these two commonly comorbid disorders.

3 Summary of Current Findings

3.1 PTSD

MRI Hippocampal atrophy has been consistently related to PTSD (Karl et al. 2006; Kitayama et al. 2005; Nelson and Tumpap 2017; Smith 2005). Notably reductions occur in the absence of treatment (Sheline et al. 2003), but volume loss is mitigated

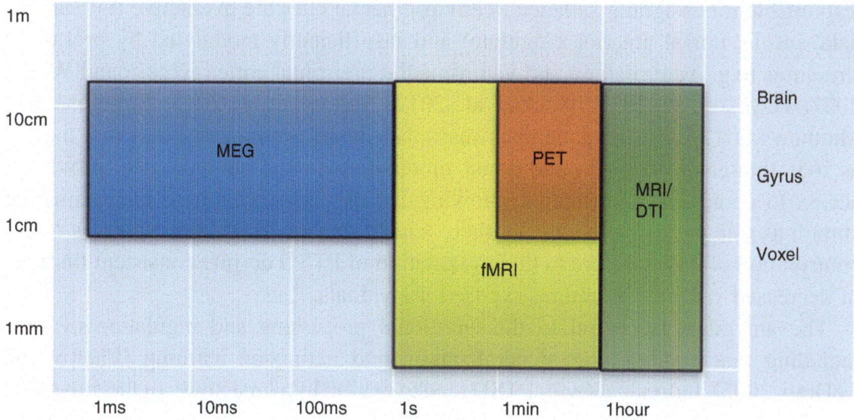

Fig. 1 Spatial vs temporal resolution of reviewed imaging methods. Adapted from Huettel et al. (2004)

Table 2 Advantages and challenges associated with reviewed imaging modalities

Imaging modality	Cost (h)	Storage	Design	PTSD[a]	TBI[a]	Advantages	Challenges
MRI	~$500–1,000	Low	Static	+	–	Availability; standardization	Sensitivity to treatment change
Diffusion imaging	~$500–1,000	Moderate-to-high	Static	–	+	Standardization; measurement of microstructure	Sensitivity to treatment change
fMRI	~$500–1,000	Moderate-to-high	Dynamic	++	+	Spatial resolution	Moderate temporal resolution
MEG	~$500–1,000	Moderate-to-high	Dynamic	+	++	Excellent temporal resolution; measurement of microstructure	Availability; moderate deep brain signal
PET	~ $2,000–4,000	Low-to-moderate	Mostly static	+	+	Targets specific neurotransmitters	Exposure to radioactive materials; low temporal resolution

Storage demands vary due to storage (raw P-files versus dicom files), scan parameters (multi-band, repetitions), and processing. Due to this variance a single file can range from less than 10 MB to over 10 GB

Ranking is subjective and anticipated to change with the issuance of further research and methodologies

[a]Potential utility in current literature

when treatment is provided (Vermetten et al. 2003; Videbech and Ravnkilde 2015). The hippocampus is critical for fear processing (Gewirtz et al. 2000; Kuhn and Gallinat 2013; Liberzon et al. 1999a; Rauch et al. 2006), in particular contextual information that modifies responses based on environmental cues (Phillips and LeDoux 1992).

fMRI A growing body of functional MRI literature supports the prevailing theory that PTSD is associated with a prefrontal–limbic imbalance wherein structures responsible for assigning salience to environmental cues are overactive (i.e., amygdala, insula, dorsal anterior cingulate) and insufficiently modulated by prefrontal structures (e.g., ventrolateral and ventromedial prefrontal cortex) (Etkin and Wager 2007; Hayes et al. 2012; Patel et al. 2012; Sartory et al. 2013; Simmons and Matthews 2012). The hippocampus may also figure prominently into this model, as it is thought to provide additional modulation to the amygdala by providing access to contextual information allowing for the classification of environmental stimuli as safe or unsafe. However, there have been few direct investigations of the contribution of this structure to the presentation of PTSD despite consistent findings of decreased volumes in trauma-exposed individuals.

The amygdala is central to the emotional processing and regulation of fear, including fear conditioning, generalization, and extinction learning (Phelps and LeDoux 2005). Individuals with PTSD tend to exhibit hyperreactivity of the amygdala to trauma-related stimuli (Hayes et al. 2012; Liberzon et al. 1999b) and greater amygdala functional connectivity with other regions of the salience network (i.e., insula) (Rabinak et al. 2011; Sripada et al. 2012a). Amygdala activity as measured by fMRI is predictive of symptom severity in PTSD patients (e.g., Liberzon et al. 1999b). It has also been hypothesized that exaggerated amygdala activity may partially account for the observed decrease in activity in the prefrontal cortex (PFC) by way of an increased feedback inhibition (Kelmendi et al. 2017). Exaggerated amygdala activity may also explain some of the core behavioral patterns in PTSD, such as hypervigilance and a failure to extinguish maladaptive fear response (Shin et al. 2006).

Evidence from functional neuroimaging studies suggests that dysfunction of the ventromedial PFC (vmPFC)–amygdala circuit may be an underlying mechanism driving PTSD symptomatology (e.g., Rauch et al. 2006). That is, in healthy individuals, volitional suppression of negative emotion and fear extinction are associated with increases in vmPFC activity and decreases in amygdala activity (e.g., Delgado et al. 2008; Milad et al. 2009). Therefore, the finding that ventral portions of medial prefrontal cortex (vmPFC) tend to be hyporesponsive (Etkin and Wager 2007; Felmingham et al. 2010; Milad et al. 2009), coupled with extensive evidence for exaggerated amygdala responsivity, argues for dysfunction of this circuitry.

The insula has figured more prominently in neural models of PTSD over the last several years. The anterior insula, which monitors internal homeostasis and the integration of homeostatic assessment (Craig 2009), has been posited as a key component in understanding PTSD (Paulus and Stein 2006). The right anterior insula responds primarily to interoceptive information, tying together cognitions, emotions, and internal body state (Craig 2003) and is a vital part of the salience network (Menon 2011).The anterior insula has been shown to encode affective distress that is associated with PTSD (Chen et al. 2009; Fonzo et al. 2010; King et al. 2009; Simmons et al. 2008, 2009; Strigo et al. 2010) and may even predict treatment response in this psychiatric samples (Dickie et al. 2011; Peres et al. 2011). Individuals with PTSD show increased activation in the insula during anticipation of an aversive image (Simmons et al. 2008), while this activation is reduced to those with marked resiliency traits (i.e., special forces operatives) (Simmons et al. 2012). New models tying the physiological and affective distress in PTSD to dysregulation in the insula have been proposed (Paulus and Stein 2006).

The exact contribution of amygdala, insula, and vmPFC to the chronology of PTSD and its clinical course has yet to be adequately studied. Recent meta-analysis suggests that PTSD may represent an excitation of the saliency network (SN) (i.e., amygdala, insula, and dorsal cingulate) and a suppression of the central executive network (CEN) (i.e., dorsal lateral prefrontal and posterior cingulate) (Patel et al. 2012). This network conceptualization of the disrupted networks overlaps and incorporates prior models of PTSD (Liberzon and Sripada 2008) and provides further evidence that relationships between these core networks may be at the heart of PTSD symptom presentation.

DTI Previous DTI studies examining the relationship between PTSD and white matter integrity in military (Schuff et al. 2011) and adult-onset PTSD civilian samples have mixed results, reporting reduced white matter integrity (lower FA) in the corpus callosum (Kitayama et al. 2007; Villarreal et al. 2004), prefrontal cortex (PFC) (Schuff et al. 2011), anterior cingulum (Kim et al. 2005; Schuff et al. 2011; Zhang et al. 2011), and posterior cingulum (Fani et al. 2012) and higher fractional anisotropy in the cingulum (Abe et al. 2006; Zhang et al. 2012) and superior longitudinal fasciculus (Zhang et al. 2011, 2012). In a recent study that compared adult-onset PTSD to controls, both exposed to a single, specific major trauma (8.0 earthquake), Li et al. (2016a) found significantly increased FA in the PTSD group in two regions of left dorsolateral prefrontal cortex (DLPFC) and in the left forceps major of the corpus callosum. Furthermore, the region of significantly decreased FA in the middle frontal gyrus was positively correlated with CAPS scores in those with PTSD (Li et al. 2016a). Results of this study are particularly interesting regarding the specific effects of PTSD on white matter, given that participants in both groups shared the traumatic event, were free of past or present psychiatric disorder, were unmedicated, and had a relatively short duration of illness (mean = 11 months). Questions remain regarding whether increased FA in DLPFC reflects a predisposition for the development of posttraumatic symptomatology rather than a consequence of the disorder or if decreased FA in the stressed control group may be associated with resilience. Differences from previous studies may also reflect sample characteristics such as chronicity, medication, psychiatric comorbidity, number of traumas, and variable methods to account for a history of TBI.

MEG MEG studies contrasting PTSD patients with healthy volunteers found hyperactivity from the amygdala, hippocampus, posterolateral orbitofrontal cortex (OFC), dorsomedial prefrontal cortex, and insular cortex in high-frequency (i.e., beta, gamma, and high-gamma) bands and hypoactivity from vmPFC, frontal pole, and dlPFC in high-frequency bands in those with PTSD. In individuals with PTSD, MEG activity in the left amygdala and posterolateral OFC correlated positively with PTSD symptom scores, whereas MEG activity in vmPFC and precuneus correlated negatively with symptom scores (Huang et al. 2014c).

MEG has also been used to delineate PTSD and healthy control subjects with >90% overall accuracy (Georgopoulos et al. 2010). Specifically, those with PTSD had differential communication between temporal and parietal and/or parieto-occipital right hemispheric areas with other brain areas (Engdahl et al. 2010).

PET Studies of posttraumatic stress disorder have shown increased amygdala activation, although this was not confirmed in resting FDG-PET studies (Molina et al. 2010).

3.2 *TBI*

MRI Brain changes in TBI emerge from an abrupt external physical force. These can result in clear regions of injury, such as in the case of an open head injury. In moderate or severe cases of TBI, even in closed head injuries, the brain often has a clear region of damage relating to the site of impact or the contralateral side of the brain. In these cases, the brain can often receive damage from the spiny ridge in the occipital bone. However, in case of mild TBI, the location of this damage cannot be detected by traditional scanning procedures. Some initial evidence suggests that hippocampal and temporoparietal damage may be related to blast injury (Wang and Huang 2013).

In a comparison of Iraq and/or Afghanistan veterans with a history of blast and blunt mTBI and without a history of mTBI, mean whole brain MPF values were lower in mTBI versus veterans without (Petrie et al. 2014). Results showed decreased values in multiple brain regions including the corpus callosum, cortical/subcortical white matter tracts, and gray matter/white matter border regions. Veterans with greater than 20 blast exposures had the lowest MPF values (Petrie et al. 2014). These findings suggest that axonal injury may be a primary marker of blast-related mTBI.

fMRI Recent meta-analyses of fMRI suggest a widely distributed network of structures that may be affected by mTBI (Eierud et al. 2014; Simmons and Matthews 2012). This is not surprising, given the heterogeneity of the impacts and sequelae that relate to mTBI. Work by Bonnelle et al. suggests that disruption in SN leads to increases in default mode network, notably the subgenual ACC (Bonnelle et al. 2012). These divergent findings may be highly dependent on the nature of impact and confounding emotional effects in this highly heterogeneous condition. Similarly, individuals who develop MDD after blast-related mTBI show maladaptive hyperactivity in emotion processing structures such as the amygdala and hypoactivity in emotional control structures such as the dorsolateral prefrontal cortex during performance of an emotional face matching task, irrespective of PTSD diagnosis (Matthews et al. 2011).

DTI Diffuse axonal injury (DAI) is caused by rapid acceleration–deceleration of the brain and has been identified as one of the most important causes of morbidity and mortality in patients with TBI (Frasure-Smith et al. 1995; Gean 1994). Such injury to white matter integrity is thought to drive observable clinical and behavioral symptoms associated with mild TBI (Arfanakis et al. 2002), although studies that test these specific relationships are sparse. Neuropathology and imaging studies of TBI have emphasized damage to white matter (Wilde et al. 2005, 2006), which is characterized by axonal stretching, disruption, and eventual separation of nerve fibers (Adams 1982). TBI alters brain tissue microstructure via widening of extracellular space secondary to glial cell shrinkage (Goetz et al. 2004), small hemorrhages within the white matter, and Wallerian degeneration (Cernak et al. 2001). These neuropathological changes lead to axonal collapse, breakdown of myelin, and possible disconnection effects (Povlishock and Katz 2005). Despite this evidence, little is known about the fundamental changes that occur in the brain of individuals who have sustained TBI from blast. Blast injuries may differ from mechanical force-related injury (e.g., motor vehicle accidents) because of different mechanisms of brain dysfunction. The mechanism of brain injury from blast results not only from DAI but also from focal injury from stroke due to air emboli that can form in blood vessels and travel to the brain (Okie 2005), and from trauma to other internal organs (i.e., the lungs or kidneys), which can affect brain function. Although results from animal studies indicate that DAI occurs even after relatively mild head injury (Frasure-Smith et al. 1995; Povlishock and Coburn 1989), diagnosis and detection of DAI are particularly challenging given that traditional neuroimaging (CT and MRI) is often insensitive to this type of white matter damage (Scheid et al. 2003). Consequently, TBI is frequently undetected or misdiagnosed, leading to inadequate treatment (Gentry et al. 1988).

Diffusion tensor analysis has revealed abnormalities in cerebellar white matter (MacDonald et al. 2011, 2013), orbitofrontal cortex (MacDonald et al. 2011), temporal regions, and callosal white matter (Petrie et al. 2014) among individuals with a history of blast injury. More generally speaking, mild TBI is associated with widespread white matter abnormalities such as lower FA and higher MD after injury (Aoki et al. 2012; Hulkower et al. 2013; Morey et al. 2013; Niogi and Mukherjee 2010), and studies of military cohorts generally echo this result. While some analyses failed to find a consistent relationship between military mTBI and specific hypothesized regions of white matter (Davenport et al. 2012; Levin et al. 2010; MacDonald et al. 2011; Sponheim et al. 2011), whole brain voxel-wise comparisons and overall measures of white matter integrity have since detected lower measures of FA associated with military mTBI (Davenport et al. 2012; Jorge et al. 2012; Morey et al. 2013; Yeh et al. 2014).

Yeh et al. (2014) employed a novel DTI "asymmetry analysis" to compare the effects of blunt-only vs blast + blunt TBI. Their results suggested that the mechanism of injury was related to the distribution of low FA clusters, as military personnel with blast + blunt TBI showed lowest FA in central superior–inferior-oriented tracts near subcortical regions (e.g., projection fibers interconnecting cortico–subcortical regions such as the superior corona radiata), while blunt trauma-only TBI subjects showed lowest FA in anterior–posterior-oriented tracts (e.g., anterior limbs of internal capsules) (Yeh et al. 2014). These results point to the potential utility of characterizing mTBI by mechanism of injury and could suggest altered sequelae with additional blast force neurotrauma.

MEG Resting-state MEG appears highly sensitive to abnormal neuronal signals resulting from brain injuries. Slow-waves, if present during wakefulness, are a sign of brain dysfunction (Kandel et al. 2013). Neurophysiological studies in animals have established a solid connection between pathological delta wave (1–4 Hz) generation in gray matter and injuries in white matter. Polymorphic delta-band slow-waves produced by white matter axonal lesions in cats were localized to the gray matter of cortex overlying the lesion (Ball et al. 1977; Gloor et al. 1977). Abnormal delta waves can also be induced by the administration of atropine in the white matter (Schaul et al. 1978). It is known that atropine is a competitive antagonist of acetylcholine (ACh) receptors and can block and/or limit ACh. These animal experiments concluded that cortical deafferentation was a key factor in abnormal delta wave production, owing to white matter lesions (i.e., axonal injury) and/or blockages/limitations in the cholinergic pathway. They also demonstrated that abnormal delta waves can directly result from axonal and/or cholinergic blockage/limitation.

Human studies in wakefulness suggest that the brains of mTBI patients generate abnormal low-frequency magnetic waves that can be measured and localized by resting-state MEG (Huang et al. 2009, 2012, 2014b; Lewine et al. 2007; Robb Swan et al. 2015). MEG may be more sensitive than conventional MRI or EEG in detecting abnormalities in mTBI patients (Lewine et al. 2007). Unlike normal resting-state MEG data, which is dominated by neuronal activity with frequencies above 8 Hz, injured neuronal tissues in many chronic neurological disorders (e.g., head trauma, brain tumors, stroke, epilepsy, Alzheimer's disease, and certain chronic neurovascular diseases) generate abnormal focal or multifocal low-frequency neuronal magnetic signals (delta-band 1–4 Hz, extending to theta-band 5–7 Hz) that can be directly measured and localized using MEG (Huang et al. 2014b). While TBI is not the only neurological disorder that generates abnormal slow-waves, in practice, brain tumor and stroke can be easily ruled out based on structural imaging (i.e., CT and MRI), whereas epilepsy, Alzheimer's disease, and other chronic neurovascular diseases (e.g., hypertension and diabetes) can be ruled out based on medical history. Using voxel-wise and ROI approaches, MEG slow-wave source imaging has been shown to detect abnormal slow-waves with ~85% sensitivity in patients with persistent post-concussive symptoms in chronic and subacute phases of mTBI (Huang et al. 2012, 2014b).

In addition, MEG source images were found to be correlated with neuropsychological exams (Robb Swan et al. 2015) and with abnormal eye movement (Diwakar et al. 2015) in individuals with mTBI. Recently, abnormal resting-state MEG functional connectivity in different frequency bands was also found in mTBI population (Engdahl et al. 2010; Huang et al. 2016b).

PET Studies have shown that patients with a history of TBI show diminished activation across wide areas of the brain as detected by PET (Kato et al. 2007; Levine et al. 2002; Nakayama et al. 2006; Shin et al. 2006; Stamatakis et al. 2002; Zhang et al. 2009) both during performance of a task and at rest. Another approach that has been successful in the differentiation of TBI from controls using PET has been the definition of clusters of contiguous voxels of outliers of FDG-PET. Notably, TBI patients showed larger clusters of deactivation that were located closer to the gray/white junction than in a healthy comparison group (Zhang et al. 2009). In an FDG-PET study of patients with TBI alone, or with PTSD + TBI compared to individuals without history of either condition, Buchsbaum et al. (2015) found that both TBI groups had larger clusters of larger, low activity, and more irregular in shape than combat controls (Buchsbaum et al. 2015).

3.3 PTSD TBI Interaction

MRI Despite the common clinical overlap (Stein and McAllister 2009), and evidence that TBI may increase the incidence and severity of PTSD (e.g., Vasterling et al. 2009), only a small number of MR studies have attempted to describe the interplay of TBI occurring in conjunction with PTSD. Examination of TBI in the presence of PTSD compared to TBI- and PTSD-only may provide some information regarding risk and resilience for this common dual diagnosis. For example, in a sample of veterans with PTSD-only ($n = 4$), TBI-only ($n = 32$; all severities), or PTSD + TBI ($n = 20$), Brenner et al. (2009) employed a standard clinical imaging approach using T2 gradient echo to evaluate TBI based upon the presence of encephalomalacia or hemosiderin deposits. TBI-only was significantly more often associated with MRI physical trauma-related findings than TBI + PTSD. There was no evidence of encephalomalacia or hemosiderin deposits in those individuals with PTSD-only (Brenner et al. 2009). One significant limitation of this study was that the TBI-only group had a higher number of participants with severe TBI ($n = 17/27$) than the TBI + PTSD group ($n = 4/32$). Negative findings signal a need for more sensitive measures of PTSD and TBI burden.

Lindemer et al. (2013) examined the interactive effects of PTSD and TBI on regional cortical thickness using T1-weighted high-resolution scans. Greater reductions in cortical thickness in bilateral superior frontal regions were found in veterans with higher cumulative lifetime burden of PTSD and mTBI compared to the PTSD-only group (Lindemer et al. 2013). Depue et al. (2014) compared comorbid PTSD + mTBI relative to PTSD-only or TBI-only groups using voxel-based and surface-based morphometry. They observed volumetric reductions in the bilateral anterior amygdala in comorbid PTSD + mTBI individuals compared to an OEF/OIF veterans without mTBI and/or PTSD (Depue et al. 2014). These observed structural alterations in frontal and limbic regions in dually diagnosed individuals point to additive effects in the central executive network and amygdala and may contribute to the putative role these structures play in the symptomatology of PTSD (Patel et al. 2012).

fMRI There have been few explicit attempts to delineate the shared and/or unique variance of PTSD and TBI using functional magnetic resonance imaging. One meta-analysis used activation likelihood estimation (ALE) to compare PTSD-only and TBI-only to control patterns of brain activation across 36 PTSD primary studies and 7 primary TBI studies. Results showed that the primary area of overlap, wherein both TBI and PTSD groups differed from control participants, was in the middle frontal gyrus (Simmons and Matthews 2012). More specifically, more activation was observed in those with PTSD and less activation in those with mTBI relative to controls in this region. Varying study and task design may have significantly contributed to the difference in direction (Simmons and Matthews 2012), rather than differences in the neural mechanisms of each disorder. Due to the small number of TBI studies available at the time, an updated analysis of this kind may illuminate additional areas of overlap.

Newsome et al. (2015) used fMRI to assess differences in brain activation during verbal working memory in Iraq/Afghanistan veterans ($n = 25$) with mild–moderate blast TBI, veterans without history of TBI ($n = 25$), civilians with a history of blunt force mild–moderate TBI ($n = 25$), and civilians with no history of TBI ($n = 25$) (Newsome et al. 2015). In regions that demonstrated a significant load effect during the encoding portion of the task (correct trials only), the military blast TBI group did not show the same monotonic increase in brain activation to increased set size (1 v 3 v 5) in the right head/body/tail of the caudate as did the three other groups, even after controlling for differences in age, education, PTSD, depression, fatigue, and pain symptoms. Of importance, military blast TBI group showed a significant and nearly significant negative relationship between caudate activations and reexperiencing and avoidance PTSD symptoms, respectively, suggesting a potential relationship between disrupted caudate response to increasing set size and these symptoms. However, as the military groups were not matched on levels of combat exposure or PTSD, and given the cross-sectional sample, this finding cannot be specifically linked to PTSD.

Interrogation of functional connectivity has been the other principal approach in evaluating the shared and dissociative contributions of TBI and PTSD to brain function. Whereas task-related connectivity describes the interaction of brain regions in the context of an external stimulus, resting-state connectivity derives from intrinsic patterns of neural synchrony. Work by Newsome et al. (2016) employed functional connectivity analyses to describe the intrinsic connectivity of subcortical gray matter structures (caudate, putamen, globus pallidus) and cortical gray matter (DLPFC) during resting state in veterans who had sustained a TBI from one or more blast exposures ($n = 17$; 15 mild, 1 moderate, 1 severe TBI) as compared to veterans who had been deployed but denied blast exposures or history of blunt TBI ($n = 15$) (Newsome et al. 2016). Notably, self-reported PTSD, neurobehavioral, and depression symptoms were significantly greater in the TBI group than combat controls. Results of functional connectivity without controlling for these group differences in PTSD and depression suggested (1) greater positive connectivity in the TBI group than the control group between the right putamen and the right angular gyrus and right lateral occipital cortex and (2) reduced connectivity in the TBI group between the right DLPFC and bilateral superior parietal lobule, supramarginal gyrus, and post-central gyrus (i.e., the control group demonstrated positive connectivity while the TBI group demonstrated an anti-correlation between regions). Previous work citing aberrant putamen connectivity in individuals with PTSD (e.g., Lei et al. 2015; Linnman et al. 2011) may suggest a specific vulnerability to the comorbid influence of TBI and PTSD in this region.

Spielberg et al. (2015) used graph theory to examine the interactive impact of PTSD and mTBI on default network connectivity in veterans with trauma exposure ($n = 208$) and mixed histories of current/past PTSD and exposures to blast and/or blunt mTBI (e.g., 96% with lifetime history of PTSD; 63% with mTBI) (Spielberg et al. 2015). Authors discovered a network including the caudate/putamen (basal ganglia) and the PFC wherein reexperiencing symptoms were related to decoupling of these regions but only in veterans with mTBI. Given previous work suggesting that the basal ganglia and PFC work interactively to provide the mechanism by which contextual information modifies working memory and distracting information (Badre and Frank 2012), authors theorized that weaker connections in this network may be associated with greater trauma-related intrusions in safe contexts (Spielberg et al. 2015). Worse caudate local efficiency, or ability to exchange information within a network, was also related to greater self-reported functional disability, underscoring the potential clinical utility of this finding (Spielberg et al. 2015). Furthermore, reexperiencing symptoms also predicted reduced insula participation, or decoupling, in the mTBI group. Given the well-established role of the insula in assessing the salience of an emotional stimulus (Simmons et al. 2013; Sripada et al. 2012b), disruption of this circuit may relate to increased errors in processing trauma-related versus safe stimuli. These findings propose specific neural mechanisms underlying a particular vulnerability to intrusive memories in the context of an mTBI. Tasks that assess these circuits in direct relation to trauma processing in the context of mild TBI are necessary to explore the potential specificity in the functional pathology associated with PTSD symptoms (Spielberg et al. 2015).

DTI Recent work reflects the growing need to examine the conjoint influence of TBI and PTSD on white matter. Jorge et al. (2012) used DTI to investigate white matter microstructure in $n = 72$ OEF/OIF veterans with blast-only mTBI and compared to 21 deployed veterans without mTBI (Jorge et al. 2012). Conventional voxel-wise tensor analysis revealed no between-group differences. However, a greater number of spatially heterogeneous areas of abnormally low fractional anisotropy, or "potholes," were identified in veterans with a history of mild TBI that were not related to age, time since trauma, presence of PTSD, mood disturbance, or alcohol misuse. A civilian comparison group with blunt-force trauma absent of psychopathology showed the greatest number of "potholes," suggesting specificity for mTBI. Across all TBI groups, the number of regions with reduced FA was also significantly associated with greater duration of posttraumatic amnesia, loss of consciousness, and poorer performance on measures of executive functioning (Jorge et al. 2012). In this sample, the presence of PTSD was not significantly related to white matter microstructure in veterans with mTBI.

Morey et al. (2013) employed voxel-wise analysis of diffusion metrics using HARDI which allowed for the quantification of primary and partial volume fractions of crossing fibers in (1) Iraq/Afghanistan veterans with mTBI with comorbid PTSD and depression ($n = 30$) and (2) non-TBI veterans without PTSD from primary control ($n = 42$) and confirmatory control ($n = 28$) groups. After controlling for age, PTSD, number of TBI events, duration of LOC, and presence of feeling dazed or confused, mTBI relative to control groups was associated with an extensive number of regions with lower white matter integrity including the corpus callosum, forceps minor and major, superior and posterior corona radiata, internal capsule, superior longitudinal fasciculus, among others (Morey et al. 2013). Interestingly, PTSD, in the context of TBI, was not associated with white matter integrity.

Bazarian et al. (2013) used conventional T1-/T2-weighted MRI and voxel-wise and ROI analysis of diffusion values (FA, MD) to investigate the relationship between measures of PTSD and mTBI acquired during deployment (Bazarian et al. 2013). In 52 OEF/OIF veterans with mixed diagnoses (i.e., PTSD and mTBI ($n = 9$), PTSD only ($n = 6$), mTBI only ($n = 21$) and neither PTSD nor mTBI ($n = 16$)), approximately 4 years post duty, self-reported PTSD severity was significantly predicted by whole brain MD values that fell into the 1st percentile of the sample distribution, as well as severity of exposure to combat events, age, time since last tour, and abnormal T1-/T2-weighted MRI. Of note, a clinical self-report measure of mTBI was not predictive of PTSD severity. They also found that blast exposure was significantly associated with 1st percentile FA, abnormal T1-/T2-weighted imaging, and severity of exposure to traumatic events. This small but well-controlled study may suggest that self-reported mTBI may not relate to neural changes obtained via brain injury and that biomechanical forces experienced during combat exposure may increase vulnerability for PTSD (Bazarian et al. 2013).

Yeh et al. (2014) used voxel-wise analysis of diffusion metrics, tract-specific analysis, asymmetrical analysis, and fiber tracking to characterize white matter microstructure in active military personnel with a history of TBI (i.e., blast + blunt and blunt-only-related concussion) and non-deployed military controls. DTI findings were examined in relation to post-concussive neuropsychological, neurobehavioral, and posttraumatic stress symptoms. Compared to military control subjects ($n = 14$), deployed individuals with a history of mild ($n = 29$), moderate ($n = 7$), and severe ($n = 1$) TBI had significantly lower FA and higher trace, or the magnitude of diffusion in a voxel, in a number of white matter tracts located principally in the frontostriatal and fronto-limbic circuits, as well as in the midbrain and brainstem regions (Yeh et al. 2014). These regions of lower FA were related to higher self-reported PTSD and neurobehavioral symptoms.

Petrie and colleagues also attempted to tease apart the influence of PTSD and TBI on white matter in Iraq- and Afghanistan-deployed veterans. They failed to find significant differences on measures of white matter anisotropy (FA) between blast-mTBI veterans ($n = 34$) with and without PTSD (Petrie et al. 2014). Importantly, there were no significant associations between PTSD symptoms and number of reported concussive events.

Davenport et al. (2015) utilized HARDI and the orientation distribution function (ODF) to study white matter in 133 OEF/OIF veterans with PTSD and/or mTBI, as compared to combat controls absent of both conditions (Davenport et al. 2015). Results indicated that PTSD was associated with high GFA in anterior thalamic radiations, corticospinal tract, inferior fronto-occipital fasciculus, and inferior and superior longitudinal fasciculus. PTSD was also associated with lower MD in the corticospinal tract. Traditional tensor measures of FA, MD, and AD proved much less sensitive to PTSD than GFA, and there were no main effects for TBI or TBI + PTSD on GFA, FA, or MD suggesting that measures that account for multiple fiber orientations may be especially sensitive to the relationship of PTSD to white matter microstructure. Authors speculated that relative elevations in FA in association with PTSD may represent preexisting vulnerabilities or symptom-related overuse in association with anxiety regulation. Lopez et al. (2017) examined individuals with mTBI ($n = 27$) or mTBI + PTSD ($n = 12$) using a DTI tensor model, volumetric imaging, and neuropsychological testing (Lopez et al. 2017). The mTBI + PTSD was found to have significantly reduced axial diffusivity (diffusion along the axon) in the right cingulum bundle, significantly *larger* volume in the right entorhinal cortex, and lower scores on tests of mental flexibility, processing speed, encoding, and retrieval. These findings indicate the addition of PTSD to mTBI may be associated with decreased white matter integrity, larger medial temporal volumes, and poorer neuropsychological performance.

In a sample of OEF/OIF veterans with a mild or moderate TBI ($n = 38$) (blunt and/or blast) and $n = 17$ veteran controls, Sorg et al. (2016) examined white matter correlates of neurocognition and PTSD symptoms (Sorg et al. 2016). After controlling for PTSD symptoms, a history of TBI significantly predicted lower FA in the genu of the corpus callosum and left cingulum bundle, and higher radial diffusivity in bilateral cingulum bundle and genu of the corpus callosum. PTSD symptoms, however, were not predictive of diffusivity values with or without controlling for mTBI. Similarly, Sorg et al. found that while TBI history was significantly associated with poorer performance on tests of memory and psychomotor speed, PTSD symptoms did not account for significant variance in test scores.

Using DTI, Dretsch and colleagues (Dretsch et al. 2017) assessed white matter integrity associated with posttraumatic symptomatology and mTBI in $n = 74$ active-duty US soldiers with posttraumatic symptoms (PTS) ($n = 16$), PTS with a comorbid history of mTBI (PTS/mTBI; $n = 28$), and a military control group ($n = 30$). They did not find significant between-group differences in the number of abnormal DTI values (i.e., >2 SDs from the mean of the control group) for FA and MD or differences in mean FA or MD across ROIs in commissure, association, and projection tracts. While the comorbid PTS/mTBI group had significantly greater traumatic stress, depression, anxiety, and post-concussive symptoms, and performed worse on neurocognitive testing compared to those with PTS and controls, more severe clinical presentation in the comorbid group did not appear to be moderated by pathological changes in white matter induced by mTBI (Dretsch et al. 2017).

Longitudinal work may help clarify conflicting findings and link them mechanistically to PTSD, TBI and their clinical course. For example, using conventional DTI, Li et al. (2016b) measured FA and MD at acute (3 days), subacute (10–20 days), and chronic stages (1–6 months) following blunt mTBI and examined whether these values prospectively relate to the development of PTSD in community participants recruited from the hospital emergency departments (Li et al. 2016b). Relative to healthy controls, mTBI was equally associated with significantly higher FA in the acute phase, regardless of eventual diagnosis. However, compared to both healthy controls and individuals who did not develop PTSD, those with onset of PTSD within 6 months also showed increased MD in subacute phase and decreased FA and increased MD in the chronic phase in several regions including the corpus callosum, inferior fronto-occipital fasciculus, uncinate fasciculus, superior longitudinal fasciculus, inferior longitudinal fasciculus, anterior thalamic radiation, and corticospinal tract. A Bayesian discriminant analysis suggested predictive classification overall accuracy of 76% using subacute MD values, while FA failed to show discriminative significance. Findings are consistent with previous reports of elevated FA within days of mTBI injury (Eierud et al. 2014) but additionally suggest that measurements of cerebral edema within weeks of the injury may help to identify blunt injury – mTBI patients with an increased risk of PTSD (Li et al. 2016b).

MEG MEG has been used to study individuals with comorbid mTBI and PTSD (Huang et al. 2016c). Abnormal MEG slow-wave generation was found in vmPFC and bilateral dlPFC areas, indicating potential injuries due to mTBI. The slow-wave generation suggests mTBI in these PFC areas. In addition, similar to the result from an earlier cohort of PTSD-only subjects (Huang et al. 2014c), these comorbid patients also showed MEG hypoactivity from vmPFC and dlPFC in high-frequency bands when compared with the HCs. In this comorbid group, the coexistence of an abnormal MEG slow-wave (mTBI component) and hypoactivity of vmPFC and dlPFC in high-frequency bands (PTSD component) suggests the mTBI injuries in PFC may result in a lack of inhibition from PFC to other areas of the PTSD neurocircuitry. These data, thus, provide evidence of abnormal slow-wave generation in these PFC areas due to mTBI and the resulting potentiation of PTSD.

PET PET has also been applied to help delineate the separate influence of TBI and PTSD on brain functioning. In an FDG-PET study of combat veterans with TBI alone, with PTSD + TBI, compared to individuals without history of either condition ($n = 15$ combat controls), Buchsbaum et al. (2015) found that comparisons between TBI and TBI + PTSD were not significant, which could be in part due to the fact that the TBI group had elevated scores on a measure of PTSD. Also, there was no PTSD comparison group, which may have provided more clarity (Buchsbaum et al. 2015). Petrie et al. (2014) also did not find significant differences in TBI vs TBI + PTSD groups on FDG-PET, which may suggest that this approach lacks in sensitivity for detecting TBI vs TBI + PTSD differences (Petrie et al. 2014).

4 Summary

In this paper, we reviewed a number of core and emerging methodologies and the findings from these modalities. In summary, we discussed how these could potentially be applied to further our understanding of the coupled disorders of PTSD and TBI. Functional imaging techniques may hold promise for detecting subtle differences in neuropathology, as these approaches (fMRI, MEG) have shown some evidence of specific variance due to PTSD and TBI in concert, while volumetric and microstructural studies have shown potentially sensitivity to the contribution of PTSD in the context of TBI.

Functional MRI provides preliminary evidence for aberrant connectivity of the basal ganglia and insula in dually diagnosed individuals (Newsome et al. 2015, 2016; Spielberg et al. 2015). Caudate/putamen and insula connectivity were related to reexperiencing symptoms in individuals with mTBI (Spielberg et al. 2015), a finding that begins to elucidate mechanistic processes driving the increased rate of PTSD in those with mTBI. Authors of a meta-analysis, agnostic for design or approach, found that the middle frontal gyrus was an overlapping area of potential vulnerability (Simmons and Matthews 2012). Results of these studies point to the continued investigation of frontal and limbic regions, and their interplay, in the increased incidence of PTSD in those with TBI. Graph theory may provide an especially powerful tool to examine the mechanistic basis of dual diagnosis that is less dependent on nodal differences. PET has yet to provide evidence of its efficacy to describe the interplay of PTSD and TBI neural correlates; however work that has looked at the shape rather than location of divergent findings in TBI may be more sensitive.

Results of MEG work in comorbid individuals also provide some additional evidence of separable contributions of TBI and PTSD to the mechanism of increased vulnerability to PTSD in those with TBI. Specifically, the confluence of slow-wave activity in ventromedial and bilateral dorsolateral frontal regions, consistent with TBI, and hypoactivity in high-frequency bands in these same regions, consistent with PTSD, suggests that mTBI injuries to the PFC may aid in the potentiation of PTSD symptoms. These findings are consistent with the hypothesis that mTBI may serve as a catalyst for the onset of PTSD (Bazarian et al. 2013).

Interrogation of macrostructural and microstructural characteristics have provided mixed results. Measurements of volume/thickness suggest structural alterations in superior frontal and amygdalar regions of dually diagnosed individuals (Depue et al. 2014; Lindemer et al. 2013) and are consistent with working models of symptomatology of both disorders, in which limbic activity is thought to be either exaggerated and thus difficult to regulate effectively or wherein limbic function is poorly regulated due to compromised frontal function. Sorg et al. (2016), Jorge et al. (2012), Morey et al. (2013), and Petrie et al. (2014) show evidence for TBI-related reductions in white matter integrity in dually diagnosed individuals relative to non-TBI controls (Jorge et al. 2012; Morey et al. 2013; Petrie et al. 2014; Sorg et al. 2016). However, the additive effect of PTSD remains unclear, as these studies failed to demonstrate a significant relationship between PTSD and white matter microstructure in the context of TBI. Furthermore, Dretsch et al. (2017) did not find significant differences in white matter across groups with posttraumatic symptomatology, mixed posttraumatic symptomatology and mTBI, and controls despite a greater preponderance of clinical symptoms in the mixed group. However, Bazarian, Davenport, Li, Lopez, and Yeh all found evidence that white matter microstructure was related to PTSD symptomatology in the context of TBI (Bazarian et al. 2013; Davenport et al. 2015; Li et al. 2016b; Lopez et al. 2017; Yeh et al. 2014). These findings may be dependent upon time of assay (e.g., Li et al. 2016b) or measurement (e.g., Davenport et al. 2015). Ultimately, multimodal approaches in which functional and microstructural relationships are dually specified, ideally in a longitudinal manner, may provide the most insight as to the nature of the observed interactions.

4.1 Limitations

Mixed etiology within group such as mechanism of injury (blast + blunt), time since event, and premorbid trauma exposure presents a major challenge in the delineation of neurobiological changes in response to these injuries and their comorbidities. Samples of generous size that allow for the examination of symptoms in the context of one another in continuous models may provide essential information regarding the neural underpinnings of these oft-shared disorders and their heterogeneous presentations. Multivariate continuous approaches may also be advantageous as categorically defined groups are difficult to match on exposures (e.g., TBI + PTSD tends to have greater trauma exposure than TBI-only or PTSD-only groups; TBI-only may have greater severity of TBI than TBI + PTSD groups). Additionally, full factorial designs are costly to image with adequate sample size. Indeed, collaborative data-sharing approaches maybe the most effective way to achieve adequate sample sizes to address these concerns (Thompson et al. 2014).

In terms of course, due to the timing of injury and trauma exposure, there may be some inherent limitations to the conclusions we can draw about the interplay of TBI and PTSD. For example, the fact that PTSD tends to follow TBI in those deployed to combat zones (e.g., Hoge et al. 2008) is of questionable utility, as many veterans (irrespective of TBI) report first noticing their PTSD symptoms when they have returned from deployments, wherein symptoms such as hypervigilance are no longer adaptive. Additionally, when there are multiple, similar, traumatic experiences, there may not be a specific event or even a finite number of events that are the singular cause of the onset of trauma-related symptomatology. Therefore, a TBI that occurred at any point throughout these exposures may appear to precede or follow a trauma (e.g., the reported worst trauma). Therefore, a purported temporal precedence may be misleading, as the posttraumatic symptomatology may be linked more strongly to the overall burden of stress than a singular event, and a reported TBI may not therefore occur as a clear risk factor for PTSD. Likewise, susceptibility to TBI sequelae following traumatic exposures may be similarly misleading. Relationships between TBI effects and trauma exposures might be more clearly delineated in a sample wherein the insults are not temporally overlapping. For example, groups of individuals with separable causal events for TBI (e.g., post-training, pre-deployment breachers) and criterion A trauma (e.g., post-training, post-deployment breachers with combat trauma exposure) and measurement at multiple time points may be helpful in further defining the mechanisms of vulnerability and resilience. The use of objective measurements of TBI presence, in advance of PTSD (or ideally trauma exposure), may be of greater utility in discerning a mechanistic contribution of TBI to the onset of PTSD. Bazarian and colleagues' finding that MR and DTI metrics predicted PTSD severity, but not clinical diagnosis of mTBI, underscores this notion (Hoge et al. 2008).

Finally, cohorts were also mostly cross-sectional, small in size, and many participants were taking medications (e.g., sedatives) at the time of participation. At this time, the literature on the comorbidity of TBI and PTSD is more or less at an exploratory stage. With improved utilization of resources, such as cross sharing of cohorts, or data sharing, the design of such studies could be greatly improved. Conventional measures in parallel with novel measures help to both validate previous results but also push the field forward include GFA, ODF, asymmetry analyses, and MPF. Additional measures of potential interest are genetic measures to assess markers of vulnerability and resilience in conjunction with neuroimaging markers and neuropsychological assessments to help understand whether neurobiological substrates may correlate with cognitive deficits that inhibit recovery from mTBI + PTSD.

Acknowledgments The authors appreciate the contributions of Michelle Liu to the literature search. Dr. Spadoni's VA salary is supported by a Career Development Award and Dr. Simmons' VA salary is supported by a Merit Award through the VA Clinical Science Research and Development Service. Additional support was generously provided by the VA Center for Excellence in Stress and Mental Health.

References

Abe O, Yamasue H, Kasai K, Yamada H, Aoki S, Iwanami A, Ohtani T, Masutani Y, Kato N, Ohtomo K (2006) Voxel-based diffusion tensor analysis reveals aberrant anterior cingulum integrity in posttraumatic stress disorder due to terrorism. Psychiatry Res Neuroimaging 146: 231–242

Adams JH (1982) Diffuse axonal injury in non-missile head injury. Injury 13:444–445

Aoki Y, Inokuchi R, Gunshin M, Yahagi N, Suwa H (2012) Diffusion tensor imaging studies of mild traumatic brain injury: a meta-analysis. J Neurol Neurosurg Psychiatry 83:870–876

Arfanakis K, Haughton VM, Carew JD, Rogers BP, Dempsey RJ, Meyerand ME (2002) Diffusion tensor MR imaging in diffuse axonal injury. Am J Neuroradiol 23(5):794–802

Ashburner J, Friston KJ (2000) Voxel-based morphometry – the methods. NeuroImage 11:805–821

Assemlal HE, Tschumperle D, Brun L, IEEE (2007) Fiber tracking on HARDI data using robust ODF fields. In: 2007 I.E. international conference on image processing, vol 1–7, pp 1261–1264

Badre D, Frank MJ (2012) Mechanisms of hierarchical reinforcement learning in cortico-striatal circuits 2: evidence from fMRI. Cereb Cortex 22:527–536

Ball GJ, Gloor P, Schaul N (1977) The cortical electromicrophysiology of pathological delta waves in the electroencephalogram of cats. Electroencephalogr Clin Neurophysiol 43:346–361

Basser PJ, Pajevic S, Pierpaoli C, Duda J, Aldroubi A (2000) In vivo fiber tractography using DT-MRI data. Magn Reson Med 44:625–632

Bazarian JJ, Donnelly K, Peterson DR, Warner GC, Zhu T, Zhong JH (2013) The relation between posttraumatic stress disorder and mild traumatic brain injury acquired during Operations Enduring Freedom and Iraqi Freedom. J Head Trauma Rehabil 28:1–12

Beaulieu C (2002) The basis of anisotropic water diffusion in the nervous system – a technical review. NMR Biomed 15:435–455

Beckmann CF, DeLuca M, Devlin JT, Smith SM (2005) Investigations into resting-state connectivity using independent component analysis. Philos Trans R Soc B-Biol Sci 360:1001–1013

Bonnelle V, Ham TE, Leech R, Kinnunen KM, Mehta MA, Greenwood RJ, Sharp DJ (2012) Salience network integrity predicts default mode network function after traumatic brain injury. Proc Natl Acad Sci U S A 109:4690–4695

Brenner LA, Ladley-O'Brien SE, Harwood JEF, Filley CM, Kelly JP, Homaifar BY, Adler LE (2009) An exploratory study of neuroimaging, neurologic, and neuropsychological findings in veterans with traumatic brain injury and/or posttraumatic stress disorder. Mil Med 174:347–352

Buchsbaum MS, Simmons AN, DeCastro A, Farid N, Matthews SC (2015) Clusters of low F-18-fluorodeoxyglucose uptake voxels in combat veterans with traumatic brain injury and post-traumatic stress disorder. J Neurotrauma 32:1736–1750

Bullmore E, Sporns O (2009) Complex brain networks: graph theoretical analysis of structural and functional systems. Nat Rev Neurosci 10:186–198

Cernak I, Wang Z, Jiang J, Bian X, Savic J (2001) Cognitive deficits following blast injury-induced neurotrauma: possible involvement of nitric oxide. Brain Inj 15:593–612

Chen S, Li L, Xu B, Liu J (2009) Insular cortex involvement in declarative memory deficits in patients with post-traumatic stress disorder. BMC Psychiatry 9:39

Cohen D (1968) Magnetoencephalography – evidence of magnetic fields produced by alpha-rhythm currents. Science 161:784–786

Craig AD (2003) Interoception: the sense of the physiological condition of the body. Curr Opin Neurobiol 13:500–505

Craig AD (2009) How do you feel – now? The anterior insula and human awareness. Nat Rev Neurosci 10:59–70

Dale AM, Fischl B, Sereno MI (1999) Cortical surface-based analysis – I. Segmentation and surface reconstruction. NeuroImage 9:179–194

Davenport ND, Lim KO, Armstrong MT, Sponheim SR (2012) Diffuse and spatially variable white matter disruptions are associated with blast-related mild traumatic brain injury. NeuroImage 59: 2017–2024

Davenport ND, Lim KO, Sponheim SR (2015) White matter abnormalities associated with military PTSD in the context of blast TBI. Hum Brain Mapp 36:1053–1064

Delgado MR, Nearing KI, LeDoux JE, Phelps EA (2008) Neural circuitry underlying the regulation of conditioned fear and its relation to extinction. Neuron 59:829–838

Depue BE, Olson-Madden JH, Smolker HR, Rajamani M, Brenner LA, Banich MT (2014) Reduced amygdala volume is associated with deficits in inhibitory control: a voxel- and surface-based morphometric analysis of comorbid PTSD/Mild TBI. Biomed Res Int 2014:691505

Dickie EW, Brunet A, Akerib V, Armony JL (2011) Neural correlates of recovery from post-traumatic stress disorder: a longitudinal fMRI investigation of memory encoding. Neuropsychologia 49: 1771–1778

Diwakar M, Harrington DL, Maruta J, Ghajar J, El-Gabalawy F, Muzzatti L, Corbetta M, Huang MX, Lee RR (2015) Filling in the gaps: anticipatory control of eye movements in chronic mild traumatic brain injury. Neuroimage Clin 8:210–223

Dretsch MN, Lange RT, Katz JS, Goodman A, Daniel TA, Deshpande G, Denney TS, Iverson GL, Robinson JL (2017) Examining microstructural white matter in active duty soldiers with a history of mild traumatic brain injury and traumatic stress. Open Neuroimaging J 11:46–57

Drevets WC, Frank E, Price JC, Kupfer DJ, Holt D, Greer PJ, Huang YY, Gautier C, Mathis C (1999) PET imaging of serotonin 1A receptor binding in depression. Biol Psychiatry 46: 1375–1387

Eierud C, Craddock RC, Fletcher S, Aulakh M, King-Casas B, Kuehl D, LaConte SM (2014) Neuroimaging after mild traumatic brain injury: review and meta-analysis. Neuroimage Clin 4: 283–294

Elsinga PH, Hatano K, Ishiwata K (2006) PET tracers for imaging of the dopaminergic system. Curr Med Chem 13:2139–2153

Engdahl B, Leuthold AC, Tan HRM, Lewis SM, Winskowski AM, Dikel TN, Georgopoulos AP (2010) Post-traumatic stress disorder: a right temporal lobe syndrome? J Neural Eng 7(6): 066005

Etkin A, Wager TD (2007) Functional neuroimaging of anxiety: a meta-analysis of emotional processing in PTSD, social anxiety disorder, and specific phobia. Am J Psychiatry 164:1476–1488

Fani N, King TZ, Jovanovic T, Glover EM, Bradley B, Choi K, Ely T, Gutman DA, Ressler KJ (2012) White matter integrity in highly traumatized adults with and without post-traumatic stress disorder. Neuropsychopharmacology 37:2740–2746

Felmingham K, Williams LM, Kemp AH, Liddell B, Falconer E, Peduto A, Bryant R (2010) Neural responses to masked fear faces: sex differences and trauma exposure in posttraumatic stress disorder. J Abnorm Psychol 119:241–247

Fischl B, Sereno MI, Dale AM (1999) Cortical surface-based analysis – II: inflation, flattening, and a surface-based coordinate system. NeuroImage 9:195–207

Fonzo GA, Simmons AN, Thorp SR, Norman SB, Paulus MP, Stein MB (2010) Exaggerated and disconnected insular-amygdalar blood oxygenation level-dependent response to threat-related emotional faces in women with intimate-partner violence posttraumatic stress disorder. Biol Psychiatry 68:433–441

Frasure-Smith N, Lesperance F, Talajic M (1995) The impact of negative emotions on prognosis following myocardial infarction: is it more than depression? Health Psychol 14:388–398

Friston KJ, Penny W (2003) Posterior probability maps and SPMs. NeuroImage 19:1240–1249

Friston KJ, Frith CD, Fletcher P, Liddle PF, Frackowiak RSJ (1996) Functional topography: multidimensional scaling and functional connectivity in the brain. Cereb Cortex 6:156–164

Friston KJ, Trujillo-Barreto N, Daunizeau J (2008) DEM: a variational treatment of dynamic systems. NeuroImage 41:849–885

Gean AD (1994) Imaging of head trauma. Raven Press, New York

Gentry LR, Godersky JC, Thompson B, Dunn VD (1988) Prospective comparative study of intermediate-field MR and CT in the evaluation of closed head trauma. AJR 150:673–682

Georgopoulos AP, Tan HRM, Lewis SM, Leuthold AC, Winskowski AM, Lynch JK, Engdahl B (2010) The synchronous neural interactions test as a functional neuromarker for post-traumatic stress disorder (PTSD): a robust classification method based on the bootstrap. J Neural Eng 7: 16011

Gewirtz JC, McNish KA, Davis M (2000) Is the hippocampus necessary for contextual fear conditioning? Behav Brain Res 110:83–95

Gloor P, Ball G, Schaul N (1977) Brain lesions that produce delta waves in the EEG. Neurology 27:326–333

Goetz P, Blamire A, Rajagopalan B, Cadoux-Hudson T, Young D, Styles P (2004) Increase in apparent diffusion coefficient in normal appearing white matter following human traumatic brain injury correlates with injury severity. J Neurotrauma 21:645–654

Haacke EM, Brown RW, Thompson MR, Venkatesan R (1999) Magnetic resonance imaging: physical principles and sequence design. Wiley-Liss, New York

Hamalainen M, Hari R, Ilmoniemi RJ, Knuutila J, Lounasmaa OV (1993) Magnetoencephalography – theory, instrumentation, and applications to noninvasive studies of the working human brain. Rev Mod Phys 65:413–497

Hayes JP, Hayes SM, Mikedis AM (2012) Quantitative meta-analysis of neural activity in posttraumatic stress disorder. Biol Mood Anxiety Disord 2:9

Hoge CW, McGurk D, Thomas JL, Cox AL, Engel CC, Castro CA (2008) Mild traumatic brain injury in U.S. soldiers returning from Iraq. N Engl J Med 358:453–463

Huang MX, Dale AM, Song T, Halgren E, Harrington DL, Podgorny I, Canive JM, Lewis S, Lee RR (2006) Vector-based spatial-temporal minimum L1-norm solution for MEG. NeuroImage 31:1025–1037

Huang MX, Theilmann RJ, Robb A, Angeles A, Nichols S, Drake A, D'Andrea J, Levy M, Holland M, Song T, Ge S, Hwang E, Yoo K, Cui L, Baker DG, Trauner D, Coimbra R, Lee RR (2009) Integrated imaging approach with MEG and DTI to detect mild traumatic brain injury in military and civilian patients. J Neurotrauma 26:1213–1226

Huang MX, Nichols S, Robb A, Angeles A, Drake A, Holland M, Asmussen S, D'Andrea J, Chun W, Levy M, Cui L, Song T, Baker DG, Hammer P, McLay R, Theilmann RJ, Coimbra R, Diwakar M, Boyd C, Neff J, Liu TT, Webb-Murphy J, Farinpour R, Cheung C, Harrington DL, Heister D, Lee RR (2012) An automatic MEG low-frequency source imaging approach for detecting injuries in mild and moderate TBI patients with blast and non-blast causes. NeuroImage 61:1067–1082

Huang MX, Huang CW, Robb A, Angeles A, Nichols SL, Baker DG, Song T, Harrington DL, Theilmann RJ, Srinivasan R, Heister D, Diwakar M, Canive JM, Edgar JC, Chen YH, Ji ZW, Shen M, El-Gabalawy F, Levy M, McLay R, Webb-Murphy J, Liu TT, Drake A, Lee RR (2014a) MEG source imaging method using fast L1 minimum-norm and its applications to signals with brain noise and human resting-state source amplitude images. NeuroImage 84: 585–604

Huang MX, Nichols S, Baker DG, Robb A, Angeles A, Yurgil KA, Drake A, Levy M, Song T, Mclay R, Theilmann RJ, Diwakar M, Risbrough VB, Ji ZW, Huang C, Chang DG, Harrington DL, Muzzatti L, Canive JM, Edgar JC, Chen YH, Lee RR (2014b) Single-subject-based whole-brain MEG slow-wave imaging approach for detecting abnormality in patients with mild traumatic brain injury. Neuroimage Clin 5:109–119

Huang MX, Yurgil KA, Robb A, Angeles A, Diwakar M, Risbrough VB, Nichols SL, McLay R, Theilmann RJ, Song T, Huang CW, Lee RR, Baker DG (2014c) Voxel-wise resting-state MEG source magnitude imaging study reveals neurocircuitry abnormality in active-duty service members and veterans with PTSD. Neuroimage Clin 5:408–419

Huang CW, Huang MX, Ji ZW, Swan AR, Angeles AM, Song T, Huang JW, Lee RR (2016a) High-resolution MEG source imaging approach to accurately localize Broca's area in patients with brain tumor or epilepsy. Clin Neurophysiol 127:2308–2316

Huang MX, Harrington DL, Robb Swan A, Angeles Quinto A, Nichols S, Drake A, Song T, Diwakar M, Huang CW, Risbrough VB, Dale A, Bartsch H, Matthews S, Huang JW, Lee RR, Baker DG (2016b) Resting-state magnetoencephalography reveals different patterns of aberrant functional connectivity in combat-related mild traumatic brain injury. J Neurotrauma 34:1412–1426

Huang MX, Risling M, Baker DG (2016c) The role of biomarkers and MEG-based imaging markers in the diagnosis of post-traumatic stress disorder and blast-induced mild traumatic brain injury. Psychoneuroendocrinology 63:398–409

Huettel SA, Song AW, McCarthy G (2004) Functional magnetic resonance imaging. Sinauer Associates, Sunderland

Hulkower MB, Poliak DB, Rosenbaum SB, Zimmerman ME, Lipton ML (2013) A decade of DTI in traumatic brain injury: 10 years and 100 articles later. Am J Neuroradiol 34:2064–2074

Jorge RE, Acion L, White T, Tordesillas-Gutierrez D, Pierson R, Crespo-Facorro B, Magnotta VA (2012) White matter abnormalities in veterans with mild traumatic brain injury. Am J Psychiatr 169:1284–1291

Josephs O, Turner R, Friston K (1997) Event-related fMRI. Hum Brain Mapp 5:243–248

Kandel ERS, James H, Jessell TM, Siggelbaum SA, Hudspeth AJ, Mack S (2013) Principles of neural science, 5th edn. McGraw-Hill, New York

Karl A, Schaefer M, Malta LS, Dorfel D, Rohleder N, Werner A (2006) A meta-analysis of structural brain abnormalities in PTSD. Neurosci Biobehav Rev 30:1004–1031

Kato T, Nakayama N, Yasokawa Y, Okumura A, Shinoda J, Iwama T (2007) Statistical image analysis of cerebral glucose metabolism in patients with cognitive impairment following diffuse traumatic brain injury. J Neurotrauma 24:919–926

Kelmendi B, Adams TG, Southwick S, Abdallah CG, Krystal JH (2017) Posttraumatic stress disorder: an integrated overview of the neurobiological rationale for pharmacology. Clin Psychol Sci Pract 24:281–297

Kim MJ, Lyoo IK, Kim SJ, Sim M, Kim N, Choi N, Jeong DU, Covell J, Renshaw PF (2005) Disrupted white matter tract integrity of anterior cingulate in trauma survivors. Neuroreport 16:1049–1053

King AP, Abelson JL, Britton JC, Phan KL, Taylor SF, Liberzon I (2009) Medial prefrontal cortex and right insula activity predict plasma ACTH response to trauma recall. NeuroImage 47: 872–880

Kitayama N, Vaccarino V, Kutner M, Weiss P, Bremner JD (2005) Magnetic resonance imaging (MRI) measurement of hippocampal volume in posttraumatic stress disorder: a meta-analysis. J Affect Disord 88:79–86

Kitayama N, Brummer M, Hertz L, Quinn S, Kim Y, Bremner JD (2007) Morphologic alterations in the corpus callosum in abuse-related posttraumatic stress disorder. J Nerv Ment Dis 195: 1027–1029

Kuhn S, Gallinat J (2013) Gray matter correlates of posttraumatic stress disorder: a quantitative meta-analysis. Biol Psychiatry 73:70–74

Le Bihan D, Breton E, Lallemand D, Grenier P, Cabanis E, Laval-Jeantet M (1986) MR imaging of intravoxel incoherent motions: application to diffusion and perfusion in neurologic disorders. Radiology 161:401–407

Leahy RM, Mosher JC, Spencer ME, Huang MX, Lewine JD (1998) A study of dipole localization accuracy for MEG and EEG using a human skull phantom. Electroencephalogr Clin Neurophysiol 107:159–173

Lei D, Li KM, Li LJ, Chen FQ, Huang XQ, Lui S, Li J, Bi F, Gong QY (2015) Disrupted functional brain connectome in patients with posttraumatic stress disorder. Radiology 276:818–827

Levin HS, Wilde E, Troyanskaya M, Petersen NJ, Scheibel R, Newsome M, Radaideh M, Wu T, Yallampalli R, Chu ZL, Li XQ (2010) Diffusion tensor imaging of mild to moderate blast-related traumatic brain injury and its sequelae. J Neurotrauma 27:683–694

Levine B, Cabeza R, McIntosh AR, Black SE, Grady CL, Stuss DT (2002) Functional reorganisation of memory after traumatic brain injury: a study with H(2)(15)0 positron emission tomography. J Neurol Neurosurg Psychiatry 73:173–181

Lewine JD, Davis JT, Bigler ED, Thoma R, Hill D, Funke M, Sloan JH, Hall S, Orrison WW (2007) Objective documentation of traumatic brain injury subsequent to mild head trauma: multimodal brain imaging with MEG, SPECT, and MRI. J Head Trauma Rehabil 22:141–155

Li L, Lei D, Huang X, Suo X, Xiao F, Kuang W, Li J, Bi F, Lui S, Kemp GJ, Sweeney JA, Gong Q (2016a) White matter abnormalities in post-traumatic stress disorder following a specific traumatic event. EBioMedicine 4:176–183

Li L, Sun G, Liu K, Li M, Li B, Qian SW, Yu LL (2016b) White matter changes in posttraumatic stress disorder following mild traumatic brain injury: a prospective longitudinal diffusion tensor imaging study. Chin Med J 129:1091–1099

Liberzon I, Sripada CS (2008) The functional neuroanatomy of PTSD: a critical review. Prog Brain Res 167:151–169

Liberzon I, Lopez JF, Flagel SB, Vazquez DM, Young EA (1999a) Differential regulation of hippocampal glucocorticoid receptors mRNA and fast feedback: relevance to post-traumatic stress disorder. J Neuroendocrinol 11:11–17

Liberzon I, Taylor SF, Amdur R, Jung TD, Chamberlain KR, Minoshima S, Koeppe RA, Fig LM (1999b) Brain activation in PTSD in response to trauma-related stimuli. Biol Psychiatry 45:817–826

Lindemer ER, Salat DH, Leritz EC, McGlinchey RE, Milberg WP (2013) Reduced cortical thickness with increased lifetime burden of PTSD in OEF/OIF veterans and the impact of comorbid TBI. Neuroimage Clin 2:601–611

Linnman C, Zeffiro TA, Pitman RK, Milad MR (2011) An fMRI study of unconditioned responses in post-traumatic stress disorder. Biol Mood Anxiety Disord 1:8

Lopez KC, Leary JB, Pham DL, Chou YY, Dsurney J, Chan L (2017) Brain volume, connectivity, and neuropsychological performance in mild traumatic brain injury: the impact of post-traumatic stress disorder symptoms. J Neurotrauma 34:16–22

MacDonald CL, Johnson AM, Cooper D, Nelson EC, Werner NJ, Shimony JS, Snyder AZ, Raichle ME, Witherow JR, Fang R, Flaherty SF, Brody DL (2011) Detection of blast-related traumatic brain injury in U.S. military personnel. N Engl J Med 364:2091–2100

MacDonald C, Johnson A, Cooper D, Malone T, Sorrell J, Shimony J, Parsons M, Snyder A, Raichle M, Fang R, Flaherty S, Russell M, Brody DL (2013) Cerebellar white matter abnormalities following primary blast injury in US military personnel. PLoS One 8:e55823

Matthews SC, Strigo IA, Simmons AN, O'Connell RM, Reinhardt LE, Moseley SA (2011) A multimodal imaging study in U.S. veterans of Operations Iraqi and Enduring Freedom with and without major depression after blast-related concussion. NeuroImage 54(Suppl 1):S69–S75

Menon V (2011) Large-scale brain networks and psychopathology: a unifying triple network model. Trends Cogn Sci 15:483–506

Milad MR, Pitman RK, Ellis CB, Gold AL, Shin LM, Lasko NB, Zeidan MA, Handwerger K, Orr SP, Rauch SL (2009) Neurobiological basis of failure to recall extinction memory in post-traumatic stress disorder. Biol Psychiatry 66:1075–1082

Mohammadian M, Roine T, Hirvonen J, Kurki T, Ala-Seppala H, Frantzen J, Katila A, Kyllonen A, Maanpaa HR, Posti J, Takala R, Tallus J, Tenovuo O (2017) High angular resolution diffusion-weighted imaging in mild traumatic brain injury. Neuroimage Clin 13:174–180

Molina ME, Isoardi R, Prado MN, Bentolila S (2010) Basal cerebral glucose distribution in long-term post-traumatic stress disorder. World J Biol Psychiatry 11:493–501

Morey RA, Haswell CC, Selgrade ES, Massoglia D, Liu CL, Weiner J, Marx CE, Cernak I, McCarthy G, Grp MW (2013) Effects of chronic mild traumatic brain injury on white matter integrity in Iraq and Afghanistan war veterans. Hum Brain Mapp 34:2986–2999

Nakayama N, Okumura A, Shinoda J, Nakashima T, Iwama T (2006) Relationship between regional cerebral metabolism and consciousness disturbance in traumatic diffuse brain injury without large focal lesions: an FDG-PET study with statistical parametric mapping analysis. J Neurol Neurosurg Psychiatry 77:856–862

Naumova AV, Akulov AE, Khodanovich MY, Yarnykh VL (2017) High-resolution three-dimensional macromolecular proton fraction mapping for quantitative neuroanatomical imaging of the rodent brain in ultra-high magnetic fields. NeuroImage 147:985–993

Nelson MD, Tumpap AM (2017) Posttraumatic stress disorder symptom severity is associated with left hippocampal volume reduction: a meta-analytic study. CNS Spectr 22:363–372

Newsome MR, Durgerian S, Mourany L, Scheibel RS, Lowe MJ, Beall EB, Koenig KA, Parsons M, Troyanskaya M, Reece C, Wilde E, Fischer BL, Jones SE, Agarwal R, Levin HS, Rao SM (2015) Disruption of caudate working memory activation in chronic blast-related traumatic brain injury. Neuroimage Clin 8:543–553

Newsome MR, Mayer AR, Lin XD, Troyanskaya M, Jackson GR, Scheibel RS, Walder A, Sathiyaraj A, Wilde EA, Mukhi S, Taylor BA, Levin HS (2016) Chronic effects of blast-related TBI on subcortical functional connectivity in veterans. J Int Neuropsychol Soc 22: 790–792

Niogi SN, Mukherjee P (2010) Diffusion tensor imaging of mild traumatic brain injury. J Head Trauma Rehabil 25:241–255

Norman KA, Polyn SM, Detre GJ, Haxby JV (2006) Beyond mind-reading: multi-voxel pattern analysis of fMRI data. Trends Cogn Sci 10:424–430

O'Doherty DCM, Chitty KM, Saddiqui S, Bennett MR, Lagopoulos J (2015) A systematic review and meta-analysis of magnetic resonance imaging measurement of structural volumes in posttraumatic stress disorder. Psychiatry Res Neuroimaging 232:1–33

Ogawa S, Lee TM, Kay AR, Tank DW (1990) Brain magnetic resonance imaging with contrast dependent on blood oxygenation. Proc Natl Acad Sci U S A 87:9868–9872

Okie S (2005) Traumatic brain injury in the war zone. N Engl J Med 352:2043–2047

Patel R, Spreng RN, Shin LM, Girard TA (2012) Neurocircuitry models of posttraumatic stress disorder and beyond: a meta-analysis of functional neuroimaging studies. Neurosci Biobehav Rev 36:2130–2142

Paulus MP, Stein MB (2006) An insular view of anxiety. Biol Psychiatry 60:383–387

Penny W, Friston K (2003) Mixtures of general linear models for functional neuroimaging. IEEE Trans Med Imaging 22:504–514

Peres JF, Foerster B, Santana LG, Fereira MD, Nasello AG, Savoia M, Moreira-Almeida A, Lederman H (2011) Police officers under attack: resilience implications of an fMRI study. J Psychiatr Res 45:727–734

Petrie EC, Cross DJ, Yarnykh VL, Richards T, Martin NM, Pagulayan K, Hoff D, Hart K, Mayer C, Tarabochia M, Raskind MA, Minoshima S, Peskind ER (2014) Neuroimaging, behavioral, and psychological sequelae of repetitive combined blast/impact mild traumatic brain injury in Iraq and Afghanistan war veterans. J Neurotrauma 31:425–436

Phelps EA, LeDoux JE (2005) Contributions of the amygdala to emotion processing: from animal models to human behavior. Neuron 48:175–187

Phillips RG, LeDoux JE (1992) Differential contribution of amygdala and hippocampus to cued and contextual fear conditioning. Behav Neurosci 106:274–285

Portnow LH, Vaillancourt DE, Okun MS (2013) The history of cerebral PET scanning from physiology to cutting-edge technology. Neurology 80:952–956

Povlishock JT, Coburn TH (1989) Morphopathological change associated with mild head injury. Oxford University Press, New York

Povlishock JT, Katz DI (2005) Update of neuropathology and neurological recovery after traumatic brain injury. J Head Trauma Rehabil 20:76–94

Rabinak CA, Angstadt M, Welsh RC, Kenndy AE, Lyubkin M, Martis B, Phan KL (2011) Altered amygdala resting-state functional connectivity in post-traumatic stress disorder. Front Psychol 2:62

Rauch SL, Shin LM, Phelps EA (2006) Neurocircuitry models of posttraumatic stress disorder and extinction: human neuroimaging research – past, present, and future. Biol Psychiatry 60:376–382

Robb Swan A, Nichols S, Drake A, Angeles A, Diwakar M, Song T, Lee RR, Huang MX (2015) Magnetoencephalography slow-wave detection in patients with mild traumatic brain injury and ongoing symptoms correlated with long-term neuropsychological outcome. J Neurotrauma 32:1510–1521

Sartory G, Cwik J, Knuppertz H, Schurholt B, Lebens M, Seitz RJ, Schulze R (2013) In search of the trauma memory: a meta-analysis of functional neuroimaging studies of symptom provocation in posttraumatic stress disorder (PTSD). PLoS One 8(3):e58150

Schaul N, Gloor P, Ball G, Gotman J (1978) The electromicrophysiology of delta waves induced by systemic atropine. Brain Res 143:475–486

Scheid R, Preul C, Gruber O, Wiggins C, von Cramon DY (2003) Diffuse axonal injury associated with chronic traumatic brain injury: evidence from T2*-weighted gradient-echo imaging at 3 T. AJNR 24:1049–1056

Schuff N, Zhang Y, Zhan W, Lenoci M, Ching C, Boreta L, Mueller SG, Wang Z, Marmar CR, Weiner MW, Neylan TC (2011) Patterns of altered cortical perfusion and diminished subcortical integrity in posttraumatic stress disorder: an MRI study. NeuroImage 54:S62–S68

Sheline YI, Gado MH, Kraemer HC (2003) Untreated depression and hippocampal volume loss. Am J Psychiatr 160(8):1516–1518

Shin YB, Kim SJ, Kim IJ, Kim YK, Kim DS, Park JH, Yeom SR (2006) Voxel-based statistical analysis of cerebral blood flow using Tc-99m ECD brain SPECT in patients with traumatic brain injury: group and individual analyses. Brain Inj 20:661–667

Simmons AN, Matthews SC (2012) Neural circuitry of PTSD with or without mild traumatic brain injury: a meta-analysis. Neuropharmacology 62:598–606

Simmons AN, Paulus MP, Thorp SR, Matthews SC, Norman SB, Stein MB (2008) Functional activation and neural networks in women with posttraumatic stress disorder related to intimate partner violence. Biol Psychiatry 64:681–690

Simmons A, Strigo IA, Matthews SC, Paulus MP, Stein MB (2009) Initial evidence of a failure to activate right anterior insula during affective set shifting in posttraumatic stress disorder. Psychosom Med 71:373–377

Simmons AN, Fitzpatrick S, Strigo IA, Potterat EG, Johnson DC, Matthews SC, Orden KF, Swain JL, Paulus MP (2012) Altered insula activation in anticipation of changing emotional states: neural mechanisms underlying cognitive flexibility in special operations forces personnel. Neuroreport 23:234–239

Simmons AN, Norman SB, Spadoni AD, Strigo IA (2013) Neurosubstrates of remission following prolonged exposure therapy in veterans with posttraumatic stress disorder. Psychother Psychosom 82:382–389

Smith ME (2005) Bilateral hippocampal volume reduction in adults with post-traumatic stress disorder: a meta-analysis of structural MRI studies. Hippocampus 15:798–807

Smith SM, Jenkinson M, Johansen-Berg H, Rueckert D, Nichols TE, Mackay CE, Watkins KE, Ciccarelli O, Cader MZ, Matthews PM, Behrens TEJ (2006) Tract-based spatial statistics: voxelwise analysis of multi-subject diffusion data. NeuroImage 31:1487–1505

Sorg SF, Schiehser DM, Bondi MW, Luc N, Clark AL, Jacobson MW, Frank LR, Delano-Wood L (2016) White matter microstructural compromise is associated with cognition but not posttraumatic stress disorder symptoms in military veterans with traumatic brain injury. J Head Trauma Rehabil 31:297–308

Spielberg JM, McGlinchey RE, Milberg WP, Salat DH (2015) Brain network disturbance related to posttraumatic stress and traumatic brain injury in veterans. Biol Psychiatry 78:210–216
Sponheim SR, McGuire KA, Kang SS, Davenport ND, Aviyente S, Bernat EM, Lim KO (2011) Evidence of disrupted functional connectivity in the brain after combat-related blast injury. NeuroImage 54:S21–S29
Sripada RK, King AP, Garfinkel SN, Wang X, Sripada CS, Welsh RC, Liberzon I (2012a) Altered resting-state amygdala functional connectivity in men with posttraumatic stress disorder. J Psychiatry Neurosci 37:241–249
Sripada RK, King AP, Welsh RC, Garfinkel SN, Wang X, Sripada CS, Liberzon I (2012b) Neural dysregulation in posttraumatic stress disorder: evidence for disrupted equilibrium between salience and default mode brain networks. Psychosom Med 74:904–911
Stamatakis EA, Wilson JT, Hadley DM, Wyper DJ (2002) SPECT imaging in head injury interpreted with statistical parametric mapping. J Nucl Med 43:476–483
Stein MB, McAllister TW (2009) Exploring the convergence of posttraumatic stress disorder and mild traumatic brain injury. Am J Psychiatry 166:768–776
Stein MB, Kessler RC, Heeringa SG, Jain S, Campbell-Sills L, Colpe LJ, Fullerton CS, Nock MK, Sampson NA, Schoenbaum M, Sun X, Thomas ML, Ursano RJ (2015) Prospective longitudinal evaluation of the effect of deployment-acquired traumatic brain injury on posttraumatic stress and related disorders: results from the Army Study to Assess Risk and Resilience in Servicemembers (Army STARRS). Am J Psychiatry 172:1101–1111
Strigo IA, Simmons AN, Matthews SC, Grimes EM, Allard CB, Reinhardt LE, Paulus MP, Stein MB (2010) Neural correlates of altered pain response in women with posttraumatic stress disorder from intimate partner violence. Biol Psychiatry 68:442–450
Tanielian TL, Jaycox L (2008) Invisible wounds of war: psychological and cognitive injuries, their consequences, and services to assist recovery. RAND Corporation, Santa Monica
Thompson PM, Stein JL, Medland SE et al (2014) The ENIGMA consortium: large-scale collaborative analyses of neuroimaging and genetic data. Brain Imaging Behav 8:153–182
Tuch DS (2004) Q-ball imaging. Magn Reson Med 52:1358–1372
Vasterling JJ, Verfaellie M, Sullivan KD (2009) Mild traumatic brain injury and posttraumatic stress disorder in returning veterans: perspectives from cognitive neuroscience. Clin Psychol Rev 29: 674–684
Vermetten E, Vythilingam M, Southwick SM, Charney DS, Bremner JD (2003) Long-term treatment with paroxetine increases verbal declarative memory and hippocampal volume in posttraumatic stress disorder. Biol Psychiatry 54:693–702
Videbech P, Ravnkilde B (2015) Hippocampal volume and depression: a meta-analysis of MRI studies. Am J Psychiatr 161(11):1957–1966
Villarreal G, Hamilton DA, Graham DP, Driscoll I, Qualls C, Petropoulos H, Brooks WM (2004) Reduced area of the corpus callosum in posttraumatic stress disorder. Psychiatry Res Neuroimaging 131:227–235
Wang EW, Huang JH (2013) Understanding and treating blast traumatic brain injury in the combat theater. Neurol Res 35:285–289
Wilde EA, Hunter JV, Newsome MR, Scheibel RS, Bigler ED, Johnson JL, Fearing MA, Cleavinger HB, Li X, Swank PR, Pedroza C, Roberson GS, Bachevalier J, Levin HS (2005) Frontal and temporal morphometric findings on MRI in children after moderate to severe traumatic brain injury. J Neurotrauma 22:333–344
Wilde EA, Chu Z, Bigler ED, Hunter JV, Fearing MA, Hanten G, Newsome MR, Scheibel RS, Li X, Levin HS (2006) Diffusion tensor imaging in the corpus callosum in children after moderate to severe traumatic brain injury. J Neurotrauma 23:1412–1426
Worsley KJ, Evans AC, Marrett S, Neelin P (1992) A 3-dimensional statistical-analysis for CBF activation studies in human brain. J Cereb Blood Flow Metab 12:900–918

Yarnykh VL, Bowen JD, Samsonov A, Repovic P, Mayadev A, Qian PQ, Gangadharan B, Keogh BP, Maravilla KR, Henson LKJ (2015) Fast whole-brain three-dimensional macromolecular proton fraction mapping in multiple sclerosis. Radiology 274:210–220

Yeh PH, Wang BQ, Oakes TR, French LM, Pan H, Graner J, Liu W, Riedy G (2014) Postconcussional disorder and PTSD symptoms of military-related traumatic brain injury associated with compromised neurocircuitry. Hum Brain Mapp 35:2652–2673

Yurgil KA, Barkauskas DA, Vasterling JJ, Nievergelt CM, Larson GE, Schork NJ, Litz BT, Nash WP, Baker DG, Team MRS (2014) Association between traumatic brain injury and risk of post-traumatic stress disorder in active-duty marines. JAMA Psychiat 71:149–157

Zhang J, Mitsis EM, Chu K, Newmark RE, Hazlett EA, Buchsbaum MS (2009) SPM and cluster counting analysis of [18F] FDG-PET imaging in traumatic brain injury. J Neurotrauma 27(1): 35–49

Zhang L, Zhang Y, Li LJ, Li ZX, Li WH, Ma N, Hou CL, Zhang ZJ, Zhang ZQ, Wang LF, Duan L, Lu GM (2011) Different white matter abnormalities between the first-episode, treatment-naive patients with posttraumatic stress disorder and generalized anxiety disorder without comorbid conditions. J Affect Disord 133:294–299

Zhang L, Li W, Shu N, Zheng H, Zhang Z, Zhang Y, He Z, Hou C, Li Z, Liu J, Wang L, Duan L, Jiang T, Li L (2012) Increased white matter integrity of posterior cingulate gyrus in the evolution of post-traumatic stress disorder. Acta Neuropsychiatrica 24:34–42

Sleep Disruption, Safety Learning, and Fear Extinction in Humans: Implications for Posttraumatic Stress Disorder

Laura D. Straus, Sean P.A. Drummond, Victoria B. Risbrough, and Sonya B. Norman

Abstract Fear learning is critical in the development and maintenance of post-traumatic stress disorder (PTSD) symptoms, and safety learning and extinction are necessary for recovery. Studies in animal models suggest that sleep disruption, and REM sleep fragmentation in particular, interfere with safety learning and extinction processes, and recently, studies are extending these findings to humans. A discussion of the human literature is presented here, which largely consists of experimental studies in healthy human control subjects. A theoretical model for the relationship between fear learning, sleep disruption, and impaired safety learning and extinction is proposed, which provides an explanatory framework for sleep disruption and its relationship to PTSD. Overall, findings suggest that sleep disruption plays a role in the development and maintenance of PTSD symptoms, and thus presents an important modifiable target in PTSD treatment.

L.D. Straus (✉)
San Diego State University/University of California, San Diego Joint Doctoral Program in Clinical Psychology, San Diego, CA, USA

Research Service, VA San Diego Healthcare System, San Diego, CA, USA
e-mail: lauradstraus@gmail.com

S.P.A. Drummond
Monash Institute of Cognitive and Clinical Neuroscience, Monash School of Psychological Sciences, Monash University, Clayton, VIC, Australia

V.B. Risbrough
Research Service, VA San Diego Healthcare System, San Diego, CA, USA

Department of Psychiatry, University of California San Diego, La Jolla, CA, USA

S.B. Norman
Research Service, VA San Diego Healthcare System, San Diego, CA, USA

Department of Psychiatry, University of California San Diego, La Jolla, CA, USA

National Center for PTSD, White River Junction, VT, USA

Curr Topics Behav Neurosci (2018) 38: 193–206
DOI 10.1007/7854_2017_31
Published Online: 23 September 2017

Keywords Fear learning • Posttraumatic stress disorder • Sleep

Contents

Epidemiological studies suggest more than two thirds of people in the general population experience at least one traumatic event over the course of their lifetime (Breslau et al. 1998). A portion of these people go on to develop Posttraumatic Stress Disorder (PTSD). PTSD is characterized by four clusters of symptoms: (1) re-experiencing the traumatic event in the form of intrusive thoughts, fearful reactions to reminders of the event, distressing dreams, and/or dissociative flashbacks; (2) avoidance of internal and external cues associated with the traumatic event; (3) negative alterations in mood and/or negative cognitions about the self, others, and the world; and (4) hyperarousal, experienced as being watchful for threats, difficulty concentrating, irritability, exaggerated startle responses, and difficulty falling or staying asleep (American Psychiatric Association 2013).

PTSD lifetime prevalence is estimated to be approximately 7% in community samples (Kessler et al. 2005), with especially high prevalence rates in victims of interpersonal violence (Resnick et al. 1993) and combat veterans (Weiss et al. 1992). PTSD is associated with negative psychiatric outcomes such as increased suicidality (Davidson et al. 1991), alcohol abuse and dependence (Stewart 1996), multiple psychiatric and medical comorbidities (Kilpatrick et al. 2003), functional impairment (Kessler 2000), and low quality of life (Mendlowicz and Stein 2014). Although evidence-based pharmacological (Brady et al. 2000) and psychosocial (Foa et al. 2007; Resick and Schnicke 1992) interventions have been developed for PTSD, not all patients respond to treatment. With regard to gold-standard psychotherapies for PTSD, it is not uncommon to observe nonresponse rates as high as 50% (Schottenbauer et al. 2008). It is, therefore, important to understand the basic biological mechanisms involved in the development and maintenance of PTSD, as well as the mechanisms involved in treatment response, in order to better predict and understand who may not respond optimally to treatment. Additionally, understanding processes that underlie PTSD symptoms will inform development of more effective interventions. This discussion highlights two such processes: impaired safety signal learning and impaired extinction memory, and examines the extant literature on these processes and their relationship to disrupted sleep. Experimental studies suggest that disrupted sleep is linked to impairments in both safety signal learning and extinction processes. A theoretical model is proposed here hypothesizing that sleep disruption plays a role in the development and maintenance of PTSD symptoms, which can be used as a framework to guide future research.

1 Fear Conditioning, Safety Learning, and Extinction Processes Are Implicated in PTSD

Fear conditioning is hypothesized to play a role in the development and maintenance of PTSD. Patients with PTSD experience intense fear reactions to cues associated with a traumatic event, which provoke strong avoidance to these cues even long after the trauma (American Psychiatric Association 2013). For example, a victim of sexual assault may experience intense anxiety when encountering someone new who physically resembles the perpetrator, or an Iraq war veteran exposed to an IED blast may experience anxiety when seeing debris on the side of the road and make the decision to change lanes to avoid it. Additionally, patients with PTSD have difficulty differentiating between dangerous and safe situations, leading to overgeneralization of avoidance and hyperarousal symptoms (Jovanovic et al. 2012). For example, a woman who experienced a sexual assault at school may completely avoid going to campus, or a combat veteran who experienced an insurgent attack while in a crowded market may avoid crowds or be watchful or "on guard" while in crowds, even after coming home from deployment. During natural recovery from trauma, fear reactions to threat cues are often reduced through extinction learning and development of extinction memories (Rothbaum and Davis 2003). Repeated exposure to these cues in a safe setting allows them to lose their predictive quality for danger (Myers and Davis 2007; Quirk and Mueller 2008).

PTSD patients show impairments in fear conditioning and extinction learning processes in experimental studies. Researchers have used classical Pavlovian conditioning paradigms to demonstrate this in the laboratory setting. In these studies, animal or human subjects are exposed to neutral cues which are repeatedly paired with an aversive stimulus (unconditioned stimulus), such as a shock or an air puff to the throat (Lonsdorf et al. 2017). Over time, subjects learn to associate the neutral stimulus with threat and become anxiously reactive in response to the stimulus alone (conditioned stimulus, or CS+). In addition to the threat signal, fear conditioning experiments often present a second neutral stimulus which is never paired with the unconditioned stimulus, so participants learn to associate this cue with safety (CS−, Jovanovic et al. 2010). During extinction learning sessions, the CS+ is then presented repeatedly without being paired with the unconditioned stimulus. Participants undergo extinction by forming new memories allowing for inhibition of the fear response, and thus they no longer respond anxiously in the presence of the CS+. Finally, some experiments (e.g., Milad et al. 2007) also test recall of extinction by presenting the CS+ during a later testing session and examining if extinction learning is retained over time.

Laboratory studies using these experimental paradigms with PTSD patients have shown abnormalities in both fear conditioning and extinction learning. For example, Orr et al. (2000) demonstrated that patients with PTSD more readily formed fear associations as compared to trauma-exposed controls, and continued to respond anxiously to the CS+ during extinction trials, suggesting that PTSD patients take longer to "un-learn" fear associations than trauma-exposed control participants. Other studies suggest impairments in extinction learning (Wessa and Flor 2007). In addition to enhanced fear conditioning responses and impaired extinction learning, PTSD patients show difficulties with retaining extinction learning over time. In a study of monozygotic twins discordant for combat exposure, combat veterans with PTSD showed impairments in extinction recall in comparison to their monozygotic twins without PTSD (Milad et al. 2008). These difficulties with extinction have been linked to re-experiencing symptoms of PTSD. PTSD patients appear to have more difficulty learning that cues previously associated with traumatic events (e.g., helicopter sounds for a Vietnam veteran, seeing trash on the side of the road for an Operation Iraqi Freedom veteran) are no longer signals for danger outside of the context of the traumatic event. These cues then continue to provoke intense fear reactions, even when repeatedly encountered in a safe setting (Mineka and Zinbarg 2006; Rothbaum and Davis 2003).

PTSD patients also show impairments in safety learning, and/or the ability to differentiate between threatening and safe cues. For example, Grillon and Morgan (1999) demonstrated that PTSD patients are less able than healthy controls to differentiate threat cues from safety cues during a fear conditioning session, showing anxious responding to cues regardless of whether they were paired with the unconditioned stimulus. In a study comparing patients with PTSD to patients with Major Depressive Disorder (MDD), PTSD patients showed difficulty with safety learning, whereas depressed patients did not (Jovanovic et al. 2010). In a large cohort study of US Marines comparing PTSD patients to controls and other clinical groups, the PTSD symptom group was the only group to show deficient discrimination between the CS+ and CS−, exhibiting larger startle responses during the safety cue during fear acquisition (Acheson et al. 2015). These studies suggest that in addition to impairments in fear extinction and retention, PTSD patients also have difficulty discriminating between cues signaling threat versus cues signaling safety. These difficulties with safety learning are also hypothesized to underlie PTSD symptoms. PTSD patients often have difficulty discriminating between threatening and safe environments, leading to overgeneralization of avoidance to both threatening and safe environments as well as hypervigilance symptoms (Acheson et al. 2012).

Both safety learning and extinction processes are critical for response to evidence-based psychosocial interventions for PTSD. Prolonged Exposure, the gold-standard behavioral PTSD treatment, involves exposing patients repeatedly to feared cues in a safe environment (Foa et al. 2007). For such treatment to be successful, it is necessary for patients to be able to distinguish between threatening and safe cues. Patients must also successfully undergo extinction learning by learning that these trauma-related cues no longer signal the presence of danger. Finally, patients must also retain extinction learning over time to maintain the gains of the treatment long-term. Although this process occurs successfully in many patients, not all patients respond, or respond completely, to treatment (Schottenbauer et al. 2008). This variability in treatment response may reflect individual vulnerabilities that impair safety learning and/or extinction processes. Indeed, in a recent study, individuals with social anxiety who demonstrated better extinction learning during a fear conditioning paradigm at baseline reported greater anxiety reduction following brief exposure therapy (Ball et al. 2017). These findings may apply more broadly to other anxiety and/or trauma-related disorders for which exposure therapy is the gold-standard treatment approach. Thus, examining factors, especially modifiable factors, that influence extinction and/or safety learning processes will identify important processes to target in PTSD treatment.

2 Rapid Eye Movement (REM) Sleep Is Disrupted in Patients with PTSD

Some of the most ubiquitous symptoms in PTSD are insomnia and nightmares (Neylan et al. 1998). As many as 90% of veterans with PTSD experience significant regular sleep disturbances (Weiss et al. 1992). Military personnel serving in a combat zone are often sleep deficient, and sleep disruption is associated with the emergence of trauma-related mental health symptoms in this population (Taylor et al. 2014). Longitudinal studies with military personnel suggest sleep problems are also the most commonly reported difficulties following deployment (McLay et al. 2010), and are associated with the later development of psychopathology, including PTSD specifically (Babson and Feldner 2010). Though previously thought to be "secondary" to other PTSD symptoms, recent reviews suggest that sleep disturbance is a core feature of the disorder (Germain et al. 2008; Germain 2013; Spoormaker and Montgomery 2008).

REM sleep, in particular, is implicated in PTSD, in part because most nightmares occur during REM sleep. Studies using polysomnography to measure sleep architecture in PTSD subjects as compared to control participants do not always show consistent differences between groups (see Germain 2013, for a brief review of this literature), perhaps in part because sleep in PTSD differs widely from individual to individual and from night to night (Straus et al. 2015). However, these types of studies often do show abnormalities in REM sleep in PTSD patients in comparison to controls or other clinical groups (Kobayashi et al. 2007). Some studies show reduced REM in PTSD compared to other groups (Mellman et al. 1997). Other studies have shown abnormalities in REM density, or the frequency of rapid eye movements during REM sleep (Mellman et al. 2002). Additionally, a growing body of research suggests that REM sleep fragmentation may be a more consistently observed biomarker of PTSD than total REM duration. In one study, REM sleep fragmentation, as measured by short REM sleep segment duration, was associated with the severity of PTSD symptoms in the early aftermath of trauma (Mellman et al. 2007). Sleep disruption has also been shown to be a risk factor for developing PTSD following trauma exposure (Babson and Feldner 2010), underscoring the importance of REM sleep and its involvement in PTSD.

3 Neural Mechanisms of Fear, Safety, and Extinction Processes and the Hypothesized Relationship with Disrupted Sleep

Despite these findings suggestive of a role of sleep in PTSD, the mechanism by which sleep, and REM sleep specifically, influences other PTSD symptoms is unknown. We hypothesize that the mechanism is a relationship between REM sleep and both safety learning and extinction processes. Studies examining biological substrates, including brain networks and neurotransmitters involved in fear conditioning and safety signal learning, provide evidence for a connection to disrupted REM sleep. Fear conditioning involves activation of the limbic system and amygdala in particular (Phillips and LeDoux 1992). Extinction and safety learning, by contrast, are "top-down" processes involving frontal regions, such as the ventromedial prefrontal cortex (vmPFC), to inhibit limbic fear response (Jovanovic and Norrholm 2011). In healthy human control subjects, REM sleep de-potentiates amygdala reactivity to emotional stimuli and strengthens vmPFC functional connectivity, suggesting that extinction and safety learning are REM-dependent processes (van der Helm et al. 2011). Safety learning also involves the hippocampus (Jovanovic and Norrholm 2011). REM sleep promotes hippocampal neurogenesis, and sleep deprivation inhibits this process (Meerlo et al. 2009). Hippocampal neurogenesis is crucial for memory consolidation (Meerlo et al. 2009; Pan et al. 2013); thus, REM sleep fragmentation may contribute to impaired discrimination between dangerous and safe cues. Fear conditioning also involves activation of adrenergic neurons in the locus coeruleus (LC), which results in elevated norepinephrine secretion (NE; Tanaka et al. 2000) and release of corticotropin releasing factor (CRF; Heinrichs and Koob 2004). This in turn may fragment REM sleep and potentially lead to nighttime PTSD symptoms, such as nightmares (Raskind et al. 2007). Fragmented REM sleep then subsequently interferes with extinction processes, so conditioned responses to trauma cues are perpetuated long-term via re-experiencing symptoms of PTSD.

4 REM Sleep Disturbance Is Related to Safety Signal Learning and Extinction Processes

See Fig. 1 for a conceptual model of the link between REM sleep disturbance and PTSD development, maintenance, and response to exposure-based treatment. Briefly, in the model, an acute stress response may be associated with safety learning impairment, which leads to REM fragmentation. REM fragmentation in turn impairs extinction processes, particularly extinction recall. This whole process makes the development of PTSD more likely. PTSD, in turn, is associated with both safety learning impairment and REM sleep disturbance, which would be expected to perpetuate or enhance extinction recall impairment, thereby exacerbating daytime symptoms and/or reducing the treatment response to exposure-based interventions. Studies testing the hypothesized links between safety learning, REM sleep, and extinction memory (yellow boxes) are underway (NIMH F31MH106209). Future studies would test the model's assumption that treating sleep prior to exposure-based PTSD treatment would facilitate response to such interventions.

Research findings from animal models and healthy humans corroborate the links we have hypothesized in this model. For example, in one study (Marshall et al. 2014), 42 healthy control participants underwent a fear conditioning/safety learning paradigm and then their sleep was monitored overnight in the laboratory prior to undergoing extinction learning. Increased safety signal learning during the acquisition phase of the study was associated with more consolidated REM sleep that night. More consolidated REM sleep, in turn, was then associated with better discrimination between the CS+ and CS− early in the extinction learning session 24 h later. This study, which was conducted in healthy human control participants, presents evidence for the logical inverse of the model proposed here, which hypothesizes that impaired safety learning is associated with more fragmented REM sleep, which is then subsequently associated with impairments in discrimination between threat and safety cues. In another study (Menz et al. 2016), healthy human participants underwent a fear and safety acquisition session, and then a split-night protocol was used to randomize participants to have a larger or smaller proportion of REM sleep the next night and prior to extinction learning (sleeping the first half of the night leads to relatively less REM sleep and sleeping only the second half of the night leads to more REM). Following a night of recovery sleep,

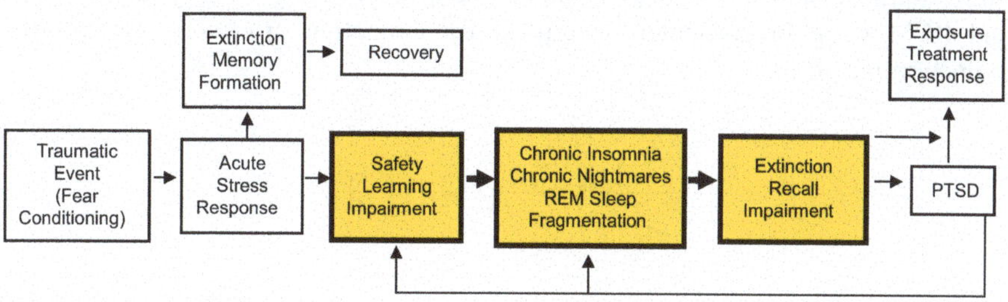

Fig. 1 Explanatory model

participants randomized to have disrupted REM sleep showed worse discrimination of threat and safety signals. This study provides additional evidence in support of the proposed model of PTSD, which postulates that impaired REM sleep is associated with worse ability to discriminate between threatening and safe cues.

In addition to studies demonstrating links between sleep and safety learning, additional research implicates sleep as a critical factor in fear extinction processes. In animal models, sleep deprivation interferes with extinction consolidation (Hussaini et al. 2009), and REM sleep deprivation in particular impairs recall of extinction learning (Fu et al. 2007; Datta and O'Malley 2013). Similar findings have been demonstrated in healthy human control participants. In one study (Pace-Schott et al. 2009), participants were randomized to two different conditions: (1) extinction learning in the evening with an extinction recall session in the morning following a normal night of sleep, or (2) extinction learning in the morning and then recall testing in the evening, without sleep in between. These researchers found that participants in the sleep condition were better able to generalize fear extinction than participants who remained awake between extinction learning and recall. In another study (Spoormaker et al. 2010), participants underwent extinction learning and then were allowed to take a nap prior to extinction recall testing. Participants were divided into two categories: (1) those who showed REM sleep during the nap, and (2) those who did not. Those participants who had REM sleep showed better extinction recall than those without REM. In an additional study in healthy human control subjects (Spoormaker et al. 2012), participants underwent fear conditioning and extinction learning, and were then randomized to normal sleep or REM deprivation overnight prior to an extinction recall session. REM-deprived participants demonstrated impairments in extinction recall relative to participants who slept normally. Finally, in another manuscript related to the Marshall et al. (2014) study cited above, 71 healthy human control participants underwent a 3-day fear acquisition, extinction learning, and extinction recall paradigm. Participants were randomized to one of three groups: (1) normal sleep throughout the study ($n = 21$), (2) total sleep deprivation for one night prior to extinction learning ($n = 25$), and (3) total sleep deprivation for one night prior to extinction recall ($n = 25$). Participants who underwent total sleep deprivation prior to extinction learning demonstrated normal extinction learning but impaired recall of extinction 1 day later (Straus et al. 2017). Additionally, REM sleep consolidation was correlated with extinction recall on the final day of testing. Taken as a whole, these studies conducted in healthy human control subjects, strongly argue sleep, and REM sleep in particular, is critical for extinction processes, particularly extinction recall.

5 Summary and Directions for Future Research

The research cited above, conducted in animals and healthy human subjects, provides evidence for the theoretical model proposed here (Fig. 1) by demonstrating that impaired safety learning is associated with REM sleep fragmentation, which then leads to subsequent difficulties discriminating between threat and safety cues. Though the exact mechanisms of how sleep modifies these PTSD-relevant processes are unknown, these studies also show that disrupted sleep, and fragmented REM sleep in particular, interferes with extinction processes, especially extinction recall.

Despite strong evidence that REM sleep disruption is linked to safety learning and extinction processes, there are some inconsistencies in the literature worth noting. For example, in the sleep deprivation study cited above, REM sleep fragmentation was linked to impairment in extinction recall, though total sleep deprivation was not (Straus et al. 2017). Some studies suggest sleep is critical in generalizing extinction learning (Pace-Schott et al. 2009), though these findings have not been replicated in other studies (Straus et al. 2017). Some researchers have postulated that these inconsistencies are due to methodological discrepancies among studies, such as differences in measurement/or and analysis of fear reactivity (Stickgold and Manoach 2017). More studies will be needed to examine consistent patterns. Additionally, future research should use direct measurement, such as functional neuroimaging of the brain systems involved in fear and safety learning to gain additional insight into the effect of sleep disruption in these processes.

To date, much of the research in this area has focused on healthy human control participants, and no studies have yet investigated the role of REM sleep in safety learning or extinction processes in PTSD patients. Such an investigation is a critical translational step in demonstrating the importance of sleep disturbance in PTSD symptom development, maintenance, and recovery. One small correlational study in veterans with PTSD (Straus et al. 2017) provides preliminary evidence that impaired safety signal learning is associated with REM sleep fragmentation, which in turn is related to subsequent difficulties with safety signal re-learning. Future studies could manipulate REM sleep via pharmacologic agents (e.g., prazosin) or behavioral treatments (e.g., Cognitive Behavioral Therapy for Insomnia) and examine the effects on safety learning and extinction processes in PTSD patients. Additionally, future research studies could also treat sleep disruption prior to initiating exposure therapy for PTSD to examine if reductions in sleep disruption are associated with greater symptom reduction in PTSD treatment. Such studies would help solidify evidence for the role that REM sleep plays in the maintenance of PTSD and will thus identify sleep as a modifiable biological process to target in treatment of PTSD.

Acknowledgements This research received support from the National Institute of Mental Health National Research Service Award No. F31MH106209 (to LDS) and the Veterans Affairs Center of Excellence for Stress and Mental Health.

VBR's work has been funded by National Institutes of Health, the Department of Defense, Navy Bureau of Medicine and Surgery, and the Department of Veterans Affairs. In the last 2 years, she has received funding from Johnson & Johnson and is a consultant for Sunovion Pharmaceuticals. SPAD's work has been funded by the National Institutes of Health, the Department of Defense, the National Science Foundation, and the National Health and Medical Research Council (Australia). Within the last 2 years, he has received compensation as Secretary Treasurer and President-elect of the Sleep Research Society and sat on an Advisory Board for Arena Pharmaceuticals. SBN's work has been funded by the National Center for PTSD, Department of Veterans Affairs and the Department of Defense.

References

Acheson DT, Gresack JE, Risbrough VB (2012) Hippocampal dysfunction effects on context memory: possible etiology for posttraumatic stress disorder. Neuropharmacology 62(2): 674–685

Acheson D, Geyer M, Baker D, Nievergelt C, Yurgil K, Risbrough V, Team M-I (2015) Conditioned fear and extinction learning performance and its association with psychiatric symptoms in active duty marines. Psychoneuroendocrinology 51:495–505

American Psychiatric Association (2013) Diagnostic and statistical manual of mental disorders, 5th edn. Washington, DC

Babson KA, Feldner MT (2010) Temporal relations between sleep problems and both traumatic event exposure and PTSD: a critical review of the empirical literature. J Anxiety Disord 24(1): 1–15

Ball TM, Knapp SE, Paulus MP, Stein MB (2017) Brain activation during fear extinction predicts exposure success. Depress Anxiety 34(3):257–266

Brady K, Pearlstein T, Asnis GM, Baker D, Rothbaum B, Sikes CR, Farfel GM (2000) Efficacy and safety of sertraline treatment of posttraumatic stress disorder: a randomized controlled trial. JAMA 283(14):1837–1844

Breslau N, Kessler RC, Chilcoat HD, Schultz LR, Davis GC, Andreski P (1998) Trauma and posttraumatic stress disorder in the community: the 1996 Detroit Area Survey of Trauma. Arch Gen Psychiatry 55(7):626–632

Datta S, O'Malley MW (2013) Fear extinction memory consolidation requires potentiation of pontine-wave activity during REM sleep. J Neurosci 33(10):4561–4569

Davidson JR, Hughes D, Blazer DG, George LK (1991) Post-traumatic stress disorder in the community: an epidemiological study. Psychol Med 21(03):713–721

Foa E, Hembree E, Rothbaum BO (2007) Prolonged exposure therapy for PTSD: emotional processing of traumatic experiences therapist guide. Oxford University Press, New York

Fu J, Li P, Ouyang X, Gu C, Song Z, Gao J et al (2007) Rapid eye movement sleep deprivation selectively impairs recall of fear extinction in hippocampus-independent tasks in rats. Neuroscience 144(4):1186–1192

Germain A (2013) Sleep disturbances as the hallmark of PTSD: where are we now? Am J Psychiatr 170(4):372–382

Germain A, Buysse DJ, Nofzinger E (2008) Sleep-specific mechanisms underlying posttraumatic stress disorder: integrative review and neurobiological hypotheses. Sleep Med Rev 12(3): 185–195

Grillon C, Morgan CA III (1999) Fear-potentiated startle conditioning to explicit and contextual cues in Gulf War veterans with posttraumatic stress disorder. J Abnorm Psychol 108(1):134

Heinrichs SC, Koob GF (2004) Corticotropin-releasing factor in brain: a role in activation, arousal, and affect regulation. J Pharmacol Exp Ther 311(2):427–440
van der Helm E, Yao J, Dutt S, Rao V, Saletin JM, Walker MP (2011) REM sleep depotentiates amygdala activity to previous emotional experiences. Curr Biol 21(23):2029–2032
Hussaini SA, Bogusch L, Landgraf T, Menzel R (2009) Sleep deprivation affects extinction but not acquisition memory in honeybees. Learn Mem 16(11):698–705
Jovanovic T, Norrholm SD (2011) Neural mechanisms of impaired fear inhibition in posttraumatic stress disorder. Front Behav Neurosci 5:44
Jovanovic T, Norrholm SD, Blanding NQ, Davis M, Duncan E, Bradley B, Ressler KJ (2010) Impaired fear inhibition is a biomarker of PTSD but not depression. Depress Anxiety 27(3): 244–251
Jovanovic T, Kazama A, Bachevalier J, Davis M (2012) Impaired safety signal learning may be a biomarker of PTSD. Neuropharmacology 62(2):695–704
Kessler RC (2000) Posttraumatic stress disorder: the burden to the individual and to society. J Clin Psychiatry 61(Suppl 5):4–12; discussion 13–14. Review
Kessler RC, Chiu WT, Demler O, Merikangas KR, Walters EE (2005) Prevalence, severity, and comorbidity of 12-month DSM-IV disorders in the National Comorbidity Survey Replication. Arch Gen Psychiatry 62(6):617–627. https://doi.org/10.1001/archpsyc.62.6.617
Kilpatrick DG, Ruggiero KJ, Acierno R, Saunders BE, Resnick HS, Best CL (2003) Violence and risk of PTSD, major depression, substance abuse/dependence, and comorbidity: results from the National Survey of Adolescents. J Consult Clin Psychol 71(4):692
Kobayashi I, Boarts JM, Delahanty DL (2007) Polysomnographically measured sleep abnormalities in PTSD: a meta-analytic review. Psychophysiology 44(4):660–669
Lonsdorf TB, Menz MM, Andreatta M, Fullana MA, Golkar A, Haaker J et al (2017) Don't fear "fear conditioning": methodological considerations for the design and analysis of studies on human fear acquisition, extinction, and return of fear. Neurosci Biobehav Rev 77:247–285
Marshall AJ, Acheson DT, Risbrough VB, Straus LD, Drummond SP (2014) Fear conditioning, safety learning, and sleep in humans. J Neurosci 34(35):11754–11760
McLay RN, Klam WP, Volkert SL (2010) Insomnia is the most commonly reported symptom and predicts other symptoms of post-traumatic stress disorder in US service members returning from military deployments. Mil Med 175(10):759
Meerlo P, Mistlberger RE, Jacobs BL, Heller HC, McGinty D (2009) New neurons in the adult brain: the role of sleep and consequences of sleep loss. Sleep Med Rev 13(3):187–194
Mellman TA, Nolan B, Hebding J, Kulick-Bell R, Dominguez R (1997) A polysomnographic comparison of veterans with combat-related PTSD, depressed men, and non-ill controls. Sleep 20(1):46–51
Mellman TA, Bustamante V, Fins AI, Pigeon WR, Nolan B (2002) REM sleep and the early development of posttraumatic stress disorder. Am J Psychiatr 159(10):1696–1701
Mellman TA, Pigeon WR, Nowell PD, Nolan B (2007) Relationships between REM sleep findings and PTSD symptoms during the early aftermath of trauma. J Trauma Stress 20(5):893–901
Mendlowicz MV, Stein MB (2000) Quality of life in individuals with anxiety disorders. Am J Psychiatr 157(5):669–682
Menz MM, Rihm JS, Büchel C (2016) REM sleep is causal to successful consolidation of dangerous and safety stimuli and reduces return of fear after extinction. J Neurosci 36(7): 2148–2160
Milad MR, Wright CI, Orr SP, Pitman RK, Quirk GJ, Rauch SL (2007) Recall of fear extinction in humans activates the ventromedial prefrontal cortex and hippocampus in concert. Biol Psychiatry 62(5):446–454
Milad MR, Orr SP, Lasko NB, Chang Y, Rauch SL, Pitman RK (2008) Presence and acquired origin of reduced recall for fear extinction in PTSD: results of a twin study. J Psychiatr Res 42(7):515–520

Mineka S, Zinbarg R (2006) A contemporary learning theory perspective on the etiology of anxiety disorders: it's not what you thought it was. Am Psychol 61(1):10
Myers K, Davis M (2007) Mechanisms of fear extinction. Mol Psychiatry 12(2):120–150
Neylan TC, Marmar CR, Metzler TJ, Weiss DS, Zatzick DF, Delucchi KL et al (1998) Sleep disturbances in the Vietnam generation: findings from a nationally representative sample of male Vietnam veterans. Sleep 155(7)
Orr SP, Metzger LJ, Lasko NB, Macklin ML, Peri T, Pitman RK (2000) De novo conditioning in trauma-exposed individuals with and without posttraumatic stress disorder. J Abnorm Psychol 109(2):290
Pace-Schott EF, Milad MR, Orr SP, Rauch SL, Stickgold R, Pitman RK (2009) Sleep promotes generalization of extinction of conditioned fear. Sleep 32(1):19
Pan YW, Storm DR, Xia Z (2013) Role of adult neurogenesis in hippocampus-dependent memory, contextual fear extinction and remote contextual memory: new insights from ERK5 MAP kinase. Neurobiol Learn Mem 105:81–92
Phillips RG, LeDoux JE (1992) Differential contribution of amygdala and hippocampus to cued and contextual fear conditioning. Behav Neurosci 106(2):274
Quirk GJ, Mueller D (2008) Neural mechanisms of extinction learning and retrieval. Neuropsychopharmacology 33(1):56–72
Raskind MA, Peskind ER, Hoff DJ, Hart KL, Holmes HA, Warren D et al (2007) A parallel group placebo controlled study of prazosin for trauma nightmares and sleep disturbance in combat veterans with post-traumatic stress disorder. Biol Psychiatry 61(8):928–934
Resick PA, Schnicke MK (1992) Cognitive processing therapy for sexual assault victims. J Consult Clin Psychol 60(5):748
Resnick HS, Kilpatrick DG, Dansky BS, Saunders BE, Best CL (1993) Prevalence of civilian trauma and posttraumatic stress disorder in a representative national sample of women. J Consult Clin Psychol 61(6):984
Rothbaum BO, Davis M (2003) Applying learning principles to the treatment of post-trauma reactions. Ann N Y Acad Sci 1008(1):112–121
Schottenbauer MA, Glass CR, Arnkoff DB, Tendick V, Gray SH (2008) Nonresponse and dropout rates in outcome studies on PTSD: review and methodological considerations. Psychiatry 71 (2):134–168
Spoormaker VI, Montgomery P (2008) Disturbed sleep in post-traumatic stress disorder: secondary symptom or core feature? Sleep Med Rev 12(3):169–184
Spoormaker V, Sturm A, Andrade K, Schröter M, Goya-Maldonado R, Holsboer F et al (2010) The neural correlates and temporal sequence of the relationship between shock exposure, disturbed sleep and impaired consolidation of fear extinction. J Psychiatr Res 44(16):1121–1128
Spoormaker VI, Schröter MS, Andrade KC, Dresler M, Kiem SA, Goya-Maldonado R et al (2012) Effects of rapid eye movement sleep deprivation on fear extinction recall and prediction error signaling. Hum Brain Mapp 33(10):2362–2376
Stewart SH (1996) Alcohol abuse in individuals exposed to trauma: a critical review. Psychol Bull 120(1):83
Stickgold R, Manoach DS (2017) The importance of sleep in fear conditioning and posttraumatic stress disorder. Biol Psychiatr Cognit Neurosci Neuroimaging 2(2):109–110
Straus LD, Drummond S, Nappi CM, Jenkins MM, Norman SB (2015) Sleep variability in military-related PTSD: a comparison to primary insomnia and healthy controls. J Trauma Stress 28(1):8–16
Straus LD, Acheson DT, Risbrough VB, Drummond SP (2017) Sleep deprivation disrupts recall of conditioned fear extinction. Biol Psychiatr Cognit Neurosci Neuroimaging 2(2):123–129
Tanaka M, Yoshida M, Emoto H, Ishii H (2000) Noradrenaline systems in the hypothalamus, amygdala and locus coeruleus are involved in the provocation of anxiety: basic studies. Eur J Pharmacol 405(1):397–406

Taylor MK, Hilton SM, Campbell JS, Beckerley SE, Shobe KK, Drummond SP (2014) Prevalence and mental health correlates of sleep disruption among military members serving in a combat zone. Mil Med 179(7):744–751
Weiss DS, Marmar CR, Schlenger WE, Fairbank JA, Kathleen Jordan B, Hough RL, Kulka RA (1992) The prevalence of lifetime and partial post-traumatic stress disorder in Vietnam theater veterans. J Trauma Stress 5(3):365–376
Wessa M, Flor H (2007) Failure of extinction of fear responses in posttraumatic stress disorder: evidence from second-order conditioning. Am J Psychiatr 164(11):1684–1692

The Future of Contextual Fear Learning for PTSD Research: A Methodological Review of Neuroimaging Studies

Daniel E. Glenn, Victoria B. Risbrough, Alan N. Simmons, Dean T. Acheson, and Daniel M. Stout

Abstract There has been a great deal of recent interest in human models of contextual fear learning, particularly due to the use of such paradigms for investigating neural mechanisms related to the etiology of posttraumatic stress disorder. However, the construct of "context" in fear conditioning research is broad, and the operational definitions and methods used to investigate contextual fear learning in humans are wide ranging and lack specificity, making it difficult to interpret findings about neural activity. Here we will review neuroimaging studies of contextual fear acquisition in humans. We will discuss the methodology associated with four broad categories of how contextual fear learning is manipulated in imaging studies (colored backgrounds, static picture backgrounds, virtual reality, and configural stimuli) and highlight findings for the primary neural circuitry involved in each paradigm. Additionally, we will offer methodological recommendations for human studies of contextual fear acquisition, including using stimuli that distinguish configural learning from discrete cue associations and clarifying how context is experimentally operationalized.

Keywords Configural learning • Context • Fear acquisition • fMRI • PTSD

Contents

D.E. Glenn, V.B. Risbrough (✉), A.N. Simmons, D.T. Acheson, and D.M. Stout
Center of Excellence for Stress and Mental Health, Veterans Affairs San Diego Healthcare System, La Jolla, CA 92093, USA

Department of Psychiatry, University of California San Diego, 9500 Gilman Dr. MC0804, La Jolla, San Diego, CA 92093, USA
e-mail: vrisbrough@ucsd.edu

Curr Topics Behav Neurosci (2018) 38: 207–228
DOI 10.1007/7854_2017_30
Published Online: 17 October 2017

1 Introduction

The ability to learn and identify threat is critical for survival; hence it is highly conserved and supported by multiple neural processes. Threat-relevant information consists not only of discrete cues (e.g., a gun) but also of the context in which a threat occurs (e.g., a gun show versus in a dark alley). Context governs the predictive value of fear and safety cues and facilitates the selection of appropriate cognitive, behavioral, and neurobiological responses. A context may act as a modulator of threat associations and/or an occasion setter for another cue and can itself also serve as a stimulus that acquires associative value (Maren et al. 2013; Urcelay and Miller 2014). Contextual information plays an important role in constraining inappropriate memory recall (Chun and Phelps 1999). Impairments in contextual fear learning during and following trauma may be involved in the etiology of posttraumatic stress disorder (PTSD) by contributing to inappropriate recall of traumatic memory (Acheson et al. 2012; Liberzon and Abelson 2016). Improved understanding of contextual fear learning may inform development of novel PTSD treatment and prevention efforts (Glenn et al. 2014; Risbrough et al. 2016), underscoring the need to better delineate the neural mechanisms associated with contextual fear.

Much of what is known about the neural mechanisms of contextual fear learning is based upon animal research (e.g., Bouton 1993; Fanselow 2000, 2010). Animal studies indicate that contextual fear may be learned through two distinct processes: elemental and configural processing of contextual information (Rudy et al. 2004). Elemental processing involves learning contextual information through separate associations with each of the salient individual elements present, which primarily requires only the amygdala. Alternatively, configural processing reflects the binding together of multimodal individual contextual features into a single gestalt representation of the entire context. The hippocampus is thought to form configural representations of a context which are subsequently used to recognize a similar context (pattern completion) or distinguish between contexts (pattern separation) (Rudy 2009) and communicate with the amygdala to control fear behavior.

Configural and elemental processes compete over encoding of contextual information such that under normal circumstances hippocampal-driven configural learning takes priority. In circumstances involving compromised hippocampal functioning, contextual information can potentially be learned through multiple amygdala-driven elemental associations.

Configural versus elemental learning of contextual information during a traumatic experience may play an important role in the etiology of PTSD. Impaired hippocampal encoding of contextual details during and following trauma could lead to the amygdala taking over, resulting in elemental encoding of the traumatic context. An important consequence of amygdala-driven elemental encoding of trauma is that subsequently encountering a single element related to the trauma may trigger recall of the traumatic memory and elicit a fear response. This differs from recall of a configural memory of trauma which is likely to be triggered only in circumstances similar to the overall traumatic context (Acheson et al. 2012; Liberzon and Abelson 2016). An example is illustrated in the following vignette: *A soldier, Joe, is driving in a long convoy through a desolate desert area of Afghanistan where prior convoys had been attacked by enemy combatants. Joe experiences the trauma of seeing a truck unexpectedly blown up in front of him by a roadside explosive.* If encoding of Joe's traumatic memory is configural, it is likely that he will only subsequently have strong recall of the trauma in situations with many of the same co-occurring contextual features (e.g., desert, roadside trash, war zone, military convoy, hot sun) but not in situations with only a single trauma reminder (e.g., seeing sand at the beach or trash by the side of the road while in the United States). Alternatively, if memory of the traumatic context was formed elementally, there is increased probability that single reminders of the event (sand, trash, hot sun, smell of automobile exhaust) will cause Joe to have increased physiological arousal related to frequent re-experiencing of the trauma. Frequent re-experiencing of traumatic memory is a core symptom of PTSD.

There is a strong theoretical rationale for researching neural mechanisms related to contextual fear learning and PTSD risk, resilience, and etiology (Acheson et al. 2012; Liberzon and Abelson 2016). However, a significant barrier to investigating these areas is a lack of clarity over what exactly the term "context" means and how to operationalize it. For example, what constitutes the context in the vignette about Joe's traumatic experience? The context shaping Joe's memory of his trauma memory likely includes a combination of the multimodal sensory details of the environment, internal affective and cognitive states, and the unpredictability of threat over an extended period of time. Context is a broad, multifaceted construct, so it is perhaps unsurprising that the operational definitions and methods used to investigate human contextual fear learning are also broad and varied. Human neuroimaging studies of contextual fear acquisition can be grouped into four broad categories in terms of the paradigms used to manipulate contextual learning: colored background, static picture background, virtual reality (VR), and configural stimuli. Across these broad paradigms, there is much variability in how context is defined and measured. Contextual fear learning may represent a behavioral phenotype with the potential to inform PTSD treatment development (Risbrough et al.

2016). In order to make progress, though, there is a need to clarify how contextual fear has been operationalized in human research and to elucidate the neural circuitry involved in distinct aspects of contextual learning.

2 Review

2.1 Review Method

The aim of the current review is to highlight the primary methods used in neuroimaging studies of human contextual fear acquisition, summarize findings regarding the neural circuitry involved, and provide methodological recommendations. There are a number of research domains related to but outside of the scope of this review, which is restricted to human neuroimaging studies with acquisition of contextual fear as a primary construct of interest. Thus, this review will not cover studies in the following areas: human neuroimaging of cued fear conditioning (e.g., Greco and Liberzon 2016), contextual fear conditioning in humans that does not involve neuroimaging (e.g., Ameli et al. 2001), contextual modulation of fear extinction in both humans and animals (e.g., Anagnostaras et al. 2001; Milad and Quirk 2012), and non-contextual fear-related hippocampal functioning (e.g., Hayes et al. 2012; Chen and Etkin 2013; Hannula and Helmstetter 2016; Nees and Pohlack 2014). This review was conducted through PubMed searches of "context fear," "contextual fear," "fMRI," and "imaging," with studies only included if they used human subjects and focused on acquisition of contextual fear. Here we will review four types of paradigms used in neuroimaging studies of contextual fear acquisition (colored background, static background, VR, configural), outline findings for the primary neural circuitry involved in each paradigm, and make methodological recommendations for human studies of contextual fear. To facilitate interpretation of the neuroimaging findings, we additionally conducted a custom meta-analysis using Neurosynth based on the studies identified and reviewed (Yarkoni et al. 2011).

2.2 Background Color or Picture

The most basic method investigators have used in imaging studies to manipulate context is by changing the background color of a screen (Armony and Dolan 2001; Barrett and Armony 2009; Cavalli et al. 2017; Lang et al. 2009; Pohlack et al. 2012). In this paradigm, the threat context (CON+) and safety context (CON−) are designated by distinct colors which correspond to whether or not an aversive unconditioned stimulus (US) will be administered. The earliest neuroimaging example of this paradigm used distinct background colors that changed to represent different contexts (Armony and Dolan 2001; Armony and Dolan 2001; Barrett and

Armony 2009). Variations of this paradigm have used background color contexts that slowly transitioned back-and-forth between CON+ and CON− (Lang et al. 2009) rather than abruptly switching between colors as in prior studies. It is noteworthy that all of the studies using a slow transition between the color of CON+ and CON− reported increased hippocampus activity for CON+ compared to CON− (see Table 1), while the distinct color background studies did not. Instead, the distinct color background studies found only amygdala and parietal activity differences in CON+ versus CON− (Armony and Dolan 2001; Barrett and Armony 2009). The absence of hippocampus activity in distinct color background studies may reflect that the simplicity of stimuli in this paradigm could be solved through elemental processing and does not require forming a configural representation of the context. This suggests that the additional perceptual component of slowly transitioning color between contexts may be sufficiently complex to require hippocampus-related processing.

Another common paradigm for studying contextual fear acquisition uses static pictures or photographs of distinct environments to serve as contexts (Marschner et al. 2008; Steiger et al. 2015). In this paradigm, pictures of two "similar but easily distinguishable rooms" serve as CON+ and CON−. The room pictures are presented for long trial durations (~60s trial), during which discrete cues (geometric shapes presented over background picture; CS+ or CS−) are presented, with timing of US presentation predictable during CS+ but unpredictable during CON+. In general, studies using this paradigm report elevated skin conductance response (SCR), US expectancy, and hippocampus activity during CON+ compared to CON− and elevated amygdala activity during discrete CS+ versus CS− trials. Most imaging studies of contextual fear use psychiatrically healthy samples, but one study using background pictures as contexts examined differences between PTSD patients ($n = 14$), trauma-exposed healthy subjects ($n = 12$), and healthy controls ($n = 11$) (Steiger et al. 2015). Self-reported arousal, negative valence, and US expectancy were higher to CON+ than CON− for all subjects, but PTSD patients had poorer contingency awareness than healthy controls and higher differential hippocampal response to CON+ versus CON− than both other groups. The authors hypothesized that the PTSD-related increase of hippocampal activity for CON+ compared to CON− may reflect compensatory neural engagement to perform a low-load contextual task for the PTSD patients. In other words, for non-PTSD subjects, the task was sufficiently easy that it required minimal hippocampus-dependent processing, while for PTSD patients task completion required greater hippocampal engagement.

In contextual fear paradigms with background colors or pictures as contexts, there are several ways in which the simplicity of the visuospatial information is a methodological strength. First, using colors or pictures as contexts may be more easily replicable across research laboratories than VR contexts due to the relative simplicity and low cost of designing the contextual stimuli. Second, neuroimaging findings from studies using background color as context are easier to interpret as being related to unpredictability of US timing rather than confounding unpredictability with learning about complex multimodal contextual features.

Table 1 Summary of neuroimaging findings for contextual fear learning paradigms

Context manipulation studies reviewed		Brain regions						
				Prefrontal cortex			Other	
		HIP	AMY	mPFC	lPFC	ACC	Cortical	Subcortical
Distinct colored background	Armony and Dolan (2001) $n = 8$; Barrett and Armony (2009) $n = 18$		✓				Parietal cortex	
Transitioning colored background	Lang et al. (2009) $n = 21$; Pohlack et al. (2012) $n = 118$	✓	✓	mPFC	IFG, SFG		INS, parietal cortex, SMG, SSC	Putamen, ventral striatum
Static back-ground picture	Marschner et al. (2008) $n = 19$; Steiger et al. (2015) $n = 37$	✓			IFG	dACC	INS, parietal cortex	
Virtual reality	Alvarez et al. (2008) $n = 13$; Alvarez et al. (2011) $n = 18$; Andreatta et al. (2015) $n = 24$; Indovina et al. (2011) $n = 23$	✓	✓	OFC, dmPFC vmPFC	IFG, SFG, MFG	dACC sgACC MCC	MTG, INS, parietal cortex	BNST, thalamus
Configural	Baeuchl et al. (2015) $n = 15$	✓	✓	mPFC	IFG, MFG		MTG, INS, parietal cortex	Thalamus, caudate

ACC anterior cingulate cortex, *AMY* amygdala, *BNST* bed nucleus of the stria terminalis, *dACC* dorsal anterior cingulate cortex, *dmPFC* dorsomedial prefrontal cortex, *HIP* hippocampus, *IFG* inferior frontal gyrus, *INS* insula, *lPFC* lateral prefrontal cortex, *MCC* mid-cingulate cortex, *MFG* middle frontal gyrus, *mPFC* medial prefrontal cortex, *MTG* middle temporal gyrus, *OFC* orbitofrontal cortex, *SFG* superior frontal gyrus, *sgACC* subgenual anterior cingulate cortex, *SMG* supramarginal gyrus, *SSC* somatosensory cortex, *vmPFC* ventromedial prefrontal cortex

Alternatively, background pictures and particularly background colors have much less ecological validity as contexts than VR environments. It is debatable to what extent different backgrounds colors or pictures actually represent different contexts versus different simple cues. Distinguishing between background colors almost certainly does not require configural processing. In studies using background pictures, without careful methodological control, distinguishing CON+ from CON− may be accomplished by solely remembering a single element present in each picture. For example, the room pictures utilized by Steiger et al. (2015) were similar in terms of overall shape and layout, but the contexts could have been differentiated by attending only to whether the left edge of each picture included a door (hallway) or books (library). It is problematic for studies ostensibly investigating hippocampus-dependent configural learning that findings of neural activity could reflect elemental rather than configural processing.

2.3 Virtual Reality

VR has been utilized in numerous studies of contextual fear conditioning (e.g., Baas et al. 2004; Grillon et al. 2006), though only a small number include neuroimaging (Andreatta et al. 2015; Alvarez et al. 2008; Alvarez et al. 2011; Indovina et al. 2011). In this paradigm, subjects passively move through VR rooms (e.g., house, airport) that serve as contexts, usually for an extended trial duration (30–40 s). Typically, the aversive US is presented unpredictably within CON+ with no US delivery during CON− or ITI. The unpredictable timing of the US in CON+ is designed to maximize conditioning to the overall environment rather than to specific features within the context. VR studies of contextual fear generally find increased SCR, US expectancy, post-training anxiety, and hippocampus and amygdala activity for CON+ versus CON−.

A modification of this paradigm adds discrete cues (e.g., 3 s auditory tone, virtual actor raising arms to ears for 4–6 s) which are presented multiple times during longer CON+ or CON− trial (Alvarez et al. 2011; Indovina et al. 2011). In these studies, the offset of the discrete CS+ is predictably paired with US administration, while timing of US presentation is unpredictable during CON+ trials. Findings using this paradigm generally demonstrate higher self-reported anxiety and SCR to the unpredictable versus predictable context and to the CS+ versus CS−. VR neuroimaging studies have found sustained activation in the bed nucleus of the stria terminalis (BNST) and in frontal and parietal regions during CON+ compared to CON− (Alvarez et al. 2011). Activation of the extended amygdala to predictable and unpredictable cues (Andreatta et al. 2015) and hippocampus and BNST activation to cues during US unpredictability have also been observed (Alvarez et al. 2011). Using a similar VR paradigm, Indovina et al. (2011) examined whether aberrant contextual fear learning is associated with individual variation in trait anxiety, a key risk factor for internalizing psychopathology (Shackman et al. 2016). Hippocampus activity during the unpredictable CON+ was increased for high trait anxious individuals, but ventromedial prefrontal cortex (vmPFC) to hippocampus functional connectivity was decreased, suggesting an anxiety-related deficit in recruiting contextual-relevant neurocircuitry. The anxiety-related enhancement of the hippocampus during the unpredictable CON+ may reflect compensatory recruitment of contextual fear learning (Steiger et al. 2015).

A major strength of using VR environments is that they represent perhaps the most ecologically valid paradigm used in fMRI studies of contextual fear acquisition. Subjects move through immersive virtual environments that mimic complex visual aspects of real-world environments. Yet, it is worth noting that VR studies of contextual fear acquisition have thus far been limited to visual elements. With continued technical advances, future studies in the VR paradigm will hopefully include other forms of sensory information such as auditory, olfactory, and tactile stimulation to create truly multimodal VR contexts as has been done in research on VR exposure therapy (e.g., Norrholm et al. 2016; Rothbaum et al. 2014).

One important methodological weakness in these VR neuroimaging studies is that configural processing was not required to differentiate contexts. In two of these studies, the different VR environments could have been distinguished solely based on the background or floor color (Andreatta et al. 2015; Indovina et al. 2011). Furthermore, none of these VR studies report having included the same elements within different contexts (i.e., feature-identical design), so contextual differentiation could be accomplished through learning the presence or location of only a few elements. It is problematic for VR imaging studies hoping to draw conclusions about hippocampus-dependent processes that contextual differentiation could be completed via hippocampal-independent means (i.e., elemental associations).

2.4 *Configural*

A novel methodological approach uses "feature-identical contexts" containing identical elements within them but which are rearranged in different contexts (Baeuchl et al. 2015). This methodology aims to require configural processing in order to differentiate contexts, i.e., a gestalt representation of each context must be learned through hippocampal-dependent processes (Rudy et al. 2004; Rudy 2009). Baeuchl et al. (2015) used two pictures of rooms as CON+ and CON−, each including seven identical elements (TV set, bookshelf, door, painting, couch, lamp, chair), four of which were rearranged differently between contexts. This design should prevent differentiating contexts through focusing only on the presence of a single element. Subjects had higher SCR, self-reported arousal, negative valence, US contingency ratings, and enhanced activity in the anterior and posterior hippocampus and the basolateral amygdala to CON+ relative to CON−. Yet, even this feature-identical design did not necessarily require learning a representation of the entire context, as contextual differentiation could have been achieved through learning only a pair of elements (e.g., *always CON+ if door is next to couch, always CON− if painting is next to couch*). For studies aiming to investigate hippocampus-dependent configural learning, it is imperative that experimental methodology ensures measurement of the configural learning, or findings cannot be definitively attributed to hippocampus-dependent processing.

2.5 General Issues in Contextual Fear Learning Paradigms

One of the earliest paradigms to investigate human contextual fear learning used variants of the no-shock, predictable shock, and unpredictable shock task (NPU task; Grillon and Davis 1997; Grillon and Morgan 1999). In these paradigms, a colored background, scene, or cue signals whether an oncoming aversive US will be administered in response to a specific cue, occur randomly, or not occur at all (Schmitz and Grillon 2012). The context, in these studies, is described as the cognitive state of certainty versus uncertainty associated with the background (Baas et al. 2004; Vansteenwegen et al. 2008). Human neuroimaging work utilizing variants of this paradigm is growing rapidly (Grupe and Nitschke 2013), but few studies continue to use the term "context" to describe the different conditions based on shock predictability. Research on predictable versus unpredictable threat has become independent of the role of contextual fear processing and more focused on differentiating sustained anxiety from phasic fear in the human brain (Shin and Liberzon 2010; Tovote et al. 2015). Notably, hippocampus-driven contextual processing does not appear to be heavily involved in threat uncertainty. Generally, investigations of threat uncertainty or unpredictability focus on the extended amygdala, primarily the central nucleus of the amygdala (CeA) and BNST (Shackman and Fox 2016; Fox et al. 2015). Most studies do not find increased hippocampal activity to uncertain threat if that is the sole manipulation (Somerville et al. 2013; Grupe et al. 2013), unless another context is manipulated (Alvarez et al. 2008, 2011) or healthy subjects are compared to psychiatric populations (Dretsch et al. 2016). This suggests that cues with temporal unpredictability (the subject knows the US is coming but not when it will occur) have their own unique underlying neurocircuitry (e.g., CeA, BNST) which may shape or interact with hippocampus-based contextual learning.

It is frequently stated that US unpredictability is necessary to elicit contextual fear conditioning (e.g., Baeuchl et al. 2015; Grillon et al. 2004), but this may only be true insofar as it is necessary to elicit sustained anxiety. Just as with a simple CS+, phasic (not sustained) fear to CON+ can be trained through pairing context offset with US onset. There is ample evidence from human and animal research that hippocampus-dependent pattern separation and pattern completion do not require stimulus uncertainty (Bakker et al. 2008; Yassa and Stark 2011; Rolls 2013). Much of the interest in contextual fear conditioning paradigms pertains to their relevance for investigating hippocampus-dependent configural processes in relation to PTSD (Acheson et al. 2012; Steiger et al. 2015), yet perhaps the most ubiquitous methodology in contextual fear studies (US unpredictability) does not depend on hippocampus activity (Shackman and Fox 2016; Somerville et al. 2013). This mismatch between methodology and neural region of interest may result from ambiguity around the term "contextual fear" which led to the conflation of two distinct contextual characteristics (temporal unpredictability versus multimodal features).

The broad definition of context has led to methodological challenges for identifying neural activity which is unique to different forms of contextual manipulation. A methodological limitation of most fMRI studies of contextual fear acquisition is that different contextual characteristics (i.e., US unpredictability, context duration, simple versus complex multimodal features) are commonly confounded with one another. For example, it is impossible to make conclusions about findings of elevated hippocampal activity during VR threat contexts (Alvarez et al. 2008; Andreatta et al. 2015) or static picture contexts (Marschner et al. 2008). The neural circuits underlying configural learning versus sustained anxiety are likely different, and they cannot be distinguished in studies such as these, which confound distinct contextual characteristics.

A general critique about imaging studies of contextual fear learning pertains to the use and reporting of psychophysiological measures. Fear-potentiated startle is one of the most commonly used and well-validated methods for measuring fear learning in humans (e.g., Grillon 2008), yet none of the studies reviewed here utilized startle. The lack of imaging studies in this area utilizing startle is likely due to the challenge of presenting auditory startle pulses over the loud background noise of an fMRI environment (>100 dB; Ravicz et al. 2000; Moelker and Pattynama 2003) or to avoid introducing electrical artifacts and noise into acquisition of brain images and vice versa (scanner pulses and changing magnetic field causing noise in EMG collection). However, simultaneous fMRI and startle measurement is possible with scanner safe equipment and additional preprocessing steps (van Well et al. 2012; Gorka et al. 2017). Relying on SCR as the sole physiological measure of fear learning is problematic because SCR and startle reactivity are distinct biological responses with different underlying neural underpinnings and potential for translation to model organisms (Davis 2006; Nagai et al. 2004; Risbrough 2010). SCR, but not startle reactivity, may be dependent on contingency awareness (Sevenster et al. 2014), and SCR may be a less sensitive measure than startle in detecting differences in fear responding related to PTSD (Acheson et al. 2014). Finally, few studies report associations (positive or negative) between physiological measures of fear and neural activity (Indovina et al. 2011; Pohlack et al. 2012). This paucity of reporting is likely related to the difficulty relating SCR and BOLD changes over long stimulus durations, as well as the broader issue in the literature of not reporting negative results. It is worth reiterating the value of reporting negative findings (e.g., Teixeira da Silva 2015) particularly for research domains such as brain activity during fear learning in which the expensive monetary cost of equipment limits both sample size and the overall number of studies.

2.6 Summary of Neuroimaging Findings

Much of the work defining the neural circuitry associated with contextual fear learning has been done in animals (Maren et al. 2013) with little neuroimaging work investigating this area in humans (Greco and Liberzon 2016). Findings from the human neuroimaging studies reviewed here are generally consistent with findings from animal models (see Table 1 for summary of imaging findings across contextual fear paradigms). The results from Marschner et al. (2008) found dissociable roles for the hippocampus and amygdala for contextual and cued fear acquisition, respectively, and replicated animal findings implicating the hippocampus as the key region for acquiring contextual/configural fear but not elemental fear (Fanselow 2010; Maren et al. 2013) and the amygdala as primary region for elemental processing (Rudy 2009). Other human studies find contextual modulation of both the hippocampus and amygdala (Andreatta et al. 2015) or the amygdala alone (Armony and Dolan 2001). One potential explanation is that the hippocampus forms a conjunctive representation of the discrete fear-related cues, which strengthens connections with the amygdala to control defensive responses (e.g., freezing; Rudy et al. 2004) and inhibits elemental contextual processing by the amygdala (Rudy 2009). This is consistent with converging findings in connectivity analyses where increased functional connectivity between the hippocampus and amygdala was observed (Baeuchl et al. 2015) and a path analysis which showed a negative association between hippocampus and the amygdala (Alvarez et al. 2008). These findings are correlational in nature and do not provide a causal explanation, but they show preliminary evidence that contextual fear learning in humans recruits similar neural circuitry as has been extensively mapped in animal models (Rudy 2009; Fanselow 2010).

Beyond the hippocampus and amygdala, many of the reviewed studies observed medial PFC (mPFC) activation, including the vmPFC and dorsal anterior cingulate cortex (dACC) (Andreatta et al. 2015; Marschner et al. 2008; Pohlack et al. 2012). Animal models demonstrate that the mPFC is an important region for encoding contextual associations (Hyman et al. 2012; Euston et al. 2012) and plays a pivotal role in fear conditioning and extinction in humans and animals (Giustino and Maren 2015). Specifically, the vmPFC has been shown to mediate fear regulation and extinction (Sehlmeyer et al. 2009; Milad and Quirk 2012) and encoding of contextual cues (Rozeske et al. 2015; Quinn et al. 2008). In contrast dorsal regions of the ACC are thought to integrate cognitive, affective, and physiological signals (Shackman et al. 2011) and be involved in the expression of fear during fear conditioning (Etkin et al. 2011; Milad et al. 2007) including contextual fear (Rozeske et al. 2015). These data suggest that subregions within the mPFC are critical to contextual fear expression.

The results of our review also indicate that a broad frontoparietal network responds to contextual fear processing. For example, Lang et al. (2009) found contextual fear was associated with sustained activation in the superior frontal gyrus, inferior frontal gyrus (IFG), frontal gyri, supramarginal gyrus, and the insula.

Marschner et al. (2008) found increased bilateral parietal cortex, bilateral insula, dACC, and orbital frontal cortex for CON+. Baeuchl et al. (2015) found configural context fear-related learning in the IFG and middle frontal gyri (MFG), bilateral parietal cortices, and bilateral insula. Such frontoparietal activity may reflect enhanced cognitive and attentional allocation to threatening contexts (Corbetta and Shulman 2002; Scolari et al. 2015; Zanto and Gazzaley 2013), or it could reflect emotional regulation processes (Etkin et al. 2015). The lateral PFC activity is consistent with research showing this region's importance for cognitively demanding fear learning such as contextual fear extinction and trace conditioning (Delgado et al. 2008; Knight et al. 2004; Gilmartin et al. 2014). The insular cortex plays an important function in anticipating aversive events (Carlson et al. 2011; Simmons et al. 2011) and in predicting cognitive control demands (Jiang et al. 2015), especially in paradigms that use unpredictable shock (Alvarez et al. 2015; Grupe and Nitschke 2013). It has also been shown to facilitate the expression of contextual fear in rodents (Alves et al. 2013). Collectively, these data suggest that contextual fear learning is associated with broad cortico-limbic circuits underlying cognitive-emotional function.

3 Contextual Fear Acquisition Meta-Analysis

To facilitate the summary of the neuroimaging findings associated with contextual fear learning, we conducted a custom meta-analysis by searching the Neurosynth database (Yarkoni et al. 2011) for the studies surveyed in our review. This strategy enabled us to aggregate the findings across multiple studies, thereby increasing the power to detect effects and visualize "core" brain circuits underlying contextual fear learning. A total of six studies were available for which we created the term "*contextual fear acquisition*." As detailed by Yarkoni et al. (2011), the Neurosynth database extracts the coordinates of brain activation maps from each of the identified studies and then creates Z-scored brain maps representing the strength of the association of the term (i.e., contextual fear acquisition) to the rest of the brain. The meta-analysis shows the probability that specific brain regions consistently activate given the custom term (e.g., contextual fear acquisition). The forward and reverse inference maps of contextual fear acquisition and a list of the six investigations used can be found online: http://www.neurosynth.org/analyses/custom/31dcd438-7888-413f/. To aid interpretation of the custom meta-analysis, we assessed how the contextual fear acquisition brain map is unique versus overlaps with the broader Neurosynth automated meta-analysis terms "fear" and "conditioning." A total of 298 studies were included for the term fear and "137" studies for conditioning as of 6/5/17. For ease of comparison between the three terms, we computed a three-way conjunction map to visualize brain regions that overlap and differ. All brain maps are visualized at a threshold of $Z > 5.0$ ($FDR < 0.01$).

As Fig. 1 shows, contextual fear acquisition (red clusters) is associated with a distributed network of cognitive and affective neural regions that differ from the

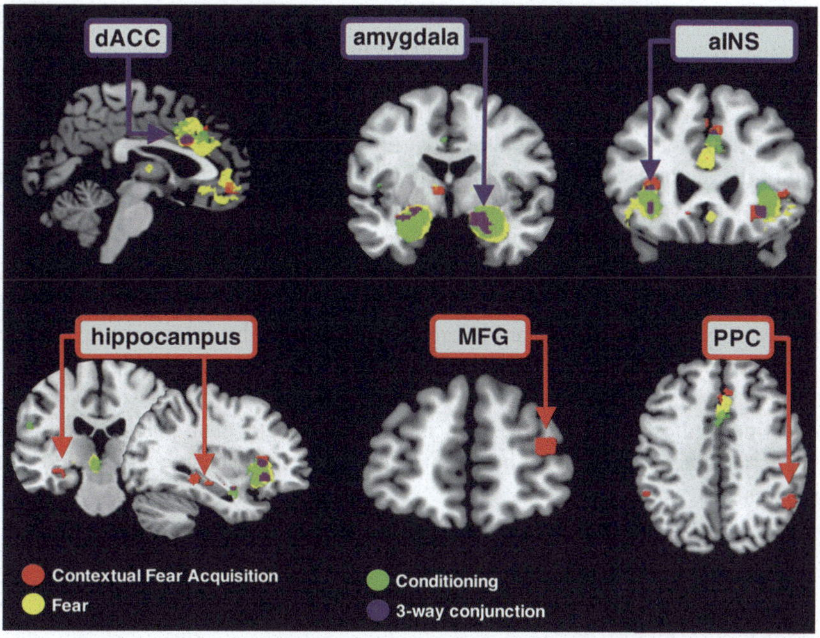

Fig. 1 Conjunction analysis showing similarities and differences between the custom meta-analysis of "contextual fear acquisition" and the automated meta-analyses of "fear" and "conditioning" using Neurosynth. Brain maps shown are forward inference maps at a threshold of $Z > 5$ and FDR corrected (FDR < 0.01)

broader constructs of fear (yellow clusters) and conditioning (green clusters). As hypothesized, studies of contextual fear acquisition consistently report activity in the hippocampus, particularly in mid to posterior subregions. The anterior hippocampus appears to share neural circuitry with contextual fear acquisition, fear, and conditioning (purple clusters). This pattern suggests that learning or expression of fear is associated with anterior portions of the hippocampus, whereas contextual aspects reliably activate mid to posterior portions. The overlap in the anterior hippocampus extends into the amygdala, where the three terms also show strong activation. These results converge with animal studies (Fanselow 2010) implicating the hippocampus and amygdala as key regions in discriminating safe versus threatening contexts.

Within the PFC contextual fear acquisition and fear overlapped in the vmPFC. All three terms (contextual fear acquisition, fear, and conditioning) converged in the dACC, with larger clusters found for fear and conditioning. Consistent MFG activation was unique to studies with contextual fear acquisition but not fear or conditioning, as was activation in the parietal cortex. These results suggest that regions important for higher-order cognition and attention may be important for contextual fear learning.

All three terms consistently show activation in the bilateral anterior insula, and contextual fear acquisition and fear overlap in the BNST. The insula and BNST have been shown to be important for sustained anxiety and modulating the anticipation of threat certainty (Fox et al. 2015; Paulus and Stein 2006; Shackman and Fox 2016). In regard to contextual fear learning, insula and BNST activity may be a result of the methodological convention to administer the US unpredictably. Future research is needed to determine the role the insula and BNST play in contextual fear paradigms that use predictable threat timing.

As this custom contextual fear acquisition meta-analysis consisted of six studies, cautious interpretation is warranted. Moreover, due to the few studies to select from, we only present the forward inference maps. Thus, the unique neural circuitry of contextual fear acquisition that is reported provides only tentative evidence that those regions may contribute to contextual fear learning. As future contextual fear acquisition studies are published, they can be added to the meta-analysis which will serve to enhance the understanding of the neural mechanisms associated with contextual fear acquisition.

4 Recommendations

4.1 Terminology

Future studies of contextual fear learning should use language that clearly identifies which types of contextual characteristics are investigated. Only using the broad term "context conditioning" or "contextual fear learning" can indicate a number of distinctly different paradigms. Just as it would be problematic to lump together without distinction studies of fear acquisition, extinction, and reinstatement all under the term "fear learning," failing to specify significant methodological differences in the study of contextual fear acquisition hinders scientific progress. Indeed, in human research the term "contextual fear learning" is vague enough that it means almost nothing when used in the absence of a specifier (e.g., *unpredictable* contextual fear, *configural* contextual fear). Until there is widespread agreement about what constitutes an associative context, researchers should specify exactly how a context is operationalized in new studies and in previous findings being cited. Further, any discussion about neural circuits involved with contextual fear learning in regard to psychiatric etiology should differentiate between distinct forms of contextual fear processes (e.g., *Is PTSD associated with dysfunction in configural processing of multimodal environmental elements, uncertainty of threat, extended duration threat, or some combination of all three?*).

4.2 Stimulus and Response Characteristics

Investigations of contextual fear should utilize experimental designs that control for rather than confound distinct contextual characteristics (US unpredictability, long stimulus duration, multimodal configuration). Such methodological considerations are particularly important for imaging studies aiming to draw conclusions about neural activity. As long as the neural circuitry underlying a given process is not fully understood, experimental designs should aim to isolate processing of only a single contextual characteristic. This strategy could include examining (a) configural learning in visually complex short duration (6–8 s) contexts with predictable US timing, (b) US unpredictability in visually simple short duration contexts, or (c) stimulus duration in visually simple contexts with predictable US timing. Once there is improved understanding of neural circuits involved in processing specific types of contextual characteristics, interactions between distinct circuits might be investigated through study designs such as a 2×2 approach with two characteristics manipulated, while the third is held constant [e.g., visuospatial complexity (simple, complex) $\times$ US timing (predictable, unpredictable) in short duration contexts].

Linking neuroimaging findings with functional indicators of fear will be critical for understanding functional relationships between circuit activity and contextual modification of fear expression. This complementary approach will greatly enhance our understanding of abnormalities in contextual fear expression and circuits in patient populations such as PTSD. Beyond SCR and self-report US expectancy/contingency ratings, additional measures of fear learning include fear-potentiated startle (van Well et al. 2012; Gorka et al. 2017) and pupil dilation (Visser et al. 2015; Korn et al. 2017), which probe different fear circuitry than SCR. Moreover, both positive and negative results regarding associations between measures of fear learning and neural activity should be reported.

4.3 Control for Cue Associations

A major reason for interest in contextual fear learning is that PTSD has been associated with hippocampal dysfunction (Acheson et al. 2012; Liberzon and Abelson 2016), and contextual fear conditioning is viewed as a useful paradigm for investigating hippocampal-dependent processes of pattern completion and pattern separation (Rudy et al. 2004; Rudy 2009). Unfortunately, much of the fMRI research on contextual fear conditioning has utilized methodology which either does not specifically probe hippocampal-dependent fear learning (i.e., configural) or which confounds complex visuospatial environmental details with threat uncertainty and extended stimulus duration. Studies investigating configural fear learning of contextual information should use feature-identical contexts which can only be distinguished by learning the overall arrangement of elements within a context. One study design that could accomplish this goal would be having a single

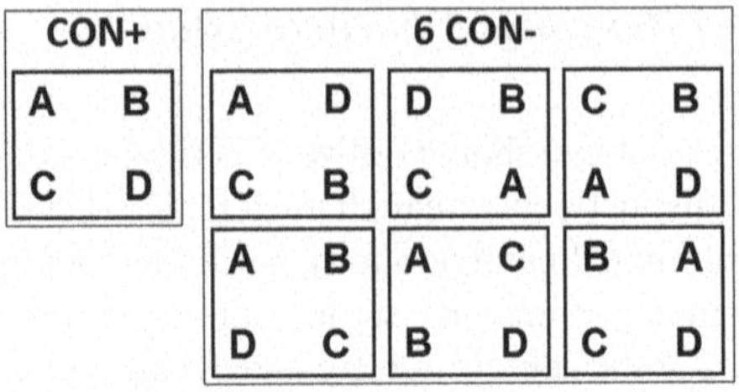

Fig. 2 Schematic design of feature-identical CON+ and CON-s with four features that require configural learning for differentiation. In order to correctly predict CON+ (which is paired with US presentation), it is necessary to learn the full configuration of ABCD in CON+ rather than learning the position of a single feature or pair of features (AB, AD, AC, BD, BC, CD)

threat context (CON+) comprised of several features and multiple safe contexts (CON−) all comprised of different arrangements of the same elements. If designed properly, such an approach can necessitate that distinguishing CON+ from CON− can only be achieved through learning the overall contextual configurations (see Fig. 2 for example). Such a feature-identical approach to contextual fear learning could also utilize VR environments.

4.4 Timing of Recall

Configural learning about contextual information is believed to require the hippocampus (Rudy 2009), but long-term encoding and retrieval of contextual information may depend upon mPFC (Quinn et al. 2008). Given that the timing of memory recall may be a critical determinant of the neural circuitry probed, imaging studies of contextual fear learning should examine short-term versus long-term recall of contextual information.

4.5 Psychiatric Samples

Given recent interest in the role of hippocampal dysfunction as a risk factor for PTSD, it is unfortunate and somewhat surprising that only a single extant neuroimaging study has examined contextual fear in individuals with PTSD (Steiger et al. 2015). The completion of more neuroimaging studies of contextual fear learning in PTSD and psychiatric samples will likely advance understanding of psychiatric risk, resilience, and etiology, particularly if improved methodology suggested above is incorporated.

5 Summary

In conclusion, there are relatively few experimental investigations of human contextual fear learning using neuroimaging (Greco and Liberzon 2016). Findings from these studies generally support the role of the hippocampus, amygdala, and mPFC in contextual fear learning as well as a broad frontoparietal network. A custom Neurosynth meta-analysis provides additional evidence that MFG, parietal cortex, and mid and posterior subregions of the hippocampus contribute to acquisition of contextual fear as compared to neural activity associated with broader constructs of "fear" and "conditioning." Unfortunately, many of the studies we reviewed have methodological confounds which limit interpretation and understanding of the neural circuitry involved. In order to make advances in understanding how humans acquire fear in complex, multimodal environments, it is necessary to increase specificity in terminology about "context" and "contextual fear." Tasks that avoid confounding distinct types of contextual characteristics (i.e., configural processing, US predictability, stimulus duration) will further refine our understanding of neural circuits underlying context modulation/mediation of learned fear. Ultimately, improved understanding of the neural circuitry involved in different aspects of human contextual fear learning may contribute to advances in characterizing risk and resilience for PTSD as well as treatment development.

Acknowledgments All authors are supported by the Center of Excellence for Stress and Mental Health, Veterans Affairs. Dr. Risbrough and Dr. Simmons receive additional funding from their respective VA Merit Awards. Writing of this manuscript was partially supported by the Office of Academic Affiliations, Advanced Fellowship Program in Mental Illness Research and Treatment, Department of Veteran Affairs.

References

Acheson DT, Geyer MA, Risbrough VB (2014) Psychophysiology in the study of psychological trauma: where are we now and where do we need to be? In: Electrophysiology and psychophysiology in psychiatry and psychopharmacology, vol 21. Springer International Publishing, Cham, pp 157–183. http://doi.org/10.1007/7854_2014_346

Acheson DT, Gresack JE, Risbrough VB (2012) Hippocampal dysfunction effects on context memory: possible etiology for posttraumatic stress disorder. Neuropharmacology 62(2): 674–685. http://doi.org/10.1016/j.neuropharm.2011.04.029

Alvarez RP, Biggs A, Chen G, Pine DS, Grillon C (2008) Contextual fear conditioning in humans: cortical-hippocampal and amygdala contributions. J Neurosci 28(24):6211–6219. http://doi.org/10.1523/JNEUROSCI.1246-08.2008

Alvarez RP, Chen G, Bodurka J, Kaplan R, Grillon C (2011) Phasic and sustained fear in humans elicits distinct patterns of brain activity. Neuroimage 55(1):389–400. http://doi.org/10.1016/j.neuroimage.2010.11.057

Alvarez RP, Kirlic N, Misaki M, Bodurka J, Rhudy JL, Paulus MP, Drevets WC (2015) Increased anterior insula activity in anxious individuals is linked to diminished perceived control. Transl Psychiatry 5(6):e591. http://doi.org/10.1038/tp.2015.84

Alves FHF, Gomes FV, Reis DG, Crestani CC, Corrêa FMA, Guimarães FS, Resstel LBM (2013) Involvement of the insular cortex in the consolidation and expression of contextual fear conditioning. Eur J Neurosci 38(2):2300–2307. http://doi.org/10.1111/ejn.12210

Ameli R, Ip C, Grillon C (2001) Contextual fear-potentiated startle conditioning in humans: replication and extension. Psychophysiology 38(3):383–390. http://doi.org/10.1111/1469-8986.3830383

Anagnostaras SG, Gale GD, Fanselow MS (2001) Hippocampus and contextual fear conditioning: recent controversies and advances. Hippocampus 11(1):8–17. http://doi.org/10.1002/1098-1063(2001)11:1%3C8::AID-HIPO1015%3E3.0.CO;2-7

Andreatta M, Glotzbach-Schoon E, Mühlberger A, Schulz SM, Wiemer J, Pauli P (2015) Initial and sustained brain responses to contextual conditioned anxiety in humans. Cortex 63: 352–363. http://doi.org/10.1016/j.cortex.2014.09.014

Armony JL, Dolan RJ (2001) Modulation of auditory neural responses by a visual context in human fear conditioning. Neuroreport 12(15):3407–3411

Baas JM, Nugent M, Lissek S, Pine DS, Grillon C (2004) Fear conditioning in virtual reality contexts: a new tool for the study of anxiety. Biol Psychiatry 55(11):1056–1060. http://doi.org/10.1016/j.biopsych.2004.02.024

Baeuchl C, Meyer P, Hoppstädter M, Diener C, Flor H (2015) Contextual fear conditioning in humans using feature-identical contexts. Neurobiol Learn Mem 121:1–11. http://doi.org/10.1016/j.nlm.2015.03.001

Bakker A, Kirwan CB, Miller M, Stark CEL (2008) Pattern separation in the human hippocampal CA3 and dentate gyrus. Science 319(5870):1640–1642. http://doi.org/10.1126/science.1152882

Barrett J, Armony JL (2009) Influence of trait anxiety on brain activity during the acquisition and extinction of aversive conditioning. Psychol Med 39(2):255–265. http://doi.org/10.1017/S0033291708003516

Bouton ME (1993) Context, time, and memory retrieval in the interference paradigms of Pavlovian learning. Psychol Bull 114(1):80–99

Carlson JM, Greenberg T, Rubin D, Mujica-Parodi LR (2011) Feeling anxious: anticipatory amygdalo-insular response predicts the feeling of anxious anticipation. Soc Cogn Affect Neurosci 6(1):74–81. http://doi.org/10.1093/scan/nsq017

Cavalli J, Ruttorf M, Pahi MR, Zidda F, Flor H, Nees F (2017) Oxytocin differentially modulates pavlovian cue and context fear acquisition. Soc Cogn Affect Neurosci. http://doi.org/10.1093/scan/nsx028

Chen AC, Etkin A (2013) Hippocampal network connectivity and activation differentiates post-traumatic stress disorder from generalized anxiety disorder. Neuropsychopharmacology 38 (10):1889–1898. http://doi.org/10.1038/npp.2013.122

Chun MM, Phelps EA (1999) Memory deficits for implicit contextual information in amnesic subjects with hippocampal damage. Nat Neurosci 2(9):844–847. http://doi.org/10.1038/12222

Corbetta M, Shulman GL (2002) Control of goal-directed and stimulus-driven attention in the brain. Nat Rev Neurosci 3(3):201–215. http://doi.org/10.1038/nrn755

Davis M (2006) Neural systems involved in fear and anxiety measured with fear-potentiated startle. Am Psychol 61(8):741–756. http://doi.org/10.1037/0003-066X.61.8.741

Delgado MR, Nearing KI, LeDoux JE, Phelps EA (2008) Neural circuitry underlying the regulation of conditioned fear and its relation to extinction. Neuron 59(5):829–838. http://doi.org/10.1016/j.neuron.2008.06.029

Dretsch MN, Wood KH, Daniel TA, Katz JS, Deshpande G, Goodman AM et al (2016) Exploring the neurocircuitry underpinning predictability of threat in soldiers with PTSD compared to deployment exposed controls. Open Neuroimag J 10(1):111–124. http://doi.org/10.2174/1874440001610010111

Etkin A, Büchel C, Gross JJ (2015) The neural bases of emotion regulation. Nat Rev Neurosci 16(11):693–700. http://doi.org/10.1038/nrn4044

Etkin A, Egner T, Kalisch R (2011) Emotional processing in anterior cingulate and medial prefrontal cortex. Trends Cogn Sci 15(2):85–93. http://doi.org/10.1016/j.tics.2010.11.004

Euston DR, Gruber AJ, McNaughton BL (2012) The role of medial prefrontal cortex in memory and decision making. Neuron 76(6):1057–1070. http://doi.org/10.1016/j.neuron.2012.12.002

Fanselow MS (2000) Contextual fear, gestalt memories, and the hippocampus. Behav Brain Res 110(1–2):73–81

Fanselow MS (2010) From contextual fear to a dynamic view of memory systems. Trends Cogn Sci 14(1):7–15. http://doi.org/10.1016/j.tics.2009.10.008

Fox AS, Oler JA, Tromp DPM, Fudge JL, Kalin NH (2015) Extending the amygdala in theories of threat processing. Trends Neurosci 38(5):319–329. http://doi.org/10.1016/j.tins.2015.03.002

Gilmartin MR, Balderston NL, Helmstetter FJ (2014) Prefrontal cortical regulation of fear learning. Trends Neurosci 37(8):455–464. http://doi.org/10.1016/j.tins.2014.05.004

Giustino TF, Maren S (2015) The role of the medial prefrontal cortex in the conditioning and extinction of fear. Front Behav Neurosci 9:298. http://doi.org/10.3389/fnbeh.2015.00298

Glenn DE, Minor TR, Vervliet B, Craske MG (2014) The effect of glucose on hippocampal-dependent contextual fear conditioning. Biol Psychiatry 75:847–854. http://doi.org/10.1016/j.biopsych.2013.09.022

Gorka SM, Lieberman L, Shankman SA, Phan KL (2017) Association between neural reactivity and startle reactivity to uncertain threat in two independent samples. Psychophysiology 54(5): 652–662. http://doi.org/10.1111/psyp.12829

Greco JA, Liberzon I (2016) Neuroimaging of fear-associated learning. Neuropsychopharmacology 41(1):320–334. http://doi.org/10.1038/npp.2015.255

Grillon C (2008) Models and mechanisms of anxiety: evidence from startle studies. Psychopharmacology (Berl) 199(3):421–437. http://doi.org/10.1007/s00213-007-1019-1

Grillon C, Davis M (1997) Fear-potentiated startle conditioning in humans: explicit and contextual cue conditioning following paired versus unpaired training. Psychophysiology 34(4):451–458

Grillon C, Morgan CA (1999) Fear-potentiated startle conditioning to explicit and contextual cues in Gulf War veterans with posttraumatic stress disorder. J Abnorm Psychol 108(1):134–142

Grillon C, Baas JMP, Cornwell B, Johnson L (2006) Context conditioning and behavioral avoidance in a virtual reality environment: effect of predictability. Biol Psychiatry 60(7):752–759. http://doi.org/10.1016/j.biopsych.2006.03.072

Grillon C, Baas JP, Lissek S, Smith K, Milstein J (2004) Anxious responses to predictable and unpredictable aversive events. Behav Neurosci 118(5):916–924. http://doi.org/10.1037/0735-7044.118.5.916

Grupe DW, Nitschke JB (2013) Uncertainty and anticipation in anxiety: an integrated neurobiological and psychological perspective. Nat Rev Neurosci 14(7):488–501. http://doi.org/10.1038/nrn3524

Grupe DW, Oathes DJ, Nitschke JB (2013) Dissecting the anticipation of aversion reveals dissociable neural networks. Cereb Cortex 23(8):1874–1883. http://doi.org/10.1093/cercor/bhs175

Hannula DE, Helmstetter FJ (2016) Hippocampal interactions with brain networks that influence learning & memory. Neurobiol Learn Mem 134 Pt A:1–4. http://doi.org/10.1016/j.nlm.2016.08.018

Hayes JP, VanElzakker MB, Shin LM (2012) Emotion and cognition interactions in PTSD: a review of neurocognitive and neuroimaging studies. Front Integr Neurosci 6:89. http://doi.org/10.3389/fnint.2012.00089

Hyman JM, Ma L, Balaguer-Ballester E, Durstewitz D, Seamans JK (2012) Contextual encoding by ensembles of medial prefrontal cortex neurons. Proc Natl Acad Sci U S A 109(13): 5086–5091. http://doi.org/10.1073/pnas.1114415109

Indovina I, Robbins TW, Núñez-Elizalde AO, Dunn BD, Bishop SJ (2011) Fear-conditioning mechanisms associated with trait vulnerability to anxiety in humans. Neuron 69(3):563–571. http://doi.org/10.1016/j.neuron.2010.12.034

Jiang J, Beck J, Heller K, Egner T (2015) An insula-frontostriatal network mediates flexible cognitive control by adaptively predicting changing control demands. Nat Commun 6:8165. http://doi.org/10.1038/ncomms9165

Knight DC, Cheng DT, Smith CN, Stein EA, Helmstetter FJ (2004) Neural substrates mediating human delay and trace fear conditioning. J Neurosci 24(1):218–228. http://doi.org/10.1523/JNEUROSCI.0433-03.2004

Korn CW, Staib M, Tzovara A, Castegnetti G, Bach DR (2017) A pupil size response model to assess fear learning. Psychophysiology 54(3):330–343. http://doi.org/10.1111/psyp.12801

Lang S, Kroll A, Lipinski SJ, Wessa M, Ridder S, Christmann C et al (2009) Context conditioning and extinction in humans: Differential contribution of the hippocampus, amygdala and prefrontal cortex. Eur J Neurosci 29:823–832. http://doi.org/10.1111/j.1460-9568.2009.06624.x

Liberzon I, Abelson JL (2016) Context processing and the neurobiology of post-traumatic stress disorder. Neuron 92(1):14–30. http://doi.org/10.1016/j.neuron.2016.09.039

Maren S, Phan KL, Liberzon I (2013) The contextual brain: implications for fear conditioning, extinction and psychopathology. Nat Rev Neurosci 14(6):417–428. http://doi.org/10.1038/nrn3492

Marschner A, Kalisch R, Vervliet B, Vansteenwegen D, Buchel C (2008) Dissociable roles for the hippocampus and the amygdala in human cued versus context fear conditioning. J Neurosci 28(36):9030–9036. http://doi.org/10.1523/JNEUROSCI.1651-08.2008

Milad MR, Quirk GJ (2012) Fear extinction as a model for translational neuroscience: ten years of progress. Annu Rev Psychol 63(1):129–151. http://doi.org/10.1146/annurev.psych.121208.131631

Milad MR, Quirk GJ, Pitman RK, Orr SP, Fischl B, Rauch SL (2007) A role for the human dorsal anterior cingulate cortex in fear expression. Biol Psychiatry 62(10):1191–1194. http://doi.org/10.1016/j.biopsych.2007.04.032

Moelker A, Pattynama PMT (2003) Acoustic noise concerns in functional magnetic resonance imaging. Hum Brain Mapp 20(3):123–141. http://doi.org/10.1002/hbm.10134

Nagai Y, Critchley HD, Featherstone E, Trimble MR, Dolan RJ (2004) Activity in ventromedial prefrontal cortex covaries with sympathetic skin conductance level: a physiological account of a "default mode" of brain function. Neuroimage 22(1):243–251. http://doi.org/10.1016/j.neuroimage.2004.01.019

Nees F, Pohlack ST (2014) Functional MRI studies of the hippocampus. Front Neurol Neurosci 34:85–94. http://doi.org/10.1159/000356427

Norrholm SD, Jovanovic T, Gerardi M, Breazeale KG, Price M, Davis M et al (2016) Baseline psychophysiological and cortisol reactivity as a predictor of PTSD treatment outcome in virtual reality exposure therapy. Behav Res Ther 82:28–37. http://doi.org/10.1016/j.brat.2016.05.002

Paulus MP, Stein MB (2006) An insular view of anxiety. Biol Psychiatry 60(4):383–387. http://doi.org/10.1016/j.biopsych.2006.03.042

Pohlack ST, Nees F, Ruttorf M, Schad LR, Flor H (2012) Activation of the ventral striatum during aversive contextual conditioning in humans. Biol Psychol 91(1):74–80. http://doi.org/10.1016/j.biopsycho.2012.04.004

Quinn JJ, Ma QD, Tinsley MR, Koch C, Fanselow MS (2008) Inverse temporal contributions of the dorsal hippocampus and medial prefrontal cortex to the expression of long-term fear memories. Learn Mem 15(5):368–372. http://doi.org/10.1101/lm.813608

Ravicz ME, Melcher JR, Kiang NY (2000) Acoustic noise during functional magnetic resonance imaging. J Acoust Soc Am 108(4):1683–1696

Risbrough V (2010) Behavioral correlates of anxiety. Curr Top Behav Neurosci 2:205–228

Risbrough VB, Glenn DE, Baker DG (2016) On the road to translation for PTSD treatment: theoretical and practical considerations of the use of human models of conditioned fear for drug development. In: Electrophysiology and psychophysiology in psychiatry and psychopharmacology, vol 28. Springer International Publishing, Cham, pp 1–24. http://doi.org/10.1007/7854_2015_5010

Rolls ET (2013) The mechanisms for pattern completion and pattern separation in the hippocampus. Front Syst Neurosci 7:74. http://doi.org/10.3389/fnsys.2013.00074

Rothbaum BO, Price M, Jovanovic T, Norrholm SD, Gerardi M, Dunlop B et al (2014) A randomized, double-blind evaluation of D-cycloserine or alprazolam combined with virtual reality exposure therapy for posttraumatic stress disorder in Iraq and Afghanistan War veterans. Am J Psychiatry 171(6):640–648. http://doi.org/10.1176/appi.ajp.2014.13121625

Rozeske RR, Valerio S, Chaudun F, Herry C (2015) Prefrontal neuronal circuits of contextual fear conditioning. Genes Brain Behav 14(1):22–36. http://doi.org/10.1111/gbb.12181

Rudy JW (2009) Context representations, context functions, and the parahippocampal-hippocampal system. Learn Mem 16(10):573–585. http://doi.org/10.1101/lm.1494409

Rudy JW, Huff NC, Matus-Amat P (2004) Understanding contextual fear conditioning: insights from a two-process model. Neurosci Biobehav Rev 28(7):675–685. http://doi.org/10.1016/j.neubiorev.2004.09.004

Schmitz A, Grillon C (2012) Assessing fear and anxiety in humans using the threat of predictable and unpredictable aversive events (the NPU-threat test). Nat Protoc 7(3):527–532. http://doi.org/10.1038/nprot.2012.001

Scolari M, Seidl-Rathkopf KN, Kastner S (2015) Functions of the human frontoparietal attention network: evidence from neuroimaging. Curr Opin Behav Sci 1:32–39. http://doi.org/10.1016/j.cobeha.2014.08.003

Sehlmeyer C, Schöning S, Zwitserlood P, Pfleiderer B, Kircher T, Arolt V, Konrad C (2009) Human fear conditioning and extinction in neuroimaging: a systematic review. PLoS One 4(6): e5865. http://doi.org/10.1371/journal.pone.0005865

Sevenster D, Beckers T, Kindt M (2014) Fear conditioning of SCR but not the startle reflex requires conscious discrimination of threat and safety. Front Behav Neurosci 8:32. http://doi.org/10.3389/fnbeh.2014.00032

Shackman AJ, Fox AS (2016) Contributions of the central extended amygdala to fear and anxiety. J Neurosci 36(31):8050–8063. http://doi.org/10.1523/JNEUROSCI.0982-16.2016

Shackman AJ, Salomons TV, Slagter HA, Fox AS, Winter JJ, Davidson RJ (2011) The integration of negative affect, pain and cognitive control in the cingulate cortex. Nat Rev Neurosci 12(3): 154–167. http://doi.org/10.1038/nrn2994

Shackman AJ, Tromp DPM, Stockbridge MD, Kaplan CM, Tillman RM, Fox AS (2016) Dispositional negativity: an integrative psychological and neurobiological perspective. Psychol Bull 142(12):1275–1314. http://doi.org/10.1037/bul0000073

Shin LM, Liberzon I (2010) The neurocircuitry of fear, stress, and anxiety disorders. Neuropsychopharmacology 35(1):169–191. http://doi.org/10.1038/npp.2009.83

Simmons AN, Stein MB, Strigo IA, Arce E, Hitchcock C, Paulus MP (2011) Anxiety positive subjects show altered processing in the anterior insula during anticipation of negative stimuli. Hum Brain Mapp 32(11):1836–1846. http://doi.org/10.1002/hbm.21154

Somerville LH, Wagner DD, Wig GS, Moran JM, Whalen PJ, Kelley WM (2013) Interactions between transient and sustained neural signals support the generation and regulation of anxious emotion. Cereb Cortex 23(1):49–60. http://doi.org/10.1093/cercor/bhr373

Steiger F, Nees F, Wicking M, Lang S, Flor H (2015) Behavioral and central correlates of contextual fear learning and contextual modulation of cued fear in posttraumatic stress disorder. Int J Psychophysiol 98(3 Pt 2):584–593. http://doi.org/10.1016/j.ijpsycho.2015.06.009

Teixeira da Silva JA (2015) Negative results: negative perceptions limit their potential for increasing reproducibility. J Negat Results Biomed 14(1):12. http://doi.org/10.1186/s12952-015-0033-9

Tovote P, Fadok JP, Lüthi A (2015) Neuronal circuits for fear and anxiety. Nat Rev Neurosci 16(6):317–331. http://doi.org/10.1038/nrn3945

Urcelay GP, Miller RR (2014) The functions of contexts in associative learning. Behav Processes 104:2–12. http://doi.org/10.1016/j.beproc.2014.02.008

van Well S, Visser RM, Scholte HS, Kindt M (2012) Neural substrates of individual differences in human fear learning: evidence from concurrent fMRI, fear-potentiated startle, and US-expectancy data. Cogn Affect Behav Neurosci 12(3):499–512. http://doi.org/10.3758/s13415-012-0089-7

Vansteenwegen D, Iberico C, Vervliet B, Marescau V, Hermans D (2008) Contextual fear induced by unpredictability in a human fear conditioning preparation is related to the chronic expectation of a threatening US. Biol Psychol 77(1):39–46. http://doi.org/10.1016/j.biopsycho.2007.08.012

Visser RM, Kunze AE, Westhoff B, Scholte HS, Kindt M (2015) Representational similarity analysis offers a preview of the noradrenergic modulation of long-term fear memory at the time of encoding. Psychoneuroendocrinology 55:8–20. http://doi.org/10.1016/j.psyneuen.2015.01.021

Yarkoni T, Poldrack RA, Nichols TE, Van Essen DC, Wager TD (2011) Large-scale automated synthesis of human functional neuroimaging data. Nat Methods 8(8):665–670. http://doi.org/10.1038/nmeth.1635

Yassa MA, Stark CEL (2011) Pattern separation in the hippocampus. Trends Neurosci 34(10):515–525. http://doi.org/10.1016/j.tins.2011.06.006

Zanto TP, Gazzaley A (2013) Fronto-parietal network: flexible hub of cognitive control. Trends Cogn Sci 17(12):602–603. http://doi.org/10.1016/j.tics.2013.10.001

The Dissociative Subtype of Post-traumatic Stress Disorder: Research Update on Clinical and Neurobiological Features

Jytte van Huijstee and Eric Vermetten

Abstract Recently, a dissociative subtype of post-traumatic stress disorder (PTSD) has been included in the DSM-5. This review focuses on the clinical and neurobiological features that distinguish the dissociative subtype of PTSD from non-dissociative PTSD. Clinically, the dissociative subtype of PTSD is associated with high PTSD severity, predominance of derealization and depersonalization symptoms, a more significant history of early life trauma, and higher levels of comorbid psychiatric disorders. Furthermore, PTSD patients with dissociative symptoms exhibit different psychophysiological and neural responses to the recall of traumatic memories. While individuals with non-dissociative PTSD exhibit an increased heart rate, decreased activation of prefrontal regions, and increased activation of the amygdala in response to traumatic reminders, individuals with the dissociative subtype of PTSD show an opposite pattern. It has been proposed that dissociation is a regulatory strategy to restrain extreme arousal in PTSD through hyperinhibition of limbic regions. In this research update, promises and pitfalls in current research studies on the dissociative subtype of PTSD are listed. Inclusion of the dissociative subtype of PTSD in the DSM-5 stimulates research on the prevalence, symptomatology, and neurobiology of the dissociative subtype of PTSD and poses a challenge to improve treatment outcome in PTSD patients with dissociative symptoms.

J. van Huijstee
Swammerdam Institute for Life Sciences, University of Amsterdam, Amsterdam, The Netherlands

E. Vermetten (✉)
Department Psychiatry, Leiden University Medical Center Utrecht, Albinusdreef 2, Leiden 2333 ZA, The Netherlands

Arq Psychotrauma Research Group, Diemen 1112 XE, The Netherlands
e-mail: e.vermetten@lumc.nl

Curr Topics Behav Neurosci (2018) 38: 229–248
DOI 10.1007/7854_2017_33
Published Online: 17 October 2017

Keywords Dissociation • Emotion regulation • Imaging • Neurobiology • PTSD • Stress

Contents

1 Introduction

The simple notion that physical/psychological trauma can have lasting psychological and physiological effects on the individual has led to an abundance of research studies into the effects of trauma on health and disease. It has become textbook knowledge that besides the effect of trauma itself, both peritraumatic dissociation and trauma-related persistent dissociation are markers for long-term psychopathology (Irwin 1994; Marmar et al. 1994; Bremner and Brett 1997; Briere 2006; Vermetten et al. 2007). Dissociative symptomatology has seen a long-standing history in both fields of psychiatry and neurology. Dissociative responses can accompany a multitude of psychiatric disorders, including post-traumatic stress disorder (PTSD), dissociative disorders, panic disorder, mood disorders, and psychoses. Moreover, dissociative responses have been reported to occur in several neurological conditions such as epilepsy, migraine headaches, cerebral vascular disease, cerebral neoplasms, and posttraumatic brain damage (Sierra and Berrios 1998).

Dissociation is a prevalent symptom of PTSD (Putnam et al. 1996; Waelde et al. 2005) and involves a disruption of or discontinuity in the usually integrated functions of consciousness, memory, identity, emotion, perception, and body awareness (American Psychiatric Association 2013). A wide range of instruments have been developed for the assessment of dissociation, e.g., the Dissociative Experience Scale (DES, Bernstein and Putnam 1986), Peritraumatic Dissociative Experiences Questionnaire (PDEQ, Marmar et al. 1997), Clinician Administered Dissociative States Scale (CADSS, Bremner et al. 1998), Somatoform Dissociation Questionnaire (SDQ-20, Nijenhuis et al. 1996), Structured Clinical Interview for DSM-IV Dissociative Disorders – Revised (SCID-DR, Steinberg 2000), and the Multidimensional Inventory of Dissociation (MID, Dell 2006), that all operationalize dissociation in different ways ranging from trait to state, normal to pathological, and psychological versus somatoform dissociation. In relation to trauma, it has been proposed that dissociation could be a psychological response to trauma exposure, providing detachment from the overwhelming emotional content of the traumatic experience (Putnam 1985; Spiegel and Cardeña 1990; van der Kolk et al. 1989). Although dissociation is a common feature of PTSD, the centrality of dissociative symptoms in PTSD has been a matter of debate (Carlson et al. 2012; Dalenberg and Carlson 2012). Some have argued that dissociative processes mediate PTSD and that PTSD should be regarded as a dissociative disorder. Others, however, have proposed that dissociation and PTSD are separate but often comorbid phenomena. Moreover, the component model conceptualizes dissociation as a component of the response to traumatic stress and proposes that, like other PTSD symptoms, dissociative symptoms do not necessarily occur in all PTSD patients but are more likely to occur if PTSD is present. Finally, the subtype model, a variant of the component model, proposes that PTSD patients with dissociative symptoms are qualitatively different from individuals with non-dissociative PTSD, showing a different pattern of symptoms, social and clinical covariates, and neurobiological substrates.

Recently, Dalenberg and Carlson (2012) investigated which model is most consistent with research findings on the relationship between dissociation and PTSD, and they found that most of the empirical evidence favors a subtype model of dissociation in PTSD. Moreover, this dissociative subtype of PTSD appears to be associated with distinct clinical and neurobiological features. On the basis of these and other results, a dissociative subtype of PTSD has been included in the DSM-5. This review will concentrate on investigations of the dissociative subtype of PTSD. Firstly, epidemiological and clinical evidence for the existence of a dissociative subtype of PTSD will be discussed. Then, we will highlight the neurobiological features that distinguish dissociative PTSD from non-dissociative PTSD. Furthermore, we will investigate why some individuals develop a clinical presentation of PTSD that is predominated by the occurrence of dissociative symptoms, while other PTSD patients do not exhibit dissociative symptoms. Finally, we will describe promises and pitfalls in the research on the dissociative subtype of PTSD.

2 Epidemiological and Clinical Evidence for the Dissociative Subtype of PTSD

Prior to the launch of the DSM-5, several studies investigated the potential existence of a dissociative subtype of PTSD using latent class, taxometric, epidemiological, and confirmatory factor analyses. For example, Putnam et al. (1996) have reported that mean dissociation scores of a PTSD sample did not reflect uniform distributions within this sample but instead resulted from a small proportion of individuals with high dissociation scores. Furthermore, taxometric analyses of dissociation in trauma-exposed Vietnam veterans revealed a taxon of highly dissociative individuals (Waelde et al. 2005). Members of the dissociative taxon were more often diagnosed with current PTSD and showed more severe PTSD symptoms than non-taxon members. However, only 32% of the veterans with a diagnosis of PTSD belonged to this dissociative taxon, suggesting that there is a subtype of severe PTSD with dissociative symptoms.

Using signal detection analysis, Ginzburg et al. (2006) identified high and low dissociation PTSD subgroups in a sample of 122 women seeking treatment for child sexual abuse. Three PTSD symptoms, namely, hypervigilance, sense of foreshortened future, and sleeping difficulties, discriminated between these high and low dissociation subgroups, indicating that there is a dissociative subtype of PTSD that is characterized by a specific constellation of symptoms. Furthermore, they found that the dissociative subtype was associated with higher levels of childhood maltreatment.

Wolf et al. (2012b) conducted a latent class analysis to examine the evidence for a dissociative subtype in PTSD in a sample of 492 trauma-exposed veterans and their partners. The latent profile analysis yielded evidence for three different classes, namely, a low PTSD severity subgroup, a high PTSD severity subgroup without dissociation, and a high severity subgroup with dissociative symptoms. This dissociative subgroup represented 12% of individuals with PTSD and was characterized by severe PTSD symptoms combined with marked elevations on items assessing flashbacks, derealization, and depersonalization. Individuals in this subgroup also reported greater exposure to childhood and adult sexual trauma. Wolf et al. (2012a) validated these findings using further latent profile analyses on PTSD and dissociation items reflecting depersonalization and derealization in a sample of 360 male veterans with PTSD and a sample of 284 female veterans and active-duty service personnel with PTSD. In both samples the latent profile analysis suggested a three-class solution. Approximately 15% of the male sample and 30% of the female sample were classified into the dissociative subtype. In contrast to their earlier finding that members of the dissociative group reported higher exposure to sexual abuse, they found no group differences related to exposure to sexual trauma. However, women in the dissociative group exhibited higher levels of comorbid personality disorder.

Another study also used latent profile analyses to investigate the existence of a dissociative subtype in a civilian sample of individuals with PTSD predominantly related to childhood abuse (Steuwe et al. 2012). Again, latent profile analyses extracted three groups, one of which was characterized by high derealization and depersonalization scores. This group accounted for 25% of the total PTSD sample. Individuals in the dissociative subgroup showed a higher number of comorbid Axis I disorders and a more significant history of childhood abuse and neglect. Confirmatory factor analyses disclosed that symptoms of derealization and depersonalization could be explained by a fifth PTSD symptom cluster. This dissociative factor was distinct from but correlated significantly with the other four core PTSD symptom clusters.

More recently, Armour et al. (2014b), using latent profile analysis, identified five classes in a sample of 432 military veterans, one of which constituted a severe PTSD class (30%) and one of which constituted a severe PTSD class with dissociative symptoms (13.7%). The dissociative subgroup could be differentiated from the severe PTSD group based on higher scores for reduced awareness, derealization and depersonalization, and sense of foreshortened future. In addition, Armour et al. (2014a) investigated whether levels of depression, anxiety, hostility, and sleeping difficulties would differentiate dissociative PTSD from a similarly severe form of PTSD without dissociation in European victims of sexual assault and rape. Utilizing latent profile analyses, they found four groups in this PTSD sample, namely, a low, moderate, and high PTSD group, and a high PTSD group with dissociative symptoms. The dissociative PTSD group encompassed 13.1% of the sample and had significantly higher mean scores on measures of depression, anxiety, hostility, and sleeping difficulties.

Importantly, all these studies providing evidence for a dissociative subtype of PTSD are based on treatment-seeking samples in Western countries, leaving it unclear whether findings regarding the prevalence of the dissociative subtype and documented associations of this subtype with factors such as history of trauma generalize to the population. Therefore, Stein et al. (2013) used epidemiologic data from 16 countries in the World Health Organization (WHO) World Mental Health (WMH) surveys to examine the cross-cultural generalizability of the dissociative subtype. Dissociative symptoms of depersonalization and derealization were present in 14.4% of respondents with PTSD, a proportion that did not differ significantly across countries. This demonstrates that the dissociative subtype is not confined to Western countries. Symptoms of dissociation in PTSD were associated with elevated PTSD symptom counts, male sex, childhood onset of PTSD, high exposure to prior traumatic events and childhood adversities, comorbid psychiatric disorders, high role impairment, and suicidality. Previous studies have not reported dissociation to be more common among men with PTSD than women with PTSD. However, the WMH findings that the dissociative subtype is associated with severity of PTSD symptoms and prior exposure to trauma are highly consistent with previous studies. Stein and colleagues argue that these results point out the value of the dissociative subtype in distinguishing a meaningful proportion of severe cases of PTSD that have distinct correlates that are roughly equivalent across countries.

Even though the total number of subgroups found in PTSD samples differed among studies, all of these studies have found evidence for the existence of a dissociative subtype of PTSD. These studies have consistently demonstrated that the dissociative PTSD subtype encompasses a subset of individuals who report high levels of PTSD symptoms, plus high levels of dissociative symptoms, in particular derealization and depersonalization. Across the different studies, a wide range of risk factors was investigated with mixed results. In a systematic review of latent class and profile analytic studies of PTSD, it was shown that childhood exposure to trauma, and more specifically childhood sexual assault and physical assault, seems to be a risk factor for dissociative PTSD (Hansen et al. 2017). Furthermore, comorbid psychopathology such as anxiety and depression have been found to be significantly associated with dissociative PTSD in multiple studies. This pattern of specific risk factors underlies the prediction of the subtype model that dissociative PTSD would differentially correlate with external social and clinical covariates.

3 Neurobiological Features That Distinguish Dissociative PTSD from Non-dissociative PTSD

3.1 Neuroimaging Studies

Neuroimaging studies have implicated decreased activity of prefrontal regions and increased activity of limbic structures in the pathophysiology of PTSD (Francati et al. 2007; Sartory et al. 2013). Interestingly, however, recent data have shown that psychobiological responses to recalling traumatic experiences can differ significantly among individuals with PTSD and that not all PTSD patients show the same brain activation patterns in response to traumatic script-driven imagery (Frewen and Lanius 2006; Hopper et al. 2007; Lanius et al. 2002). Lanius et al. (2002) found that while approximately 70% of PTSD patients reported to relive their traumatic experience in response to traumatic scripts and showed an increase in heart rate, the other 30% of PTSD subjects reported experiences of derealization, depersonalization, and a feeling of emotional detachment and did not show a significant increase in heart rate when exposed to a traumatic script. Thus, similarly to the epidemiological studies discussed above, neurobiological studies have identified a dissociative subtype of PTSD that can be distinguished from non-dissociative PTSD. Moreover, these non-dissociative and dissociative response types have been associated with opposite patterns of brain activation in brain regions that are involved in arousal and emotion regulation. Patients who relived their traumatic experiences upon exposure to traumatic scripts showed less activation of the mPFC, ACC, and thalamus during script-driven recall of traumatic events (Lanius et al. 2001). This is consistent with the decreased prefrontal activation that most PTSD neuroimaging studies to date have reported. However, PTSD patients in a dissociative state showed more activation in the mPFC, ACC, medial frontal gyrus, inferior frontal gyrus, superior and middle temporal gyri, occipital lobe, and parietal lobe during symptom provocation compared to control subjects (Lanius et al. 2002). Thus, although patients with re-experiencing/hyperaroused PTSD exhibit lower activation of prefrontal regions that are implicated in arousal modulation and emotion regulation, dissociative PTSD patients show higher activation of these prefrontal regions.

On the basis of these findings, Lanius et al. (2010b) proposed a model in which the two different response types are viewed as two distinct types of emotion dysregulation. In this model, re-experiencing/hyperaroused reactivity is thought to reflect emotional undermodulation, mediated by failure of prefrontal inhibition of limbic regions. The dissociative PTSD patients can, on the other hand, be conceptualized as experiencing emotional overmodulation, mediated by increased activation of prefrontal regions and hyperinhibition of limbic regions. Although this model would predict that re-experiencing/hyperaroused PTSD patients would show heightened amygdala activation and that dissociative PTSD subjects would show decreased amygdala activation in comparison with controls, Lanius and colleagues did not observe amygdala activation in either the re-experiencing/hyperaroused subjects nor the dissociative subjects in the abovementioned studies. However, they have reported a case study of two survivors of the same motor vehicle accident who displayed different responses to traumatic script-driven imagery. While one of them exhibited a re-experiencing/hyperaroused response and increased amygdala activation, the other person showed a dissociative response with no amygdala activation (Lanius et al. 2003).

Further support for the emotion dysregulation model has been provided by Hopper et al. (2007) who assessed the relationship between severity of state re-experiencing, avoidance, and dissociation symptom responses to script-driven imagery and neural activation patterns. They found that state re-experiencing severity was positively associated with right anterior insula activity, an area involved in the neural representation of somatic aspects of emotional states, including acute sympathetic arousal (Craig 2002). Furthermore, re-experiencing correlated negatively with right rACC activity. Since the rACC is known to inhibit the amygdala, the finding of a negative association between re-experiencing and rACC activation is consistent with the hypothesis that re-experiencing/hyperaroused reactivity can be viewed as a form of emotional undermodulation, mediated by failed prefrontal inhibition of limbic regions. In contrast, Hopper and colleagues showed that dissociation correlated positively with activity in the left medial prefrontal cortex and negatively with activation of the right anterior insula. The finding of a positive association between dissociation and medial prefrontal activation provides support for the hypothesized hyperinhibition of limbic regions by medial prefrontal areas during dissociative states. In addition, Hopper and co-workers found that state dissociation severity correlated positively with activity in the right superior temporal cortex and negatively with activation in the left superior temporal cortex. These positive and negative correlations of dissociation with activation in the right and left temporal lobe structures, respectively, have also been reported by previous studies (Lanius et al. 2002, 2005). Interestingly, investigations of temporal lobe epilepsy have reported that patients exhibit dissociative symptoms during seizures (Lanius et al. 2006). In addition, it has been reported that patients exhibit dissociative symptoms in response to stimulation of the superior and middle temporal gyrus during neurosurgery. These findings suggest a role for the temporal lobe in dissociation. However, the functional significance of superior temporal activations during dissociative responses, including laterality effects, still has to be investigated.

A study by Felmingham et al. (2008) has provided further support for the corticolimbic inhibition model of dissociation. Using fMRI, these researchers compared brain activation patterns of two groups of PTSD patients, one with high dissociation scores and one with low dissociation scores, during the processing of consciously and nonconsciously perceived fear stimuli. During conscious fear processing, patients with high dissociation scores showed enhanced activation of the vmPFC in comparison with patients with low dissociation scores. This result supports the assumption that dissociative PTSD is characterized by emotional overmodulation accompanied by increased activation of medial prefrontal structures. During nonconscious fear processing, however, dissociative PTSD was associated with enhanced activation in limbic regions, such as the amygdala, insula, and left thalamus, compared to non-dissociative PTSD. These authors therefore suggested that dissociation is a regulatory strategy that can be applied during conscious fear processing in order to cope with extreme arousal in PTSD through hyperinhibition of limbic regions.

The pain neurobiology literature provides additional evidence for the hyperinhibition of limbic regions during dissociative responses. Röder et al. (2007) demonstrated that healthy individuals exhibited decreased activation in sensory and affective pain-related areas, including the amygdala, in response to painful stimuli during hypnosis-induced states of depersonalization. Furthermore, Mickleborough et al. (2011) found that trait dissociation correlated negatively with activity in the right amygdala and left putamen during traumatic script-induced analgesia in individuals with PTSD. In addition, Ludäscher et al. (2010) found increased activity in the left cingulate gyrus and bilateral insula in patients with borderline personality disorder and comorbid PTSD during script-driven imagery-induced dissociation, providing evidence for increased cortical inhibition during dissociative responses. Of note, the finding of increased insular cortex activity during dissociative states is not consistent with the results of Hopper et al. (2007), who reported increased insula activation during re-experiencing/hyperaroused states and decreased insula activation during dissociative states. Thus, exactly how insula activation is involved in the distinct re-experiencing and dissociative response types requires further investigation.

In addition to the research describing differential patterns of neural activation during symptom provocation, evidence has also been provided for baseline functional differences between patients with dissociative PTSD and those with re-experiencing PTSD. A study comparing resting-state functional connectivity between PTSD patients with and without the dissociative subtype found that individuals with dissociative PTSD exhibited greater resting-state functional connectivity between the amygdala and the medial frontal gyrus than subjects with non-dissociative PTSD (Nicholson et al. 2015). This finding is in line with the proposed increased PFC inhibition of limbic regions in dissociative PTSD (Lanius et al. 2010a, b).

The studies reviewed above have investigated which functional brain alterations distinguish dissociative PTSD from non-dissociative PTSD. However, information about which structural alterations are related to dissociation in PTSD samples is scarce. Recently though, Nardo et al. (2013) investigated brain structural alterations associated with trait dissociation in a sample of traumatized subjects with and without PTSD. Their results show that PTSD and dissociation are related to opposite volumetric patterns in the prefrontal cortex. While the volume of the medial PFC and lateral PFC was decreased in PTSD subjects compared to traumatized controls, PFC volume was increased as a function of trait dissociation. Moreover, trait dissociation was related to increased gray matter volume in prefrontal, orbitofrontal, parahippocampal, temporal polar, and inferior parietal cortices. Another study investigating the differences in volumetric brain morphology between patients with dissociative and non-dissociative PTSD found that individuals with the dissociative subtype exhibited greater gray matter volume in the right precentral and fusiform gyri and less gray matter volume in the right inferior temporal gyrus than patients with non-dissociative PTSD (Daniels et al. 2016). Furthermore, they found that dissociation severity was positively correlated with gray matter volume in the right middle frontal gyrus, a region known to subserve the downregulation of emotional arousal. To the extent that volumetric alterations can be interpreted as the structural counterpart of functional alterations, the finding of a positive correlation between dissociation and frontal gyrus volume is consistent with the model proposed by Lanius et al. (2010a, b), according to which hyperaroused PTSD subjects exhibit functional reduction in prefrontal regions (emotional undermodulation), while dissociative PTSD subjects exhibit increased prefrontal inhibition of limbic regions (emotional overmodulation).

In sum, neuroimaging studies have identified two distinct response types in PTSD subjects: a re-experiencing/hyperaroused subtype and a dissociative subtype. These response types are characterized by different psychophysiological and neural responses to the recall of traumatic memories (Table 1). While individuals with re-experiencing/hyperaroused PTSD appear to exhibit an increased heart rate, lower activation of prefrontal regions, and increased activation of the amygdala,

Table 1 Overview of the features that distinguish dissociative PTSD from non-dissociative PTSD

	Dissociative PTSD	Non-dissociative PTSD
Epidemiology	– More severe PTSD symptoms, complex, chronic	– More often associated with trauma in adult life
	– Associated with early life trauma/repetitive trauma	– Less cumulative trauma
	– Higher level of comorbid psychiatric disorders	
Reaction after exposure to traumatic stimuli	– Dissociation, numbing	– Fear/anxiety-driven
	– Suppression of autonomic response	– Increased autonomic arousal
	– No increase in heart rate	– Increased heart rate
	– No increase in skin conductance	– Increased skin conductance
Phenomenology	Dissociation	Re-experiencing
Imaging	Anterior cingulate cortex ↑	Anterior cingulate cortex ↓
	Medial prefrontal cortex ↑	Medial prefrontal cortex ↓
	Amygdala ↓	Amygdala ↑
	Right anterior insula ↓	Right anterior insula ↑
Concept	Emotional overmodulation	Emotional undermodulation

evidence is accumulating that individuals with dissociative PTSD do not show an increase in heart rate and exhibit higher activation of prefrontal regions and no increase in activation of limbic regions during the recall of traumatic memories. The corticolimbic model of dissociation postulates that once a threshold of anxiety is reached, the medial prefrontal cortex inhibits emotional processing in limbic structures, resulting in dissociative symptoms. Furthermore, there are indications that dissociation is associated with activity in the temporal lobes and insular cortex and an increase in PFC volume.

3.2 Baseline and Functional Neuroendocrinological Studies

As discussed in this update, individuals with dissociative PTSD show a different psychophysiological response and different activation patterns of the mPFC and amygdala during the recall of traumatic memories than individuals with non-dissociative PTSD. However, PTSD is not only associated with alterations in stress-regulating brain regions but also with neuroendocrinological alterations, such as exaggerated cortisol reactivity and increased NE levels in response to stressors (de Kloet et al. 2006; Southwick et al. 1999). It is not yet fully clear whether these neuroendocrinological changes also apply to individuals with dissociative PTSD or whether these patients exhibit different neuroendocrine characteristics than patients without dissociative symptoms. Although a growing number of studies have examined cortisol response as a function of dissociative symptoms, results are inconclusive. Higher (Simeon et al. 2007), lower (Simeon et al. 2008), and no difference (Koopman et al. 2003) in baseline cortisol levels in high versus low dissociators have been observed. Furthermore, some studies have reported that increased cortisol reactivity in response to a stressor was positively related to dissociative symptoms (Giesbrecht et al. 2007; Powers et al. 2006), whereas other studies have reported negative correlations of dissociation scores and cortisol levels after exposure to acute stress (Morgan et al. 2001; Simeon et al. 2007) or did not find a difference in cortisol reactivity between high and low dissociators (Simeon et al. 2008). Koopman et al. (2003) observed increased salivary cortisol in individuals reporting greater dissociative symptoms 24 h after being interviewed about traumatic life events, but not immediately or 48 h after the interview, suggesting that differences in cortisol reactivity between individuals with and without dissociative symptoms might depend on the time between a stressful event and the collection of samples. Thus, the relation between cortisol levels and dissociative symptoms remains unclear. Moreover, the relation between NE levels and dissociation has, to our knowledge, not yet been investigated.

To address the inconsistencies in findings on the function of the hypothalamic–pituitary–adrenal (HPA) axis in patients suffering from PTSD, Zaba et al. (2015) compared a sample of female PTSD patients with either early life trauma, adult trauma, or combined early life and adult trauma to healthy controls in the Trier social stress test. They identified and characterized for the first time two different HPA axis reactivity subgroups in PTSD. One of these subgroups showed a blunted HPA axis response and distinct clinical characteristics such as a higher prevalence of dissociative symptoms and of combined early life and adult trauma. This subgroup largely drove the diminished HPA axis response of the total PTSD cohort. The existence of different HPA axis reactivity endophenotypes in PTSD could offer an explanation for the inconsistent reports on HPA axis function in PTSD patients. Furthermore, these findings suggest that individuals with the dissociative subtype might display a different HPA axis reactivity than patients with non-dissociative PTSD.

4 Developmental Pathways to Emotional Dysregulation

Both neurobiological and clinical research have shown that PTSD has a re-experiencing/hyperaroused subtype and a dissociative subtype. The remaining question is why some individuals develop dissociative symptoms, while others do not. As discussed previously, PTSD patients with a history of prolonged trauma, such as childhood abuse, often show a clinical syndrome that is characterized by dissociative symptoms, as opposed to patients who have suffered from more acute forms of trauma (Ginzburg et al. 2006; Stein et al. 2013; Steuwe et al. 2012; Wolf et al. 2012b). Support for these findings comes from a study that demonstrated that individuals who experienced early-onset abuse showed greater levels of dissociative symptoms than disaster survivors and individuals who experienced late-onset abuse (van der Kolk et al. 1996). It could thus be that prolonged traumatic experiences, particularly during childhood, may promote the use of dissociation as a defense strategy to cope with aversive experiences.

Lanius et al. (2010a) have proposed two distinct pathways to emotional dysregulation and brain dysfunction in PTSD. The first pathway proposes that fear reactions experienced during the aftermath of an acute traumatic event, followed by repeated re-experiencing of the traumatic event, can lead to sensitization of the fear and stress response. This sensitization can result in a progressive augmentation of PTSD symptoms and general emotional dysregulation. It is thought that a combination of the individual's distress, physiological reactivity, and related neurotransmitter and neurohormonal responses that in turn have an effect on a variety of brain regions, including the mPFC, amygdala, and hippocampus, underlies this process of sensitization. Lanius and co-workers hypothesized that individuals following this pathway would most likely be adults who experienced adult-onset trauma and do not have a history of childhood abuse. In contrast, the second pathway describes that an impoverished childhood environment and/or childhood maltreatment can lead to inadequate neurodevelopment of emotion and arousal regulatory systems, resulting in general emotional dysregulation and the inability to regulate physiological arousal to traumatic events. As a consequence of this inability, exposure to (a) traumatic event(s) later in life would lead to an exacerbation of emotional dysregulation, including the development of PTSD. Interestingly, empirical research has indeed shown that different types of childhood trauma exposure are related to distinct anomalies in HPA axis functioning (Kuhlman et al. 2015), providing evidence for the idea that the type and timing of trauma influence the type of stress-related pathology an individual develops.

To summarize, it is proposed that individuals with a history of severe or repetitive trauma exposure follow a different pathway to the development of PTSD once exposed to trauma, than individuals that are trauma naïve and exposed to trauma in adulthood. Furthermore, individuals with a history of prolonged trauma are more likely to develop greater levels of dissociative symptoms. However, it cannot be stated that all individuals with a significant trauma history will develop the dissociative subtype of PTSD and that trauma survivors with no prior history of trauma will not develop dissociative symptoms. Re-experiencing and dissociative response types are not completely distinct, and individuals with PTSD may show both response types at different time points. Nevertheless, prolonged and early traumatic experiences do present a risk factor for developing the dissociative subtype of PTSD.

5 Pitfalls and Promises in Current Research Studies on the Dissociative Subtype of PTSD

5.1 Limitations of Current Research Studies

Although during the past three decades there have been major advances in our understanding of the neurobiology of PTSD, neurobiological findings in PTSD are not unequivocal. Reports on alterations in cortisol and norepinephrine (NE) responses, as well as alterations in the activity of the amygdala, mPFC, and insular cortex in PTSD, have been inconsistent. This could be due to methodological differences, such as differences in task design, time of sampling of cortisol and NE levels, and imaging methods or resolution. Meta-analyses of existing studies that take into account these methodological differences could possibly reconcile some of these inconsistent findings. Importantly, inconsistent findings could also be explained by the heterogeneity of symptomatic and biological responses to traumatic reminders in PTSD. As we have discussed in this review, individuals with the re-experiencing and the dissociative subtype of PTSD exhibit different physiological and neurobiological responses to trauma-related stimuli. This suggests that grouping PTSD subjects with different symptom patterns may hamper our understanding of the pathophysiology of PTSD. Therefore, the heterogeneity within PTSD samples should be addressed in the designs of futures PTSD studies.

The majority of PTSD studies have been cross-sectional, comparing PTSD patients to healthy controls who were either exposed to trauma but did not develop the disorder or were not exposed. Therefore, it has been difficult to determine whether the observed neurobiological differences between PTSD patients and controls are causes or consequences of the pathophysiology of PTSD. In a recent review, Admon et al. (2013) have attempted to allocate PTSD neural characteristics to either predisposing or acquired factors, by gathering evidence from genetic, environmental, twin, and prospective studies. They have suggested that abnormal structure of the amygdala and dACC, and their heightened responsivity to negative stimuli, represent predisposing neural abnormalities. Since the amygdala and dACC are mediators of fear generation and expression, these neural predispositions may make individuals prone to express heightened fear upon trauma exposure, reducing their chances to cope adequately with the traumatic experience. In contrast, they have suggested that reduced volume of the mPFC, as well as reduced vmPFC structural and functional connectivity with the hippocampus, may represent acquired neural characteristics. Compromised vmPFC structure and connectivity with the hippocampus may impair an individual's ability to extinguish fear, which is an essential need for recovery from traumatic stress. However, this causal model is based on only a few studies, and the authors emphasize that more genetic, twin, and longitudinal studies are needed to validate these findings. Furthermore, it would be interesting to investigate whether predisposing factors are different for people that develop dissociative PTSD than for individuals that do not develop dissociative symptoms. More knowledge about predisposing factors for dissociative/non-dissociative PTSD could facilitate the accurate identification of vulnerable individuals and would enable the implementation of early clinical intervention to prevent the development of PTSD and dissociative symptoms. In addition, longitudinal studies can give us more insight in the trajectories of neurobiological changes in dissociative/non-dissociative PTSD, advancing our understanding of the pathophysiology of PTSD. Future studies will also need to explore genome-wide associations as well as specific genetic polymorphisms to dissociation and stability and change over intergenerational transmission of trauma.

Another limitation of PTSD research is that most studies have focused on neurobiological systems involved in stress and fear responses and how these responses are altered in PTSD, conceptualizing PTSD as a fear disorder. Although the amygdala, hippocampus, ACC, mPFC, and insula are indeed involved in fear processing, the functions of these regions extend beyond fear expression and inhibition. The mPFC and hippocampus, for instance, are not only implicated in fear inhibition and extinction but also contribute to PTSD deficits in attention and declarative memory, deficits that have not received as much attention. Furthermore, the amygdala, ACC, and mPFC are known to mediate important aspects of general emotion regulation and are involved in a variety of emotional states such as anger, guilt, and shame, emotions that are frequently reported by PTSD patients. Therefore, Lanius et al. (2010a) have suggested that PTSD is associated with general emotion dysregulation and that fear is not the prevailing emotion but rather one of several components implicated in a dysfunctional emotional system that also mediates problems regulating anger, guilt, shame, dissociation, and numbing. Furthermore, the role of neural systems that are not primarily implicated in the stress and fear response should be considered. For instance, the reward circuitry, of which the nucleus accumbens is a core element, has generally been overlooked in PTSD research. However, recent studies suggest that PTSD might be associated with reduced nucleus accumbens reward responsivity (Elman et al. 2009; Sailer et al. 2008). This reduced reward responsivity may contribute to PTSD symptoms of diminished motivation and restricted affect range of positive emotions. Thus, we suggest that we could further our understanding of PTSD and its underlying neurocircuitry by looking beyond impairments in fear expression and fear inhibition.

Recently, researchers have pointed out some shortcomings in the current definition of the dissociative subtype of PTSD in the DSM-5. In the DSM-5, the dissociative subtype of PTSD is defined by dissociative symptoms of derealization and depersonalization. This is based on the empirical evidence showing that a group of individuals with PTSD who have elevations of these symptoms can clearly be distinguished. However, dissociative symptoms in PTSD do not only include derealization and depersonalization but comprise a variety of dissociative symptoms, including several positive (e.g., intrusions such as flashbacks and jarring trauma-related physical pain) and negative dissociative symptoms (e.g., amnesia, emotional numbing, derealization, and depersonalization). Remarkably, symptoms of intrusions and amnesia are also present in the so-called non-dissociative subtype of PTSD. Thus, PTSD is associated with several dissociative symptoms, but only if individuals exhibit predominant derealization and/or depersonalization symptoms, the DSM-5 states they have the dissociative subtype of PTSD. Therefore, Dorahy and van der Hart (2015) have argued that limiting the dissociative subtype of PTSD to depersonalization and derealization symptoms creates a false dichotomy between dissociative symptoms in PTSD and between a PTSD that is dissociative and one that apparently is not. Considerable evidence suggests that dissociation is generally heightened in PTSD, indicating that the relevance of dissociation for PTSD is not limited to the dissociative subtype. Carlson et al. (2012) reviewed 13 studies that have assessed dissociation in PTSD and trauma-exposed non-PTSD samples and found that dissociation levels were significantly higher in participants with PTSD in 11 of the 13 studies. Importantly, scatter plots suggested that this relation between dissociation and PTSD was not created by a small subset of participants with extremely high dissociation scores but rather represented a linear pattern of data. Similarly, two taxometric studies found that dissociation was on a continuum rather than being a categorical variable (Forbes et al. 2005; Ruscio et al. 2002). For these reasons, Dorahy and van der Hart argue that even though there are differences between the two subtypes of PTSD, the definition of the dissociative subtype in the DSM-5 is problematic, since PTSD generally involves dissociation, and the symptoms of derealization and depersonalization which characterize DSM-5's dissociative subtype are a restricted and nonrepresentative reflection of dissociative symptoms in PTSD. Future studies of dissociation in PTSD need to examine the degree to which the different positive and negative dissociative symptoms may vary among PTSD patients to allow a greater elucidation of dissociation in PTSD. An overview of the described limitations in PTSD research can be found in Table 2.

Table 2 Limitations in PTSD research

• Neurobiological findings are not unequivocal and inconsistencies across studies have not yet been resolved
• The majority of studies have been cross-sectional, making it difficult to follow developmental trajectories of neurobiological changes in PTSD and to discriminate predisposing from acquired neural characteristics
• The majority of studies have focused on neurobiological systems involved in stress and fear responses and have generally not paid attention to the function of these brain regions beyond fear expression and inhibition
• Studies of the dissociative subtype of PTSD have focused on the dissociative symptoms of depersonalization and derealization and have not investigated how other dissociative symptoms may vary among PTSD patients

5.2 *Benefits of Research Studies on the Dissociative Subtype of PTSD*

Even though the current definition of the dissociative subtype of PTSD might be criticized, the addition of the dissociative subtype to the DSM-5 comes with great benefits. First of all, it boosts empirical and theoretical efforts to further understand the links between dissociation and PTSD. Furthermore, the inclusion of the dissociative subtype in the DSM-5 improves clinician familiarity with and screening for dissociative symptoms. Individuals with the dissociative subtype are an important group to identify in the clinical context, since this group of patients generally exhibits more severe PTSD symptoms, greater role impairment, and increased suicidality (Stein et al. 2013). Therefore, we would like to highlight thc importance of the development of an advanced assessment tool for the diagnosis of dissociative symptoms in PTSD. Recently, a brief assessment has been proposed that assesses dissociative symptoms in line with the CAPS-5 (Eidhof et al. 2017). The National Center of PTSD is also developing a novel screener for the assessment of the dissociative subtype (Wolf et al. 2017). The accurate assessment of dissociative symptoms in PTSD is also essential for the treatment of patients with dissociative PTSD, since it has been shown that individuals with the dissociative subtype of PTSD exhibit a different response to conventional PTSD treatments than patients with non-dissociative PTSD (Lanius et al. 2012). Specifically, Jaycox et al. (1998) have proposed that dissociation can reduce the effectiveness of exposure therapy by preventing sustained emotional engagement with trauma-related information. Therefore, it has been suggested that PTSD patients with dissociative symptoms require a different type of treatment (Cloitre et al. 2012; Resick et al. 2012). Cloitre et al. (2012) have suggested that a "phase-based" intervention, where a preparatory phase is introduced to focus on reducing dissociation before engaging in exposure to traumatic memories, will result in a more successful treatment outcome. However, a recent study challenges the perception that conventional exposure therapy would not be effective for individuals with the dissociative subtype of PTSD (Wolf et al. 2016). Wolf and colleagues found that although female veterans and active-duty service members with dissociative PTSD did not respond as well to prolonged exposure and present-centered therapy as those without the dissociative subtype, both PTSD and dissociation symptoms did improve in the group with the dissociative subtype. These findings indicate that more research is needed to investigate the need for distinct approaches in the treatment of dissociative and non-dissociative PTSD.

Overall, acknowledging the existence of a dissociative subtype of PTSD brings the importance of trauma-related dissociation into clinical and empirical focus and will stimulate careful analysis of the prevalence, symptomatology, neurobiology, and treatment of individuals with the dissociative subtype of PTSD.

6 Conclusion

There is abundant evidence that there is a dissociative subtype of PTSD with neurobiological and clinical features that distinguish it from non-dissociative PTSD. Individuals with the dissociative subtype of PTSD report high levels of PTSD symptoms and high levels of dissociative symptoms, in particular derealization and depersonalization. Furthermore, the dissociative subtype of PTSD is associated with a more significant history of early life trauma and higher levels of comorbid psychiatric disorders. In addition, PTSD patients with dissociative symptoms exhibit different psychophysiological and neural responses to the recall of traumatic memories. While individuals with re-experiencing/hyperaroused PTSD appear to exhibit an increased heart rate, lower activation of prefrontal regions, and increased activation of the amygdala in response to traumatic reminders, individuals with dissociative PTSD do not show an increase in heart rate and exhibit higher activation of prefrontal regions and no increase in activation of limbic regions during the recall of traumatic memories. Furthermore, there are indications that dissociation is associated with activity in the temporal lobes and insular cortex and an increase in PFC volume, although more research is required to validate these findings. It has been proposed that dissociation is a regulatory strategy invoked to cope with extreme arousal in PTSD through hyperinhibition of limbic regions. Future studies will need to explore genome-wide associations and specific genetic polymorphisms to dissociation as well as stability and change over intergenerational transmission of trauma. As first reports are coming out, it needs to be understood what the implications are for the treatment of the dissociative subtype (Dutra and Wolf 2017). Although there is some criticism on the current definition of the dissociative subtype of PTSD in the DSM-5, acknowledgment of this subtype stimulates research on the prevalence, symptomatology, and neurobiology of the dissociative subtype of PTSD and may improve treatment outcome in PTSD patients with dissociative symptoms.

References

Admon R, Milad MR, Hendler T (2013) A causal model of post-traumatic stress disorder: disentangling predisposed from acquired neural abnormalities. Trends Cogn Sci 17 (7):337–347

American Psychiatric Association (2013) Diagnostic and statistical manual of mental disorders: DSM-V, 5th edn. American Psychiatric Publishing, Arlington, VA

Armour C, Elklit A, Lauterbach D, Elhai JD (2014a) The DSM-5 dissociative-PTSD subtype: can levels of depression, anxiety, hostility, and sleeping difficulties differentiate between dissociative-PTSD and PTSD in rape and sexual assault victims? J Anxiety Disord 28 (4):418–426

Armour C, Karstoft KI, Richardson JD (2014b) The co-occurrence of PTSD and dissociation: differentiating severe PTSD from dissociative-PTSD. Soc Psychiatry Psychiatr Epidemiol 49 (8):1297–1306

Bernstein EM, Putnam FW (1986) Development, reliability, and validity of a dissociation scale. J Nerv Ment Dis 174(12):727–735

Bremner JD, Brett E (1997) Trauma-related dissociative states and long-term psychopathology in posttraumatic stress disorder. J Trauma Stress 10(1):37–49

Bremner JD, Krystal JH, Putnam FW, Southwick SM, Marmar C, Charney DS, Mazure CM (1998) Measurement of dissociative states with the clinician-administered dissociative states scale (CADSS). J Trauma Stress 11(1):125–136

Briere J (2006) Dissociative symptoms and trauma exposure: specificity, affect dysregulation, and posttraumatic stress. J Nerv Ment Dis 194(2):78–82

Carlson EB, Dalenberg C, McDade-Montez E (2012) Dissociation in posttraumatic stress disorder part I: definitions and review of research. Psychol Trauma Theor Res Pract Pol 4(5):479–489

Cloitre M, Petkova E, Wang J (2012) An examination of the influence of a sequential treatment on the course and impact of dissociation among women with PTSD related to childhood abuse. Depress Anxiety 29(8):709–717

Craig AD (2002) How do you feel? Interoception: the sense of the physiological condition of the body. Nat Rev Neurosci 3(8):655–666

Dalenberg C, Carlson EB (2012) Dissociation in posttraumatic stress disorder part II: how theoretical models fit the empirical evidence and recommendations for modifying the diagnostic criteria for PTSD. Psychol Trauma Theor Res Pract Pol 4(6):551

Daniels JK, Frewen P, Theberge J, Lanius RA (2016) Structural brain aberrations associated with the dissociative subtype of post-traumatic stress disorder. Acta Psychiatr Scand 133 (3):232–240

Dell PF (2006) The multidimensional inventory of dissociation (MID): a comprehensive measure of pathological dissociation. J Trauma Dissociation 7(2):77–106

Dorahy MJ, van der Hart O (2015) DSM–5's posttraumatic stress disorder with dissociative symptoms: challenges and future directions. J Trauma Dissociation 16(1):7–28

Dutra SJ, Wolf EJ (2017) Perspectives on the conceptualization of the dissociative subtype of PTSD and implications for treatment. Curr Opin Psychol 14:35–39

Eidhof M, ter Heide JJ, Vermetten E (2017) The dissociative subtype of PTSD-interview. Arq Diemen, The Netherlands

Elman I, Lowen S, Frederick BB, Chi W, Becerra L, Pitman RK (2009) Functional neuroimaging of reward circuitry responsivity to monetary gains and losses in posttraumatic stress disorder. Biol Psychiatry 66(12):1083–1090

Felmingham K, Kemp AH, Williams L, Falconer E, Olivieri G, Peduto A, Bryant R (2008) Dissociative responses to conscious and non-conscious fear impact underlying brain function in post-traumatic stress disorder. Psychol Med 38(12):1771–1780

Forbes D, Haslam N, Williams BJ, Creamer M (2005) Testing the latent structure of posttraumatic stress disorder: a taxometric study of combat veterans. J Trauma Stress 18(6):647–656

Francati V, Vermetten E, Bremner JD (2007) Functional neuroimaging studies in posttraumatic stress disorder: review of current methods and findings. Depress Anxiety 24(3):202–218
Frewen PA, Lanius RA (2006) Toward a psychobiology of posttraumatic self-dysregulation. Ann N Y Acad Sci 1071(1):110–124
Giesbrecht T, Smeets T, Merckelbach H, Jelicic M (2007) Depersonalization experiences in undergraduates are related to heightened stress cortisol responses. J Nerv Ment Dis 195 (4):282–287
Ginzburg K, Koopman C, Butler LD, Palesh O, Kraemer HC, Classen CC, Spiegel D (2006) Evidence for a dissociative subtype of post-traumatic stress disorder among help-seeking childhood sexual abuse survivors. J Trauma Dissociation 7(2):7–27
Hansen M, Ross J, Armour C (2017) Evidence of the dissociative PTSD subtype: a systematic literature review of latent class and profile analytic studies of PTSD. J Affect Disord 213:59–69
Hopper JW, Frewen PA, van der Kolk BA, Lanius RA (2007) Neural correlates of reexperiencing, avoidance, and dissociation in PTSD: symptom dimensions and emotion dysregulation in responses to script-driven trauma imagery. J Trauma Stress 20(5):713–725
Irwin HJ (1994) Proneness to dissociation and traumatic childhood events. J Nerv Ment Dis 182 (8):456–460
Jaycox LH, Foa EB, Morral AR (1998) Influence of emotional engagement and habituation on exposure therapy for PTSD. J Consult Clin Psychol 66(1):185–192
de Kloet CS, Vermetten E, Geuze E, Kavelaars A, Heijnen CJ, Westenberg HGM (2006) Assessment of HPA-axis function in posttraumatic stress disorder: pharmacological and non-pharmacological challenge tests, a review. J Psychiatr Res 40(6):550–567
van der Kolk BA, Pelcovitz D, Roth S, Mandel FS (1996) Dissociation, somatization, and affect dysregulation: the complexity of adaption to trauma. Am J Psychiatry 153(Suppl):83–93
Koopman C, Sephton S, Abercrombie HC, Classen C, Butler LD, Gore-Felton C et al (2003) Dissociative symptoms and cortisol responses to recounting traumatic experiences among childhood sexual abuse survivors with PTSD. J Trauma Dissociation 4(4):29–46
Kuhlman KR, Geiss EG, Vargas I, Lopez-Duran NL (2015) Differential associations between childhood trauma subtypes and adolescent HPA-axis functioning. Psychoneuroendocrinology 54:103–114
Lanius RA, Williamson PC, Densmore M, Boksman K, Gupta MA, Neufeld RW et al (2001) Neural correlates of traumatic memories in posttraumatic stress disorder: a functional MRI investigation. Am J Psychiatry 158(11):1920–1922
Lanius RA, Williamson PC, Boksman K, Densmore M, Gupta M, Neufeld RW et al (2002) Brain activation during script-driven imagery induced dissociative responses in PTSD: a functional magnetic resonance imaging investigation. Biol Psychiatry 52(4):305–311
Lanius RA, Hopper JW, Menon RS (2003) Individual differences in a husband and wife who developed PTSD after a motor vehicle accident: a functional MRI case study. Am J Psychiatry 160(4):667–669
Lanius RA, Williamson PC, Bluhm RL, Densmore M, Boksman K, Neufeld RWJ et al (2005) Functional connectivity of dissociative responses in posttraumatic stress disorder: a functional magnetic resonance imaging investigation. Biol Psychiatry 57(8):873–884
Lanius RA, Bluhm R, Lanius U, Pain C (2006) A review of neuroimaging studies in PTSD: heterogeneity of response to symptom provocation. J Psychiatr Res 40(8):709–729
Lanius RA, Frewen PA, Vermetten E, Yehuda R (2010a) Fear conditioning and early life vulnerabilities: two distinct pathways of emotional dysregulation and brain dysfunction in PTSD. Eur J Psychotraumatol 1
Lanius RA, Vermetten E, Loewenstein RJ, Brand B, Schmahl C, Bremner JD, Spiegel D (2010b) Emotion modulation in PTSD: clinical and neurobiological evidence for a dissociative subtype. Am J Psychiatry 167(6):640–647
Lanius RA, Brand B, Vermetten E, Frewen PA, Spiegel D (2012) The dissociative subtype of posttraumatic stress disorder: rationale, clinical and neurobiological evidence, and implications. Depress Anxiety 29(8):701–708

Ludäscher P, Valerius G, Stiglmayr C, Mauchnik J, Lanius RA, Bohus M, Schmahl C (2010) Pain sensitivity and neural processing during dissociative states in patients with borderline personality disorder with and without comorbid posttraumatic stress disorder: a pilot study. J Psychiatry Neurosci 35(3):177–184

Marmar CR, Weiss DS, Schlenger WE, Fairbank JA, Jordan BK, Kulka RA, Hough RL (1994) Peritraumatic dissociation and posttraumatic stress in male Vietnam theater veterans. Am J Psychiatry 151(6):902–907

Marmar CR, Weiss DS, Metzler TJ (1997) The peritraumatic dissociative experiences questionnaire. Assessing psychological trauma and PTSD, 2nd edn., pp 144–168

Mickleborough MJ, Daniels JK, Coupland NJ, Kao R, Williamson PC, Lanius UF et al (2011) Effects of trauma-related cues on pain processing in posttraumatic stress disorder: an fMRI investigation. J Psychiatry Neurosci 36(1):6–14

Morgan CA III, Wang S, Rasmusson A, Hazlett G, Anderson G, Charney DS (2001) Relationship among plasma cortisol, catecholamines, neuropeptide Y, and human performance during exposure to uncontrollable stress. Psychosom Med 63(3):412–422

Nardo D, Högberg G, Lanius R a, Jacobsson H, Jonsson C, Hällström T, Pagani M (2013) Gray matter volume alterations related to trait dissociation in PTSD and traumatized controls. Acta Psychiatr Scand 128(3):222–233

Nicholson AA, Densmore M, Frewen PA, Théberge J, Neufeld RW, McKinnon MC, Lanius RA (2015) The dissociative subtype of posttraumatic stress disorder: unique resting-state functional connectivity of basolateral and centromedial amygdala complexes. Neuropsychopharmacology 40(10):2317–2326

Nijenhuis ER, Spinhoven P, Van Dyck R, Van Der Hart O, Vanderlinden J (1996) The development and psychometric characteristics of the somatoform dissociation questionnaire (SDQ-20). J Nerv Ment Dis 184(11):688–694

Powers SI, Gunlicks M, Laurent H, Balaban S, Bent E, Sayer A (2006) Differential effects of subtypes of trauma symptoms on couples' hypothalamus-pituitary-adrenal (HPA) axis reactivity and recovery in response to interpersonal stress. Ann N Y Acad Sci 1071:430–433

Putnam FW (1985) Dissociation as a response to extreme trauma. In: Cluft RP (ed) Childhood antecedents of multiple personality. American Psychiatric Association, Washington, DC, pp 65–97

Putnam FW, Carlson EB, Ross CA, Anderson G, Clark P, Torem M et al (1996) Patterns of dissociation in clinical and nonclinical samples. J Nerv Ment Dis 184(11):673–679

Resick PA, Suvak MK, Johnides BD, Mitchell KS, Iverson KM (2012) The impact of dissociation on PTSD treatment with cognitive processing therapy. Depress Anxiety 29(8):718–730

Röder CH, Michal M, Overbeck G, van de Ven VG, Linden DE (2007) Pain response in depersonalization: a functional imaging study using hypnosis in healthy subjects. Psychother Psychosom 76(2):115–121

Ruscio AM, Ruscio J, Keane TM (2002) The latent structure of posttraumatic stress disorder: a taxometric investigation of reactions to extreme stress. J Abnorm Psychol 111(2):290

Sailer U, Robinson S, Fischmeister FPS, König D, Oppenauer C, Lueger-Schuster B et al (2008) Altered reward processing in the nucleus accumbens and mesial prefrontal cortex of patients with posttraumatic stress disorder. Neuropsychologia 46(11):2836–2844

Sartory G, Cwik J, Knuppertz H, Schürholt B, Lebens M, Seitz RJ, Schulze R (2013) In search of the trauma memory: a meta-analysis of functional neuroimaging studies of symptom provocation in posttraumatic stress disorder (PTSD). PLoS One 8(3):e58150

Sierra M, Berrios GE (1998) Depersonalization: neurobiological perspectives. Biol Psychiatry 44 (9):898–908

Simeon D, Knutelska M, Yehuda R, Putnam F, Schmeidler J, Smith LM (2007) Hypothalamic-pituitary-adrenal axis function in dissociative disorders, post-traumatic stress disorder, and healthy volunteers. Biol Psychiatry 61(8):966–973

Simeon D, Yehuda R, Knutelska M, Schmeidler J (2008) Dissociation versus posttraumatic stress: cortisol and physiological correlates in adults highly exposed to the World Trade Center attack on 9/11. Psychiatry Res 161(3):325–329

Southwick SM, Bremner JD, Rasmusson A, Morgan CA III, Arnsten A, Charney DS (1999) Role of norepinephrine in the pathophysiology and treatment of posttraumatic stress disorder. Biol Psychiatry 46(9):1192–1204

Spiegel D, Cardeña E (1990) Dissociative mechanisms in posttraumatic stress disorder. In: Wolf ME, Mosnaim AD (eds) Posttraumatic stress disorder: etiology, phenomenology, and treatment. American Psychiatric Publishing, Arlington, VA, pp 23–34

Stein DJ, Koenen KC, Friedman MJ, Hill E, McLaughlin KA, Petukhova M et al (2013) Dissociation in posttraumatic stress disorder: evidence from the world mental health surveys. Biol Psychiatry 73(4):302–312

Steinberg M (2000) Advances in the clinical assessment of dissociation: the SCID-DR. Bull Menninger Clin 64(2):146

Steuwe C, Lanius RA, Frewen PA (2012) Evidence for a dissociative subtype of PTSD by latent profile and confirmatory factor analyses in a civilian sample. Depress Anxiety 29(8):689–700

Van der Kolk BA, Brown P, Van der Hart O (1989) Pierre Janet on post-traumatic stress. J Trauma Stress 2(4):365–378

Vermetten E, Dorahy MJ, Spiegel D (eds) (2007) Traumatic dissociation: neurobiology and treatment Amercian Psychiatric Press, Washington, DC

Waelde LC, Silvern L, Fairbank JA (2005) A taxometric investigation of dissociation in Vietnam veterans. J Trauma Stress 18(4):359–369

Wolf EJ, Lunney CA, Miller MW, Resick PA, Friedman MJ, Schnurr PP (2012a) The dissociative subtype of PTSD: a replication and extension. Depress Anxiety 29(8):679–688

Wolf EJ, Miller MW, Reardon AF, Ryabchenko KA, Castillo D, Freund R (2012b) A latent class analysis of dissociation and posttraumatic stress disorder: evidence for a dissociative subtype. Arch Gen Psychiatry 69(7):698–705

Wolf EJ, Lunney CA, Schnurr PP (2016) The influence of the dissociative subtype of posttraumatic stress disorder on treatment efficacy in female veterans and active duty service members. J Consult Clin Psychol 84(1):95–100

Wolf EJ, Mitchell KS, Sadeh N, Hein C, Fuhrman I, Pietrzak RH, Miller MW (2017) The dissociative subtype of PTSD scale: initial evaluation in a national sample of trauma-exposed veterans. Assessment 24(4):503–516

Zaba M, Kirmeier T, Ionescu IA, Wollweber B, Buell DR, Gall-Kleebach DJ, Schubert CF, Novak B, Huber C, Köhler K, Holsboer F (2015) Identification and characterization of HPA-axis reactivity endophenotypes in a cohort of female PTSD patients. Psychoneuroendocrinology 55:102–115

Does Anhedonia Presage Increased Risk of Posttraumatic Stress Disorder?

Adolescent Anhedonia and Posttraumatic Disorders

Victoria B. Risbrough, Laura M. Glynn, Elysia P. Davis, Curt A. Sandman, Andre Obenaus, Hal S. Stern, David B. Keator, Michael A. Yassa, Tallie Z. Baram, and Dewleen G. Baker

V. B. Risbrough (✉) and D. G. Baker
Department of Psychiatry, University of California, San Diego, La Jolla, CA, USA

Center of Excellence for Stress and Mental Health, San Diego Veterans Administration, La Jolla, CA, USA
e-mail: vrisbrough@ucsd.edu

L. M. Glynn
Department of Psychology, Chapman University, Orange, CA, USA

Department of Psychiatry and Human Behavior, University of California, Irvine, CA, USA

E. P. Davis
Department of Psychiatry and Human Behavior, University of California, Irvine, CA, USA

Department of Psychology, University of Denver, Denver, CO, USA

C. A. Sandman
Department of Psychiatry and Human Behavior, University of California, Irvine, CA, USA

A. Obenaus
Department of Pediatrics, University of California, Irvine, CA, USA

H. S. Stern
Department of Statistics, University of California, Irvine, CA, USA

D. B. Keator
Department of Psychiatry and Human Behavior, University of California, Irvine, CA, USA

Department of Computer Science, University of California, Irvine, CA, USA

Department of Neurology, University of California, Irvine, CA, USA

M. A. Yassa
Department of Neurobiology and Behavior, University of California, Irvine, CA, USA

Center for the Neurobiology of Learning and Memory, University of California, Irvine, CA, USA

T. Z. Baram
Department of Pediatrics, University of California, Irvine, CA, USA

Department of Neurology, University of California, Irvine, CA, USA

Department of Anatomy/Neurobiology, University of California, Irvine, CA, USA

Curr Topics Behav Neurosci (2018) 38: 249–266
DOI 10.1007/7854_2018_51
Published Online: 22 May 2018

Abstract Anhedonia, the reduced ability to experience pleasure, is a dimensional entity linked to multiple neuropsychiatric disorders, where it is associated with diminished treatment response, reduced global function, and increased suicidality. It has been suggested that anhedonia and the related disruption in reward processing may be critical precursors to development of psychiatric symptoms later in life. Here, we examine cross-species evidence supporting the hypothesis that early life experiences modulate development of reward processing, which if disrupted, result in anhedonia. Importantly, we find that anhedonia may confer risk for later neuropsychiatric disorders, especially posttraumatic stress disorder (PTSD). Whereas childhood trauma has long been associated with increased anhedonia and increased subsequent risk for trauma-related disorders in adulthood, here we focus on an additional novel, emerging direct contributor to anhedonia in rodents and humans: fragmented, chaotic environmental signals ("FRAG") during critical periods of development. In rodents, recent data suggest that adolescent anhedonia may derive from aberrant pleasure/reward circuit maturation. In humans, recent longitudinal studies support that FRAG is associated with increased anhedonia in adolescence. Both human and rodent FRAG exposure also leads to aberrant hippocampal function. Prospective studies are underway to examine if anhedonia is also a marker of PTSD risk. These preliminary cross-species studies provide a critical construct for future examination of the etiology of trauma-related symptoms in adults and for and development of prophylactic and therapeutic interventions. In addition, longitudinal studies of reward circuit development with and without FRAG will be critical to test the mechanistic hypothesis that early life FRAG modifies reward circuitry with subsequent consequences for adolescent-emergent anhedonia and contributes to risk and resilience to trauma and stress in adulthood.

Keywords Anhedonia • Brain circuits • Corticotropin releasing factor (CRF) • Depression • Early life adversity • Posttraumatic stress disorder (PTSD) • Reward circuits • Rodent • Unpredictability and fragmented sensory signals (FRAG)

Contents

1 Anhedonia, A Dimensional Construct with Transdiagnostic Relevance

Mental illness, including PTSD, depression, and suicide, afflict >20% of adolescents and young adults, with significant social and fiscal costs (Insel 2011; Merikangas et al. 2010). The origins of mental illness are complex, involving genetic and environmental contributions, specifically during sensitive developmental periods (Bale et al. 2010; Lupien et al. 2009; Nelson et al. 2007; Osborne and Monk 2013). Anhedonia, defined as a reduction in pleasure and appetitive/reward seeking behaviors, is a dimensional construct that is transdiagnostic, cutting across mood, anxiety, and substance use disorders, with clearly delineated operational measures and underlying neurocircuitry [for review, see (Pizzagalli 2014; Rizvi et al. 2016)]. Across mood and anxiety disorders, anhedonia correlates with poor treatment response, suicidality, and diminished global function (Pizzagalli 2014; Rizvi et al. 2016). Because of its association with poor treatment outcomes, it is clear that understanding mechanisms underlying anhedonia and its role in both development and maintenance of neuropsychiatric disorders may be critical for effective treatment strategies. Here, we will discuss a neurodevelopmental hypothesis of anhedonia and its contribution to risk and resilience to stress and trauma in adulthood. Specifically, we will review the role of anhedonia and its underlying reward circuitry in trauma-related disorders, discuss its known association with early life adversity, and its link to a newly described form of early life adversity, fragmented and chaotic maternal signals (FRAG) during critical phases of development. We will describe recent cross-species evidence for this form of fragmented and chaotic maternal care to mediate anhedonia in adolescence/early adulthood (Molet et al. 2016a; Bolton et al. 2018) and discuss how this newly identified form of developmental stress might shape reward circuit development and neuropsychiatric risk. Throughout, we will identify knowledge gaps and research needed to understand how FRAG may affect reward processes and reward circuitry, and how it subsequently affects susceptibility to trauma disorders in later life.

2 Anhedonia and Disrupted Reward Circuits as a Marker of Heterogeneity in Trauma Disorders

Posttraumatic stress (PTS) symptoms manifest after severe trauma exposure, affecting 7–8% of the US population (Seedat and Stein 2001) and up to 20% of armed service members (Thomas et al. 2010). Besides the extensive mental health service utilization required for treatment, trauma-related disorders, such as posttraumatic stress disorder (PTSD), are associated with greater overall medical service utilization due to higher rates of chronic physical illness experienced by these patients (Baker et al. 2009; O'Donnell et al. 2013). PTSD is phenotypically and etiologically heterogeneous, posing a significant challenge to identifying its biological mechanisms, creation of objective, non-symptom-based nosological categories that cut across current diagnostic boundaries, and development of novel therapeutics. Hence, the current greater focus in psychiatric research is towards identifying fundamental underlying dimensional constructs that contribute significantly to mortality, global function, and treatment response, such as anhedonia.

Anhedonia and Reward Abnormalities in Trauma-Related Disorders Self-reported anhedonia contributes to symptom heterogeneity in PTSD; it is associated with increased risk for suicide (Spitzer et al. 2018) and social withdrawal (Cao et al. 2016) as well as reduced reward responsiveness (Nawijn et al. 2016). PTS-related anhedonic symptoms in combat veterans increase the risk for comorbidity with substance use, depression, and anxiety disorders (>50% of veterans with PTSD have at least one of these comorbidities) (Kashdan et al. 2006). Theoretically, anhedonia is derived from both reward motivation and consumption ("wanting" and "liking," respectively). A comprehensive recent review by Nawijn et al. suggests that reward abnormalities most robustly observed in trauma disorders are in reward motivation or "wanting" components of the hedonic process such a reward response and effortful-approach behavior, while arousal and valence ratings of rewarding stimuli are not different from those of controls (Nawijn et al. 2015). PTSD patients report significant reductions in the experience of pleasure, i.e., anhedonia (Vujanovic et al. 2017). PTSD patients exhibit anhedonia with or without co-occurring major depressive disorder (MDD), suggesting that anhedonia in trauma-exposed patients is not purely a function of comorbid major depression (Franklin and Zimmerman 2001). Importantly, trauma exposure alone is not related to reduced reward functioning (Nawijn et al. 2015).

Reward Circuit Abnormalities Associated with Anhedonia and Trauma-Related Disorders Anhedonia has been examined across multiple neuropsychiatric disorders including mood, trauma-related, and psychotic disorders. Evidence clearly indicates overlap in the circuitry mediating anhedonia and reward abnormalities across disorders, supporting that anhedonia spans classical diagnostic categories (Sharma et al. 2017; Zhang et al. 2016). Symptoms of anhedonia are robustly associated with the brain's reward processing pathways and reduced responsiveness to reward information (Berghorst et al. 2013; Bogdan and Pizzagalli 2006; Pizzagalli et al. 2008).

The brain's reward circuits consist of the ventral tegmental area (VTA) which sends dopaminergic projections to the nucleus accumbens (NAc) in the ventral striatum. VTA neurons also innervate the prefrontal cortex (PFC), the amygdala, and the hippocampus. Most of the extant literature on the reward pathway in humans comes from fMRI studies that implicate decreased activity levels in the VTA and NAc in anhedonia (Drevets et al. 1992; Lee et al. 2012; Mayberg et al. 2000; Russo and Nestler 2013). Consistent with these imaging findings, deep brain stimulation within the NAc reduces anhedonia in treatment-resistant depression (Bewernick et al. 2010, 2012; Schlaepfer et al. 2008). There is also some evidence that amygdala activation is associated with anhedonia (Kumar et al. 2014; Stuhrmann et al. 2013). Amygdala hyperactivity is one of the most robust phenotypes in PTSD (Koch et al. 2016; Etkin and Wager 2007) and may be a pre-trauma risk factor for PTSD

(Stevens et al. 2017). However, there is no reliable evidence as to the level of amygdala excitability in the context of anhedonia. It is possible that amygdala hyperactivation is a distinct feature of emotional processing aberrations found in mood and trauma disorders and not reward processing per se.

In general, fewer studies have examined anhedonia in individuals with PTSD compared to depression and schizophrenia. The circuit activation patterns are similar across these disorders in relation to anhedonic symptoms, with altered BOLD fMRI activity levels in reward circuitry including the ventral striatum/NAc, the amygdala, and the PFC (Nawijn et al. 2016; Frewen et al. 2012). Reductions in ventral striatum/NAc activity in response to rewards is reliably associated with increased PTS symptoms or PTSD diagnosis (Admon et al. 2013a, b; Elman et al. 2009; Sailer et al. 2008). Interestingly, CT-based lesion analysis also revealed a link between anhedonia and injury to the ventrolateral PFC in Vietnam combat veterans (Lewis et al. 2015). This finding poses a potential complication in studying populations with comorbid head injury and PTSD as regional vulnerabilities in the PFC may manifest in both and may be linked to anhedonic symptoms.

3 Anhedonia, Potential Risk Factor or Marker of Symptom State in Posttraumatic Stress Disorder?

Evidence for Anhedonia as a Preexisting Risk Factor for Depression and PTSD Substantial evidence supports anhedonia as a robust phenotype associated with reward circuit disruption in PTSD patients. However, whether anhedonia and reward dysfunction are precursors to clinical dysfunction precipitated by environmental challenges such as stress and trauma or only develop after trauma/chronic stress exposure and symptom development is unknown. Evidence for anhedonia as a risk factor is quite preliminary and still circumstantial at this stage, consisting primarily of correlational and cross-sectional findings. First, depressed and PTSD subjects continue to show deficient reward learning even after symptom remission (Whitton et al. 2016; Kalebasi et al. 2015) (note that this PTSD study is very small, $N = 12$/group). PTSD subjects also show reward processing and response abnormalities when controlling for psychoactive medication (Elman et al. 2005; Hopper et al. 2008). Potential interpretations of these findings are that anhedonia may be a more "fixed" trait that increases risk for development of PTSD, or that anhedonia is less responsive than other symptoms to current treatments. To determine if anhedonia and reward processing abnormalities are pre-trauma risk factors, twin studies and prospective longitudinal studies will be required. No studies have yet examined this question in trauma-related disorders; however, one small prospective study in adolescent girls reported that low reward sensitivity predicts later adult depression (Bress et al. 2013). To answer this question, we have used the Marine Resiliency Study (MRS) database to test a prospective role of multiple risk factors for PTSD. MRS is a prospective, longitudinal study of psychological, physiological, and biological risk factors for development of combat-related PTSD, in which infantry Marine participants (average age range 18–22) were comprehensively assessed both before leaving for a combat deployment to Iraq or Afghanistan, and 3 and 6 months after their return from deployment (Baker et al. 2012). We previously reported that childhood trauma is associated with PTSD in this population (Agorastos et al. 2014). We are currently examining if in healthy participants at pre-deployment, self-reported levels of anhedonia (as measured by the anhedonia subscale of the Beck Depression Inventory) predict increased risk for their developing PTSD after deployment. Our findings are in preparation for publication and were reported at the 2017 American College of Neuropharmacology Annual Meeting (Risbrough et al. 2017). The findings suggest that anhedonia is a significant predictor of risk, and this association is independent of self-reported depression symptoms and deployment trauma exposure. Studies such as this ongoing analysis in MRS will be required to support the hypothesis that anhedonia in late-adolescence/early adulthood may contribute to pre-trauma risk for PTS.

Might Anhedonia Mediate the Relationship Between PTSD Vulnerability and Early Life Adversity? During early life, the fetus and infant are vulnerable to the consequences of adversity with lasting consequences for infant, toddler, child, and adolescent outcomes (Kim et al. 2014, 2016, 2017; Sandman et al. 2012, 2015; Davis et al. 2011a, b; Davis and Sandman 2006, 2010, 2012; Davis et al. 2007; Glynn et al. 2007; Howland et al. 2016; Stout et al. 2015). The evidence that early experiences including maternal mental health (Goodman and Gotlib 1999; Oberlander et al. 2008; Monk 2001; Monk et al. 2011; Plant et al. 2015), quality of maternal care (Belsky and Fearon 2002; NICHD Early Child Care Research Network 1999, 2006; Masur et al. 2005; Hane et al. 2010; Feldman 2007, 2010, 2015), trauma (Treadway et al. 2009; Rao et al. 2010; Copeland et al. 2007; Mulvihill 2005; Pynoos et al. 1999; Young et al. 2017; Nelson 2013; McLaughlin et al. 2010), and poverty (Kim et al. 2013; Evans and Kim 2007, 2012; Barch et al. 2016; Javanbakht et al. 2016; Noble et al. 2015; Johnson et al. 2016; Does amount of time spent in child care predict socioemotional adjustment during the transition to kindergarten? 2003) profoundly influence later mental health is undisputed. There is also emerging research implicating exposure to early life adversity in dysfunction of reward-related circuitry (Pechtel and Pizzagalli 2011), particularly in response to adversity during early childhood (Hanson et al. 2016). Developmental trajectories of reward-related ventral striatum (VS) activity mediate the relationship between early life stress and mood disorders in adolescence (Hanson et al. 2015). Childhood trauma, for example, is a well-known risk factor for increased depression, PTSD, and physical health problems (Agorastos et al. 2014), yet it can also increase resilience in some populations (Liu et al. 2017). Others have shown that the strongest correlation of aspects of early life trauma to subsequent PTSD (Lowe et al. 2016) was indication of safety or lack thereof. Childhood trauma is also linked to anhedonia and aberrant reward processing in later life [e.g., (Frewen et al. 2012; Pechtel and Pizzagalli 2013)]. But, trauma exposure is not the only factor that can alter reward processing in development; both children and adults exposed to early social deprivation also exhibit altered reward processing and reduced activity in corticostriatal circuits (Dillon et al. 2009; Mehta et al. 2010; Goff et al. 2013; Guyer et al. 2006; Hanson et al. 2017). Crucially, the scope and trajectory of differential developmental effects of early life experiences and mechanisms associated with resilience vs. risk are still being elucidated. We and others have shown that maternal signals alone can have profound developmental effects on neural circuit development, including small-world network architecture, rich-club organization, and distinct modular structure (Kim et al. 2014, 2016, 2017; Sandman et al. 2012, 2015; Davis et al. 2007, 2011a, b; Davis and Sandman 2006, 2010, 2012; Glynn et al. 2007; Howland et al. 2016; Stout et al. 2015). We have recently identified a novel source of early life adversity across rodent models and humans: unpredictable and fragmented sensory signals (FRAG), particularly those derived from maternal care and the home environment. We have proposed that this relatively unexamined component of early developmental experience may be critical in shaping neural circuits underlying risk for neuropsychiatric disorders in adulthood (Baram et al. 2012). In humans, FRAG is assessed in two broad manners. Prenatally, inconsistency and fragmentation of maternal mood is examined via questionnaires and analyzed by applying Shannon's entropy to the distribution of self-reported mood over multiple time points both in pregnancy and after birth (Glynn et al. 2017). In addition, unpredictability of maternal care sequences is examined using behavioral observations of the mother's interactions with her infant (Molet et al. 2016a; Davis et al. 2017). In animals, fragmentation of maternal care behaviors is measured via assessment of duration of stereotyped maternal behavior bouts (e.g., nursing and grooming). In addition, unpredictability of the sequences of maternal care behaviors is examined via observations, and analyzed using Shannon's entropy, as done for humans (Guyer et al. 2006; Hanson et al. 2017). This new construct, predictability of maternal and environmental signals during early development, may be a vital factor that influences synapse strengthening and circuit maturation of reward processing and related circuitry.

4 Anhedonia Develops in Adolescent/Young Adult Rodents as a Result of FRAG

Because of the links between poverty, neglect, abuse, and other forms of childhood trauma with neuropathology and psychiatric symptoms in adulthood, numerous animal models of early life adversity have been developed. However, effects of early life stress on anhedonia in animal models have been conflicting, potentially because of their variable effects on the prime determinant of adversity: maternal care. Maternal care has been well recognized as influencing offspring outcome: Gene expression within the brain and the long-lasting emotional and cognitive consequences of early life experience are governed primarily by sensory signals derived from *active* maternal nurturing behaviors (Baram et al. 2012;

Champagne et al. 2003; Raineki et al. 2012; Weaver et al. 2004; Pena et al. 2014; Eghbal-Ahmadi et al. 1999; Suchecki et al. 1993). Most studies have manipulated the *quantity* of maternal care using well-established models of maternal separation (Stanton and Levine 1990; Aisa et al. 2007; Colorado et al. 2006; Hill et al. 2014; Kundakovic et al. 2013; Matthews et al. 1996; Michaels and Holtzman 2006, 2007). Reduced quantity of maternal care has often led in the offspring to measures of depressive-like behavior in the forced swim test (Stanton and Levine 1990; Aisa et al. 2007; Shalev and Kafkafi 2002; Dimatelis et al. 2012; Lee et al. 2007) as well as to anxiety-like behaviors (Shalev and Kafkafi 2002; Lee et al. 2007). However, the development of anhedonia does not seem to depend on quantity of care: Reduced quantity of maternal care through intermittent deprivation has been reported to increase (Hill et al. 2014; Kundakovic et al. 2013; Matthews et al. 1996; Michaels and Holtzman 2006), reduce (Stanton and Levine 1990), or not change sucrose preference, a measure of anhedonia.

To develop a robust and reproducible model of early life adversity, the Baram Laboratory developed a model of simulated poverty, which involves normal quantity of maternal care but *changes the patterns of care*. Specifically, limiting bedding and nesting material in the cages induces fragmented and unpredictable patterns of maternal care. This paradigm, adopted widely around the world (Walker et al. 2017), provokes profound anhedonia in adolescent male rats (Molet et al. 2016a; Bolton et al. 2017, 2018; Der-Avakian and Markou 2012). Anhedonia is manifest as both reduced preference for sucrose (a highly rewarding stimulus in rodents) and reduced social play without effecting other social behaviors. More recently, rats exposed to early life fragmentation also exhibit reduced pleasure from sweet rich foods and reduced response to cocaine have been established (Molet et al. 2016a). These data suggest that some types of early life adversity, specifically those associated with disruption of maternal care patterns (e.g., chaotic household, evictions, and change of caretakers) might predispose to pathology specifically in reward response.

5 Anhedonia Reflects Aberrant Brain Circuits Subsequent to Early Life Adversity in Rodents

Cognitive and emotional brain functions involve coordinated activities of brain circuits that integrate molecular, cellular, synaptic, and network signaling (Khazipov et al. 2004; Caspi et al. 2003). And as previously noted, mental disorders may arise from dysfunction (failure) of crucial brain circuits originating from genetic risk and environmental influences, particularly during sensitive developmental periods (Khazipov et al. 2004; Caspi et al. 2003; Heim and Binder 2012). Environment-derived sensory signals clearly influence development of brain circuits (Khazipov et al. 2004; Espinosa and Stryker 2012; Dulac et al. 2014) and, in some cases, may drive aberrant circuit maturation that can promote mental and cognitive problems.

Specifically, we found that a type of early life adversity characterized by unpredictable sensory input early in life reduced sucrose preference and social play during adolescence (Molet et al. 2016a). Both of these behaviors depend on an intact dopaminergic pleasure/reward circuitry (Dym et al. 2009; Kraft et al. 2015; Muscat and Willner 1989; Siviy et al. 2011; Siviy and Panksepp 2011; Trezza et al. 2010; Vanderschuren et al. 1997). The dopaminergic pleasure/reward system is not fully mature until the third postnatal week in rodents (Voorn et al. 1988) and is sensitive to the influence of early life experiences (Pena et al. 2014; Ventura et al. 2013). It has been reported that predictable sequences of events engage and shape the reward system (Berns et al. 2001; Rutledge et al. 2014). These observations lead us to speculate that predictable patterns of maternal care provide crucial cues for maturation of the pleasure/reward system (Pena et al. 2014; Khazipov et al. 2004; Berns et al. 2001). In the absence of such input, the ability to experience reward from pleasurable sensations including the sweetness of sucrose or the joy of play with peers might be (Pena et al. 2014; Ventura et al. 2013) generating anhedonia (Romer Thomsen et al. 2015). Indeed, recent data demonstrate aberrant interactions of reward and fear/anxiety circuits after early life FRAG (Der-Avakian and Markou 2012; Glenn et al. 2017).

Sensory input early in life governs neuronal activity, which influences circuit maturation (Khazipov et al. 2004; Espinosa and Stryker 2012; Evans et al. 2005) as demonstrated in visual, tactile, and olfactory circuits. We propose that patterns of maternal-derived sensory input influence the maturation of both fear and pleasure–reward emotional systems within the developing brain. The pleasure/reward circuit in rodents is remarkably similar to that found in the human (Russo and Nestler 2013; Der-Avakian and Markou 2012; Nestler 2015), comprised of connectivity between the VTA, NAc, PFC, amygdala, and the hippocampus. The FRAG model appears to disrupt this circuit at a number of nodes, including the amygdala, hippocampus, frontal cortex, and NAc. First, we have recently reported that FRAG produces altered connectivity in adolescent rats between the medial PFC and amygdala as measured by diffusion tensor imaging (DTI) (Bolton et al. 2018). The anhedonia phenotype in the FRAG model is reversed by transcriptional suppression of corticotropin releasing hormone (CRH) signaling in the amygdala, suggesting that augmented CRH release from amygdala-centered cell bodies is required for anhedonia associated with FRAG exposure. Another group recently examined FRAG effects on resting state fMRI of the reward circuit in adults, reporting reduced connectivity between the PFC and caudate putamen in FRAG exposed rats compared to controls (Yan et al. 2017). Finally, in the FRAG model of early life adversity, there is a significant loss of hippocampal dendritic arborization and volumes as well as increased fractional anisotropy (FA – a measure of structural integrity derived from DTI) in the hippocampus (Molet et al. 2016b). These hippocampal effects of FRAG also have real functional consequences. We have recently reported that FRAG across both humans and rodents, as measured by entropy of maternal interactions with offspring, is associated with poor performance in hippocampus-dependent memory tasks in later life (Davis et al. 2017; Molet et al. 2016b). Importantly, low hippocampal volume is a robust phenotype in PTSD, and both twin and prospective studies indicate that reductions in hippocampal function and structure are risk factors for development of PTSD [for review, see (Acheson et al. 2012; Logue et al. 2018; van Rooij et al. 2017)]. In rodents, high hippocampal FA correlates with susceptibility to chronic social defeat stress (Anacker et al. 2016), suggesting that across species, developmental effects on hippocampal circuits and functions may play a role in trauma susceptibility.

6 Conclusions and Future Research Directions

There is clear evidence for anhedonia and reward disruption in PTSD, but if these symptoms develop after trauma or contribute to pre-trauma risk is still very unclear. There are hints from the depression field and our own preliminary data in the MRS that adolescent anhedonia may be predictive of later risk for depression and PTSD. Childhood adversity is robustly linked to PTSD risk and anhedonia, but the specific early life experiences that contribute to this risk are not well understood. The FRAG model of early life adversity in both animals and humans suggests that anhedonia in adolescence/early adulthood can be a consequence of chaotic sensory and maternal signals during development. First, entropy of maternal mood during pregnancy influences mood and anxiety of her children at ages 10–13 (Glynn et al. 2017). In rodents, FRAG is robustly linked to increased anhedonia-like behaviors in adolescence with concurrent alterations in reward circuits (Bolton et al. 2018; Glenn et al. 2017). As discussed above, mood in adolescence is important an predictor of neuropsychiatric disorders in adulthood (Carlson and Pataki 2016; Hofstra et al. 2002; Copeland et al. 2014). Second, FRAG, as assessed by observing entropy of maternal interactions with infants in both rodents and humans, is associated with reductions in hippocampal function in later life (Davis et al. 2017). Hippocampal function has emerged as an important risk factor for the development of trauma disorders (Acheson et al. 2012; Glenn et al. 2017) and may provide an additional mechanism through which FRAG could contribute to PTSD risk. Clearly, research is required to understand the developmental trajectory of FRAG effects on circuit development, and most importantly, its contribution as an independent risk factor for trauma disorders. To begin to answer these questions, our group has developed novel measures in humans to operationalize FRAG: (1) in utero via entropy of maternal mood (Glynn et al. 2017), (2) in early life via observational studies of predictability of maternal care (Davis et al. 2017), and (3) in later life via a retrospective self-report instrument to document past FRAG exposure for use in adult populations such as in the MRS. These complementary measures of FRAG will support efforts to integrate studies across age groups and high-risk populations to identify the developmental trajectory of FRAG effects on reward circuits and subsequent anhedonia, and their contribution to adult trauma disorders.

Acknowledgements Support for this work includes NIMH P50MH096889 (Drs. Baram, Glynn, Davis, Sandman, Stern, Keator, and Baram) and MH73136 (Baram), a VA Merit Award and NIH

R01AA026560 (Risbrough), project No. SDR 09-0128 (Drs. Baker and Risbrough) from the Veterans Administration Health Service Research and Development, the US Marine Corps and Navy Bureau of Medicine and Surgery (Drs. Baker, and Risbrough).

References

Acheson DT, Gresack JE, Risbrough VB (2012) Hippocampal dysfunction effects on context memory: possible etiology for posttraumatic stress disorder. Neuropharmacology 62(2):674–685

Admon R et al (2013a) Imbalanced neural responsivity to risk and reward indicates stress vulnerability in humans. Cereb Cortex 23(1):28–35

Admon R, Milad MR, Hendler T (2013b) A causal model of post-traumatic stress disorder: disentangling predisposed from acquired neural abnormalities. Trends Cogn Sci 17(7):337–347

Agorastos A et al (2014) The cumulative effect of different childhood trauma types on self-reported symptoms of adult male depression and PTSD, substance abuse and health-related quality of life in a large active-duty military cohort. J Psychiatr Res 58:46–54

Aisa B et al (2007) Cognitive impairment associated to HPA axis hyperactivity after maternal separation in rats. Psychoneuroendocrinology 32(3):256–266

Anacker C et al (2016) Neuroanatomic differences associated with stress susceptibility and resilience. Biol Psychiatry 79(10):840–849

Baker DG et al (2009) Trauma exposure, branch of service, and physical injury in relation to mental health among U.S. veterans returning from Iraq and Afghanistan. Mil Med 174(8):773–778

Baker DG et al (2012) Predictors of risk and resilience for posttraumatic stress disorder among ground combat Marines: methods of the Marine Resiliency Study. Prev Chronic Dis 9:E97

Bale TL et al (2010) Early life programming and neurodevelopmental disorders. Biol Psychiatry 68(4):314–319

Baram TZ et al (2012) Fragmentation and unpredictability of early-life experience in mental disorders. Am J Psychiatry 169(9):907–915

Barch D et al (2016) Effect of hippocampal and amygdala connectivity on the relationship between preschool poverty and school-age depression. Am J Psychiatry 173(6):625–634

Belsky J, Fearon RM (2002) Early attachment security, subsequent maternal sensitivity, and later child development: does continuity in development depend upon continuity of caregiving? Attach Hum Dev 4(3):361–387

Berghorst LH et al (2013) Acute stress selectively reduces reward sensitivity. Front Hum Neurosci 7:133

Berns GS et al (2001) Predictability modulates human brain response to reward. J Neurosci 21(8):2793–2798

Bewernick BH et al (2010) Nucleus accumbens deep brain stimulation decreases ratings of depression and anxiety in treatment-resistant depression. Biol Psychiatry 67(2):110–116

Bewernick BH et al (2012) Long-term effects of nucleus accumbens deep brain stimulation in treatment-resistant depression: evidence for sustained efficacy. Neuropsychopharmacology 37(9):1975–1985

Bogdan R, Pizzagalli DA (2006) Acute stress reduces reward responsiveness: implications for depression. Biol Psychiatry 60(10):1147–1154

Bolton JL et al (2017) New insights into early-life stress and behavioral outcomes. Curr Opin Behav Sci 14:133–139

Bolton JL et al (2018) Anhedonia following early-life adversity involves aberrant interaction of reward and anxiety circuits and is reversed by partial silencing of amygdala corticotropin-releasing hormone gene. Biol Psychiatry 83(2):137–147

Bress JN et al (2013) Blunted neural response to rewards prospectively predicts depression in adolescent girls. Psychophysiology 50(1):74–81

Cao X et al (2016) DSM-5 posttraumatic stress disorder symptom structure in disaster-exposed adolescents: stability across gender and relation to behavioral problems. J Abnorm Child Psychol

Carlson GA, Pataki C (2016) Understanding early age of onset: a review of the last 5 years. Curr Psychiatry Rep 18(12):114

Caspi A et al (2003) Influence of life stress on depression: moderation by a polymorphism in the 5-HTT gene. Science 301(5631):386–389

Champagne FA et al (2003) Variations in maternal care in the rat as a mediating influence for the effects of environment on development. Physiol Behav 79(3):359–371

Colorado RA et al (2006) Effects of maternal separation, early handling, and standard facility rearing on orienting and impulsive behavior of adolescent rats. Behav Processes 71(1):51–58

Copeland WE et al (2007) Traumatic events and posttraumatic stress in childhood. Arch Gen Psychiatry 64(5):577–584

Copeland WE et al (2014) Longitudinal patterns of anxiety from childhood to adulthood: the Great Smoky Mountains Study. J Am Acad Child Adolesc Psychiatry 53(1):21–33

Davis EP, Sandman CA (2006) Prenatal exposure to maternal stress and stress hormones influences child development. Infants Young Child 19(3):246–259

Davis EP, Sandman CA (2010) The timing of prenatal exposure to maternal cortisol and psychosocial stress is associated with human infant cognitive development. Child Dev 81(1):131–148

Davis EP, Sandman CA (2012) Prenatal psychobiological predictors of anxiety risk in preadolescent children. Psychoneuroendocrinology 37(8):1224–1233

Davis EP et al (2007) Prenatal exposure to maternal depression and cortisol influences infant temperament. J Am Acad Child Adolesc Psychiatry 46(6):737–746

Davis EP et al (2011a) Children's brain development benefits from longer gestation. Front Psychol 2:1

Davis EP et al (2011b) Prenatal maternal stress programs infant stress regulation. J Child Psychol Psychiatry 52(2):119–129

Davis EP et al (2017) Exposure to unpredictable maternal sensory signals influences cognitive development across species. Proc Natl Acad Sci U S A 114(39):10390–10395

Der-Avakian A, Markou A (2012) The neurobiology of anhedonia and other reward-related deficits. Trends Neurosci 35(1):68–77

Dillon DG et al (2009) Childhood adversity is associated with left basal ganglia dysfunction during reward anticipation in adulthood. Biol Psychiatry 66(3):206–213

Dimatelis JJ, Stein DJ, Russell VA (2012) Behavioral changes after maternal separation are reversed by chronic constant light treatment. Brain Res 1480:61–71

Does amount of time spent in child care predict socioemotional adjustment during the transition to kindergarten? (2003) Child Dev 74(4):976–1005

Drevets WC et al (1992) A functional anatomical study of unipolar depression. J Neurosci 12(9):3628–3641

Dulac C, O'Connell LA, Wu Z (2014) Neural control of maternal and paternal behaviors. Science 345(6198):765–770

Dym CT et al (2009) Genetic variance contributes to dopamine receptor antagonist-induced inhibition of sucrose intake in inbred and outbred mouse strains. Brain Res 1257:40–52

Eghbal-Ahmadi M et al (1999) Differential regulation of the expression of corticotropin-releasing factor receptor type 2 (CRF2) in hypothalamus and amygdala of the immature rat by sensory input and food intake. J Neurosci 19(10):3982–3991

Elman I et al (2005) Probing reward function in post-traumatic stress disorder with beautiful facial images. Psychiatry Res 135(3):179–183

Elman I et al (2009) Functional neuroimaging of reward circuitry responsivity to monetary gains and losses in posttraumatic stress disorder. Biol Psychiatry 66(12):1083–1090

Espinosa JS, Stryker MP (2012) Development and plasticity of the primary visual cortex. Neuron 75(2):230–249

Etkin A, Wager TD (2007) Functional neuroimaging of anxiety: a meta-analysis of emotional processing in PTSD, social anxiety disorder, and specific phobia. Am J Psychiatry 164(10):1476–1488

Evans GW, Kim P (2007) Childhood poverty and health: cumulative risk exposure and stress dysregulation. Psychol Sci 18(11):953–957

Evans GW, Kim P (2012) Childhood poverty and young adults' allostatic load: the mediating role of childhood cumulative risk exposure. Psychol Sci 23(9):979–983

Evans GW et al (2005) The role of chaos in poverty and children's socioemotional adjustment. Psychol Sci 16(7):560–565

Feldman R (2007) Parent-infant synchrony and the construction of shared timing; physiological precursors, developmental outcomes, and risk conditions. J Child Psychol Psychiatry 48(3–4):329–354

Feldman R (2010) The relational basis of adolescent adjustment: trajectories of mother-child interactive behaviors from infancy to adolescence shape adolescents' adaptation. Attach Hum Dev 12(1–2):173–192

Feldman R (2015) Mutual influences between child emotion regulation and parent-child reciprocity support development across the first 10 years of life: implications for developmental psychopathology. Dev Psychopathol 27(4 Pt 1):1007–1023

Franklin CL, Zimmerman M (2001) Posttraumatic stress disorder and major depressive disorder: investigating the role of overlapping symptoms in diagnostic comorbidity. J Nerv Ment Dis 189(8):548–551

Frewen PA, Dozois DJ, Lanius RA (2012) Assessment of anhedonia in psychological trauma: psychometric and neuroimaging perspectives. Eur J Psychotraumatol 3. https://doi.org/10.3402/ejpt.v3i0.8587

Glenn DE et al (2017) The future of contextual fear learning for PTSD research: a methodological review of neuroimaging studies. Curr Top Behav Neurosci. https://doi.org/10.1007/7854_2017_30

Glynn LM et al (2007) Postnatal maternal cortisol levels predict temperament in healthy breastfed infants. Early Hum Dev 83(10):675–681

Glynn LM et al (2017) Prenatal maternal mood patterns predict child temperament and adolescent mental health. J Affect Disord 228:83–90

Goff B et al (2013) Reduced nucleus accumbens reactivity and adolescent depression following early-life stress. Neuroscience 249:129–138

Goodman SH, Gotlib IH (1999) Risk for psychopathology in the children of depressed mothers: a developmental model for understanding mechanisms of transmission. Psychol Rev 106(3):458–490

Guyer AE et al (2006) Behavioral alterations in reward system function: the role of childhood maltreatment and psychopathology. J Am Acad Child Adolesc Psychiatry 45(9):1059–1067

Hane AA et al (2010) Ordinary variations in human maternal caregiving in infancy and biobehavioral development in early childhood: a follow-up study. Dev Psychobiol 52(6):558–567

Hanson JL, Hariri AR, Williamson DE (2015) Blunted ventral striatum development in adolescence reflects emotional neglect and predicts depressive symptoms. Biol Psychiatry 78(9):598–605

Hanson JL et al (2016) Cumulative stress in childhood is associated with blunted reward-related brain activity in adulthood. Soc Cogn Affect Neurosci 11(3):405–412

Hanson JL et al (2017) Early adversity and learning: implications for typical and atypical behavioral development. J Child Psychol Psychiatry 58(7):770–778

Heim C, Binder EB (2012) Current research trends in early life stress and depression: review of human studies on sensitive periods, gene-environment interactions, and epigenetics. Exp Neurol 233(1):102–111

Hill RA et al (2014) Sex-specific disruptions in spatial memory and anhedonia in a "two hit" rat model correspond with alterations in hippocampal brain-derived neurotrophic factor expression and signaling. Hippocampus 24(10):1197–1211

Hofstra MB, Van Der Ende JAN, Verhulst FC (2002) Child and adolescent problems predict DSM-IV disorders in adulthood: a 14-year follow-up of a Dutch epidemiological sample. J Am Acad Child Adolesc Psychiatry 41(2):182–189
Hopper JW et al (2008) Probing reward function in posttraumatic stress disorder: expectancy and satisfaction with monetary gains and losses. J Psychiatr Res 42(10):802–807
Howland MA et al (2016) Fetal exposure to placental corticotropin-releasing hormone is associated with child self-reported internalizing symptoms. Psychoneuroendocrinology 67:10–17
Insel TR (2011) The global cost of mental illness. [Cited 28 September 2011]. https://www.nimh.nih.gov/about/directors/thomas-insel/blog/2011/the-global-cost-of-mental-illness.shtml
Javanbakht A et al (2016) Sex-specific effects of childhood poverty on neurocircuitry of processing of emotional cues: a neuroimaging study. Behav Sci (Basel) 6(4)
Johnson SB, Riis JL, Noble KG (2016) State of the art review: poverty and the developing brain. Pediatrics 137(4). pii: e20153075. https://doi.org/10.1542/peds.2015-3075
Kalebasi N et al (2015) Blunted responses to reward in remitted post-traumatic stress disorder. Brain Behav 5(8):e00357
Kashdan TB, Elhai JD, Frueh BC (2006) Anhedonia and emotional numbing in combat veterans with PTSD. Behav Res Ther 44(3):457–467
Khazipov R et al (2004) Early motor activity drives spindle bursts in the developing somatosensory cortex. Nature 432(7018):758–761
Kim P et al (2013) Effects of childhood poverty and chronic stress on emotion regulatory brain function in adulthood. Proc Natl Acad Sci U S A 110(46):18442–18447
Kim DJ et al (2014) Longer gestation is associated with more efficient brain networks in preadolescent children. Neuroimage 100:619–627
Kim DJ et al (2016) Children's intellectual ability is associated with structural network integrity. Neuroimage 124(Pt A):550–556
Kim DJ et al (2017) Prenatal maternal cortisol has sex-specific associations with child brain network properties. Cereb Cortex 27(11):5230–5241
Koch SB et al (2016) Aberrant resting-state brain activity in posttraumatic stress disorder: a meta-analysis and systematic review. Depress Anxiety 33(7):592–605
Kraft TT et al (2015) Dopamine D1 and opioid receptor antagonist-induced reductions of fructose and saccharin intake in BALB/c and SWR inbred mice. Pharmacol Biochem Behav 131:13–18
Kumar P et al (2014) Differential effects of acute stress on anticipatory and consummatory phases of reward processing. Neuroscience 266:1–12
Kundakovic M et al (2013) Sex-specific and strain-dependent effects of early life adversity on behavioral and epigenetic outcomes. Front Psych 4:78
Lee JH et al (2007) Depressive behaviors and decreased expression of serotonin reuptake transporter in rats that experienced neonatal maternal separation. Neurosci Res 58(1):32–39
Lee JS et al (2012) Hippocampus and nucleus accumbens activity during neutral word recognition related to trait physical anhedonia in patients with schizophrenia: an fMRI study. Psychiatry Res 203(1):46–53
Lewis JD et al (2015) Anhedonia in combat veterans with penetrating head injury. Brain Imaging Behav 9(3):456–460
Liu H et al (2017) Association of DSM-IV posttraumatic stress disorder with traumatic experience type and history in the World Health Organization World Mental Health Surveys. JAMA Psychiat 74(3):270–281
Logue MW et al (2018) Smaller hippocampal volume in posttraumatic stress disorder: a multisite ENIGMA-PGC study: subcortical volumetry results from posttraumatic stress disorder consortia. Biol Psychiatry 83(3):244–253
Lowe SR et al (2016) Childhood trauma and neighborhood-level crime interact in predicting adult posttraumatic stress and major depression symptoms. Child Abuse Negl 51:212–222
Lupien SJ et al (2009) Effects of stress throughout the lifespan on the brain, behaviour and cognition. Nat Rev Neurosci 10(6):434–445

Masur EF, Flynn V, Eichorst DL (2005) Maternal responsive and directive behaviours and utterances as predictors of children's lexical development. J Child Lang 32(1):63–91
Matthews K, Wilkinson LS, Robbins TW (1996) Repeated maternal separation of preweanling rats attenuates behavioral responses to primary and conditioned incentives in adulthood. Physiol Behav 59(1):99–107
Mayberg HS et al (2000) Regional metabolic effects of fluoxetine in major depression: serial changes and relationship to clinical response. Biol Psychiatry 48(8):830–843
McLaughlin KA et al (2010) Childhood adversities and adult psychiatric disorders in the national comorbidity survey replication II: associations with persistence of DSM-IV disorders. Arch Gen Psychiatry 67(2):124–132
Mehta MA et al (2010) Hyporesponsive reward anticipation in the basal ganglia following severe institutional deprivation early in life. J Cogn Neurosci 22(10):2316–2325
Merikangas KR et al (2010) Lifetime prevalence of mental disorders in U.S. adolescents: results from the national comorbidity survey replication–adolescent supplement (NCS-A). J Am Acad Child Adolesc Psychiatry 49(10):980–989
Michaels CC, Holtzman SG (2006) Neonatal stress and litter composition alter sucrose intake in both rat dam and offspring. Physiol Behav 89(5):735–741
Michaels CC, Holtzman SG (2007) Enhanced sensitivity to naltrexone-induced drinking suppression of fluid intake and sucrose consumption in maternally separated rats. Pharmacol Biochem Behav 86(4):784–796
Molet J et al (2016a) Fragmentation and high entropy of neonatal experience predict adolescent emotional outcome. Transl Psychiatry 6:e702
Molet J et al (2016b) MRI uncovers disrupted hippocampal microstructure that underlies memory impairments after early-life adversity. Hippocampus 26(12):1618–1632
Monk C (2001) Stress and mood disorders during pregnancy: implications for child development. Psychiatry Q 72(4):347–357
Monk C, Fitelson EM, Werner E (2011) Mood disorders and their pharmacological treatment during pregnancy: is the future child affected? Pediatr Res 69(5 Pt 2):3R–10R
Mulvihill D (2005) The health impact of childhood trauma: an interdisciplinary review, 1997-2003. Issues Compr Pediatr Nurs 28(2):115–136
Muscat R, Willner P (1989) Effects of dopamine receptor antagonists on sucrose consumption and preference. Psychopharmacology (Berl) 99(1):98–102
Nawijn L et al (2015) Reward functioning in PTSD: a systematic review exploring the mechanisms underlying anhedonia. Neurosci Biobehav Rev 51:189–204
Nawijn L et al (2016) Intranasal oxytocin enhances neural processing of monetary reward and loss in post-traumatic stress disorder and traumatized controls. Psychoneuroendocrinology 66:228–237
Nelson CA (2013) Biological embedding of early life adversity. JAMA Pediatr 167(12):1098–1100
Nelson CA 3rd et al (2007) Cognitive recovery in socially deprived young children: the Bucharest Early Intervention Project. Science 318(5858):1937–1940
Nestler EJ (2015) Role of the brain's reward circuitry in depression: transcriptional mechanisms. Int Rev Neurobiol 124:151–170
NICHD Early Child Care Research Network (1999) Chronicity of maternal depressive symptoms, maternal sensitivity, and child functioning at 36 months. Dev Psychol 35(5):1297–1310
NICHD Early Child Care Research Network (2006) Infant-mother attachment classification: risk and protection in relation to changing maternal caregiving quality. Dev Psychol 42:38–58
Noble KG et al (2015) Family income, parental education and brain structure in children and adolescents. Nat Neurosci 18(5):773–778
O'Donnell ML et al (2013) Disability after injury: the cumulative burden of physical and mental health. J Clin Psychiatry 74(2):e137–e143
Oberlander TF et al (2008) Prenatal exposure to maternal depression, neonatal methylation of human glucocorticoid receptor gene (NR3C1) and infant cortisol stress responses. Epigenetics 3(2):97–106

Osborne LM, Monk C (2013) Perinatal depression – the fourth inflammatory morbidity of pregnancy?: Theory and literature review. Psychoneuroendocrinology 38(10):1929–1952

Pechtel P, Pizzagalli DA (2011) Effects of early life stress on cognitive and affective function: an integrated review of human literature. Psychopharmacology (Berl) 214(1):55–70

Pechtel P, Pizzagalli DA (2013) Disrupted reinforcement learning and maladaptive behavior in women with a history of childhood sexual abuse: a high-density event-related potential study. JAMA Psychiat 70(5):499–507

Pena CJ et al (2014) Effects of maternal care on the development of midbrain dopamine pathways and reward-directed behavior in female offspring. Eur J Neurosci 39(6):946–956

Pizzagalli DA (2014) Depression, stress, and anhedonia: toward a synthesis and integrated model. Annu Rev Clin Psychol 10:393–423

Pizzagalli DA et al (2008) Reduced hedonic capacity in major depressive disorder: evidence from a probabilistic reward task. J Psychiatr Res 43(1):76–87

Plant DT et al (2015) Maternal depression during pregnancy and offspring depression in adulthood: role of child maltreatment. Br J Psychiatry 207(3):213–220

Pynoos RS, Steinberg AM, Piacentini JC (1999) A developmental psychopathology model of childhood traumatic stress and intersection with anxiety disorders. Biol Psychiatry 46(11):1542–1554

Raineki C et al (2012) Effects of early-life abuse differ across development: infant social behavior deficits are followed by adolescent depressive-like behaviors mediated by the amygdala. J Neurosci 32(22):7758–7765

Rao U et al (2010) Hippocampal changes associated with early-life adversity and vulnerability to depression. Biol Psychiatry 67(4):357–364

Risbrough V et al (2017) Pre-deployment anhedonia is a risk factor for posttraumatic psychopathology after combat trauma. In: ACNP 56th annual meeting. Neuropsychopharmacology. Palm Springs, CA, pp S111–S293

Rizvi SJ et al (2016) Assessing anhedonia in depression: potentials and pitfalls. Neurosci Biobehav Rev 65:21–35

Romer Thomsen K, Whybrow PC, Kringelbach ML (2015) Reconceptualizing anhedonia: novel perspectives on balancing the pleasure networks in the human brain. Front Behav Neurosci 9:49

Russo SJ, Nestler EJ (2013) The brain reward circuitry in mood disorders. Nat Rev Neurosci 14(9):609–625

Rutledge RB et al (2014) A computational and neural model of momentary subjective well-being. Proc Natl Acad Sci U S A 111(33):12252–12257

Sailer U et al (2008) Altered reward processing in the nucleus accumbens and mesial prefrontal cortex of patients with posttraumatic stress disorder. Neuropsychologia 46(11):2836–2844

Sandman CA, Davis EP, Glynn LM (2012) Prescient human fetuses thrive. Psychol Sci 23(1):93–100

Sandman CA et al (2015) Fetal exposure to maternal depressive symptoms is associated with cortical thickness in late childhood. Biol Psychiatry 77(4):324–334

Schlaepfer TE et al (2008) Deep brain stimulation to reward circuitry alleviates anhedonia in refractory major depression. Neuropsychopharmacology 33(2):368–377

Seedat S, Stein MB (2001) Post-traumatic stress disorder: a review of recent findings. Curr Psychiatry Rep 3(4):288–294

Shalev U, Kafkafi N (2002) Repeated maternal separation does not alter sucrose-reinforced and open-field behaviors. Pharmacol Biochem Behav 73(1):115–122

Sharma A et al (2017) Common dimensional reward deficits across mood and psychotic disorders: a connectome-wide association study. Am J Psychiatry 174(7):657–666

Siviy SM, Panksepp J (2011) In search of the neurobiological substrates for social playfulness in mammalian brains. Neurosci Biobehav Rev 35(9):1821–1830

Siviy SM et al (2011) Dysfunctional play and dopamine physiology in the Fischer 344 rat. Behav Brain Res 220(2):294–304

Spitzer EG et al (2018) Posttraumatic stress disorder symptom clusters and acquired capability for suicide: a reexamination using DSM-5 criteria. Suicide Life Threat Behav 48(1):105–115

Stanton ME, Levine S (1990) Inhibition of infant glucocorticoid stress response: specific role of maternal cues. Dev Psychobiol 23(5):411–426

Stevens JS et al (2017) Amygdala reactivity and anterior cingulate habituation predict posttraumatic stress disorder symptom maintenance after acute civilian trauma. Biol Psychiatry 81(12):1023–1029

Stout SA et al (2015) Fetal programming of children's obesity risk. Psychoneuroendocrinology 53:29–39

Stuhrmann A et al (2013) Mood-congruent amygdala responses to subliminally presented facial expressions in major depression: associations with anhedonia. J Psychiatry Neurosci 38(4):249–258

Suchecki D, Rosenfeld P, Levine S (1993) Maternal regulation of the hypothalamic-pituitary-adrenal axis in the infant rat: the roles of feeding and stroking. Brain Res Dev Brain Res 75(2):185–192

Thomas JL et al (2010) Prevalence of mental health problems and functional impairment among active component and National Guard soldiers 3 and 12 months following combat in Iraq. Arch Gen Psychiatry 67(6):614–623

Treadway MT et al (2009) Early adverse events, HPA activity and rostral anterior cingulate volume in MDD. PLoS One 4(3):e4887

Trezza V, Baarendse PJ, Vanderschuren LJ (2010) The pleasures of play: pharmacological insights into social reward mechanisms. Trends Pharmacol Sci 31(10):463–469

van Rooij SJH et al (2017) The role of the hippocampus in predicting future posttraumatic stress disorder symptoms in recently traumatized civilians. Biol Psychiatry. pii: S0006-3223(17) 31989-3. https://doi.org/10.1016/j.biopsych.2017.09.005

Vanderschuren LJ, Niesink RJ, Van Ree JM (1997) The neurobiology of social play behavior in rats. Neurosci Biobehav Rev 21(3):309–326

Ventura R et al (2013) Postnatal aversive experience impairs sensitivity to natural rewards and increases susceptibility to negative events in adult life. Cereb Cortex 23(7):1606–1617

Voorn P et al (1988) The pre- and postnatal development of the dopaminergic cell groups in the ventral mesencephalon and the dopaminergic innervation of the striatum of the rat. Neuroscience 25(3):857–887

Vujanovic AA et al (2017) Reward functioning in posttraumatic stress and substance use disorders. Curr Opin Psychol 14:49–55

Walker CD et al (2017) Chronic early life stress induced by limited bedding and nesting (LBN) material in rodents: critical considerations of methodology, outcomes and translational potential. Stress 20(5):421–448

Weaver IC et al (2004) Epigenetic programming by maternal behavior. Nat Neurosci 7(8):847–854

Whitton AE et al (2016) Blunted neural responses to reward in remitted major depression: a high-density event-related potential study. Biol Psychiatry 1(1):87–95

Yan CG et al (2017) Aberrant development of intrinsic brain activity in a rat model of caregiver maltreatment of offspring. Transl Psychiatry 7(1):e1005

Young A et al (2017) The effects of early institutionalization on emotional face processing: evidence for sparing via an experience-dependent mechanism. Br J Dev Psychol 35(3):439–453

Zhang B et al (2016) Mapping anhedonia-specific dysfunction in a transdiagnostic approach: an ALE meta-analysis. Brain Imaging Behav 10(3):920–939

The manufacturer's authorised representative in the EU is Springer Nature Customer Service Centre GmbH, Europaplatz 3, 69115 Heidelberg, Germany. If you have any concerns regarding our products, please contact ProductSafety@springernature.com

Printed and bound by CPI Group (UK) Ltd, Croydon, CR0 4YY
25/09/2026
02222891-0004